Basic Civil Engineering

Basic Civil Engineering

Satheesh Gopi

ISBN 978-81-317-2988-5

First Impression, 2010
Twentieth Impression, 2022
Twenty first Impression

Published by Pearson India Education Services Pvt. Ltd, CIN: U72200TN2005PTC057128.

Head Office: 15th Floor, Tower-B, World Trade Tower, Plot No. 1, Block-C, Sector 16,
Noida 201 301, Uttar Pradesh, India.
Registered Office: 7th Floor, SDB2, ODC 7, 8 & 9, Survey No. 01 ELCOT IT/ ITES-SEZ,
Sholinganallur, Chennai – 600 119, Tamilnadu, India.
Phone: 044-66540100
Website: in.pearson.com; Email: companysecretary.india@pearson.com

Printed in India at : Saurabh Printers Pvt Ltd.

contents

Chapter 10 Paints and Varnishes

Chapter 11 Miscellaneous Building Materials

Part II Building Construction

Chapter 12 Component Parts of a Building

Chapter 13 Foundation

Chapter 14 Mortar

preface

The aim of compiling this book has been to give a working knowledge of the important details of civil construction, materials used in civil engineering, including the source of raw materials, their characteristics, the process of manufacture, their defects, structure and uses in the industry, and the basics of surveying and levelling and several other major topics in civil engineering to all engineering students in a systematic way. The book is written in a clear and easy-to-read style, presenting fundamentals of surveying at a level that can be quickly grasped by a beginner. The basic surveying topics deal with modern instruments such as total station, GPS and digital levels, reflecting modern field procedures.

A book on a technical subject is hardly complete without illustrations, and one of the special aims of this book is to present a number of diagrams, which are presented mainly with a view to emphasize the important features of the manufacture, so as not to burden the students with unimportant details. The first part of the book consists of 11 chapters and gives a description of all the materials used for different constructions in the field of civil engineering and the processes involved in the manufacture of the same. The second part of the book discusses briefly about building construction and maintenance of buildings and explains the different stages of construction. It contains 14 chapters. The third part of the book contains two chapters and gives a clear view of basic surveying concepts. The fourth part of the book has five chapters and provides a clear idea about the major topics in civil engineering, viz., geotechnical engineering, transportation engineering, irrigation and water resources engineering and environmental engineering. The fourth part also includes a chapter on Computer-Aided Design (CAD).

The author wishes to thank Dr Sathikumar and Shri N. Madhu, Professors, College of Engineering, Trivandrum, for their help during the writing of this book.

about the author

Satheesh Gopi has over 19 years of experience as a hydrographer and over five years of experience as a civil engineer and is currently the deputy director in the Hydrographic Survey Wing of the Kerala Port Department. He graduated in civil engineering from the College of Engineering, Trivandrum, and also holds a Master's Degree in Information Technology.

He is the author of books such as *Advanced Surveying* and *Global Positioning System, Principles and Applications*. He has published research papers on surveying, GPS and hydrography. He is a life member of the Indian Cartographic Association. His areas of interest include marine construction, advanced surveying, GPS and hydrography.

Part-I

Materials for Construction

chapter 1

Stones

Stones form one of the most important building materials in civil engineering. Stones are derived from rocks, which form the earth's crust and have no definite shape or chemical combination but are mixtures of two or more minerals. The mineral is a substance which is formed by the natural inorganic process and possesses a definite chemical combination and molecular structure. They are strong, durable and descent in appearance.

The main uses of stone as a building material are:

1. As a principal material for foundation of civil engineering works, and for the construction of walls, arches, abutments and dams.
2. In stone masonry in places where it is naturally available.
3. As coarse aggregate in cement concrete (crushed form of rock).
4. As a roofing material in the form of slates.
5. As a flag or thin slab for paving.
6. As a soling material in the construction of highways and runways.
7. As ballast for railway tracks.
8. As a veneer for decorative front and interior of buildings.
9. Limestone for construction of important buildings like temples, churches and mosques.
10. Limestone for the manufacture of cement and as a flux in blast furnace.

Numerous examples of magnificent buildings made partly or wholly of stones can be given from different parts of the world. The Taj Mahal, Red Fort and temples of Jagannath, Puri and Mahabalipuram are best-known buildings of our country made up entirely of stones.

1.1 SOURCES OF STONES

Stones are obtained from rocks. A rock represents a definite portion of the earth's surface. They occur almost everywhere in mountainous and hilly areas. It has no definite chemical composition and shape. Stones are available in large quantities in different parts of our country. It is necessary to know the availability of good stones in and around the construction site to make maximum use of this naturally available and cheap building material.

It is known as monomineralic rock, if it contains only one mineral and it is known as polymineralic rock, if it contains several minerals. Quartz sand, chemically pure gypsum, magnesite, etc. are examples of monomineralic rocks and basalt, granite, etc. are examples of polymineralic rocks.

Minerals are the units of which the rocks are made up of. A mineral indicates a substance having definite chemical composition and molecular structure. It is formed by natural inorganic processes. The properties of a rock are governed by the properties of minerals present in the structure. The common rock forming minerals are quartz, feldspars, calcite and mica, hornblende, etc.

1.2 CLASSIFICATION OF ROCKS

Rocks are classified in four different ways:

1. Geological classification
2. Chemical classification
3. Physical classification
4. Practical classification

1.2.1 Geological classification

According to the mode of origin, rocks are divided into three principal classes or groups, namely:

a. Igneous rocks
b. Sedimentary rocks
c. Metamorphic rocks

1.2.1.1 *Igneous rocks*

The molten material present in the inside portion of the earth's surface is known as magma and this magma occasionally tries to come out to the earth's surface through cracks or weak portions. The rocks which are formed by the cooling of magma are called igneous rocks. The portion of lava which comes outside the surface cools quickly and forms a rock of non-crystalline nature called as trap or basalt. The rest which remains inside the earth undergoes cooling at a slow rate and results in the formation of a rock of crystalline variety known as granite. The igneous rocks are classified into the following three types.

i. *Plutonic rocks:* They are formed by the cooling of magma at a considerable depth from the earth's surface. The cooling is slow and the rocks possess coarsely grained crystalline structure. This rock is mostly used for construction purposes. Granite is the leading example of this type of rock.

ii. *Hypabyssal rocks:* They are formed by the cooling of magma at a relatively shallow depth from the earth's surface. The cooling is quick and, hence, the rocks possess a finely grained crystalline structure. Dolerite is an example of this type of rock.

iii. *Volcanic rocks:* In the case of these rocks, solidification of magma takes place on or near the surface of the earth. The cooling is very rapid as compared to the previous two cases. Hence, the rocks are extremely fine grained in structure. Basalt is an example of this type of rock.

1.2.1.2 *Sedimentary rocks*

These rocks are formed by the weathering action of natural elements on the original rock and subsequent transportation by air, river, glacier and sea and deposition at a different locality. The following four types of deposits occur:

i. *Residual deposits:* Some portion of the products of weathering remains at the site of origin. Such deposit is known as a residual deposit.

ii. *Sedimentary deposits:* The insoluble products of weathering are carried away in suspension, and when such products are deposited, they give rise to sedimentary rocks.

iii. *Chemical deposits:* Some material that is carried away in solution may be deposited by some physio-chemical process such as evaporation and precipitation. It gives rise to chemical deposits.

iv. *Organic deposits:* Some portion of the product of weathering gets deposited through the agency of organisms. Such deposits are known as organic deposits.

Examples of sedimentary rocks are sandstone, limestone, gypsum, lignite, etc.

1.2.1.3 Metamorphic rocks

These rocks are formed by the change in character of the pre-existing rocks. Igneous as well as sedimentary rocks change in character when they are subject to great heat and pressure. The process of change is known as metamorphism. Table 1.1 gives the names of the original and metamorphic rocks. Mineral composition and texture of a rock represent a system which is in equilibrium with its physio-chemical surroundings. Increase of temperature and pressure upsets this equilibrium and metamorphism results from an effort to re-establish a new equilibrium. In this process, original constituent minerals, which are unstable under the changed conditions, are converted into newer ones, which are more stable under the changed conditions. These minerals are arranged in a manner, that is more suitable to the new environment. It should, however, be noted that changes produced by weathering and sedimentation are not included in metamorphism.

There are three agents of metamorphism, namely heat, pressure and chemically acting fluids. Heat may be supplied by the general rise of temperature with depth or by igneous magma. Pressure may be developed due to the load of rocks or movement of the earth. Chemically acting fluids play a passive role only and they do not take active part in the process of metamorphism. Pressure may be uniform or directed. Uniform pressure may be applied to solids and liquids. Directed pressure or stress can exist only in solids and it is converted into uniform pressure if applied to liquids. Following are the four types of metamorphism that occur with various combinations of heat, uniform pressure and directed pressure.

i. *Thermal metamorphism:* Heat is the predominant factor in this type of metamorphism.

ii. *Cataclastic metamorphism:* At the surface of the earth, temperature is low and metamorphism is brought about by directed pressure only. Such metamorphism is known as cataclastic metamorphism.

iii. *Dynamo-thermal metamorphism:* There will always be a rise in temperature with an increase in depth. Hence, heat in combination with stress brings about the changes in the rock. Such metamorphism is known as dynamo-thermal metamorphism.

iv. *Plutonic metamorphism:* Stress is effective only up to a certain depth. This is due to the fact that rocks become plastic in nature at certain depths. At great depths, a stage is reached when stress cannot exist as it is converted into uniform pressure because of the plasticity of rocks. Metamorphic changes at great depths are, therefore, brought about by uniform pressure and heat. Such metamorphism is known as plutonic metamorphism.

Various types of metamorphic rocks that originated from various types of rocks are given in Table 1.1.

Table 1.1 Various Types of Metamorphic Rock and Its Origin

Name of the original rock		
Igneous	Sedimentary	Name of the metamorphic rock
		Gneiss
Granite	Limestone	Marble
	Sandstone	Quartzite
	Clay	Slate

1.2.2 Chemical classification

On the basis of dominant chemical composition, the building stone may fall into any of the following three groups:

a. Silicious rocks: In these rocks, silica predominates. These rocks are hard and durable. They are not easily affected by the weathering agencies. Silica, however, in combination with weaker minerals, may disintegrate easily. It is therefore necessary that these rocks should contain maximum amount of free silica for making them hard and durable. Granites, quartzite, etc. are examples of silicious rocks.

b. Argillaceous rocks: In these rocks, clay predominates. Such rocks may be dense and compact or they may be soft. Slates, laterites, etc. are examples of silicious rocks.

c. Calcarious rocks: In these rocks, calcium carbonate predominates. The durability of these rocks will depend upon the constituents present in the surrounding atmosphere. Limestone, marbles, etc. are examples of calcarious rocks.

Classification of Rocks According to Their Chemical Composition

Chemical classification	Composition	Name of the rock
1. Silicious rock	Predominance of silica	Granite, sandstone, basalt
2. Argillaceous rock	Predominance of clay	Slate, laterite, schist
3. Calcareous rock	Predominance of lime	Limestone, marbles, dolomite

1.2.3 Physical classification

This classification is based on the general structure of rocks. According to this classification, the rocks are divided into three types.

a. Stratified rocks: These rocks possess planes of stratification or cleavage and such rocks can easily be split up along these planes. Sedimentary rocks are distinctly stratified rocks.

b. Unstratified rocks: These rocks are unstratified. The structure may be crystalline granular or compact granular. Igneous rocks of volcanic agency and sedimentary rocks affected by movements of the earth are of this type of rocks.

c. Foliated rocks: These rocks have a tendency to be split up in a definite direction only. Foliated structure is very common in case of metamorphic rocks.

Classification of Rocks According to Their Structure

Physical classification	Characteristics	Typical name
1. Stratified rock	Has many strata	Slate
2. Unstratified rock	Does not have strata	Granite
3. Foliated rocks	Has foliated structure	Gneiss

1.2.4 Practical classification

Practical classification is based on the usage. Practically stones have been classified as granite, basalt, laterite, marble, limestone, sandstone and slate.

1.3 DRESSING OF STONES

A place where exposed surfaces of good quality natural rocks are abundantly available is known as 'quarry' and the process of taking out stones from the natural bed is known as 'quarrying'. This is done with the help of hand tools like pickaxe, chisels, etc., or with the help of machines. Blasting using explosives is another method used in quarrying.

The stones after being quarried are to be cut into suitable sizes and with suitable surfaces. This process is known as the dressing of stones and it is carried out for the following purposes:

1. To make the transport from the quarry easy and economical.
2. To suit the requirements of stone masonry.
3. To get the desired appearance for the stonework.

The different stages of dressing are:

1. *Sizing:* This is reducing the irregular blocks to the desired dimensions by removing extra portions. It is done with help of hand hammers and chisels.
2. *Shaping:* This follows sizing and involves removing of the sharp projections. Many stones are used in common construction after shaping.
3. *Plaining:* This is rather an advanced type of dressing in which the stone is cleared off all the irregularities from the surface.
4. *Finishing:* This is done in case of specially dressed stones only and consists of rubbing of the surface of stones with suitable abrasive materials such as silicon carbide.
5. *Polishing:* This is the last name in dressing and is done only on marbles, limestone and granite.

1.3.1 Types of dressing

a. *Hammer dressing:* A hammer-dressed stone has no sharp and irregular corners and has a comparatively even surface to fit well in the masonry (Figure 1.1).

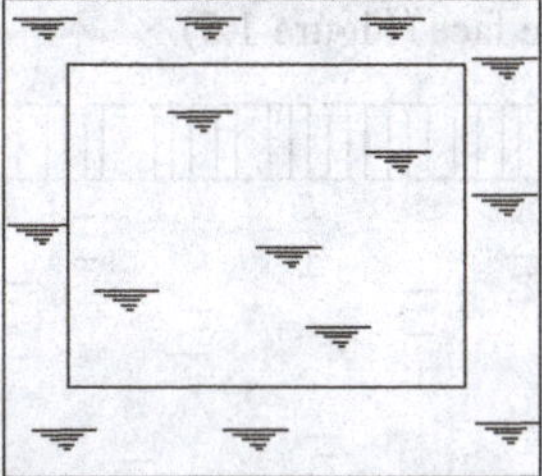

Figure 1.1 Hammer dressing

b. *Chisel drafting:* In this method drafts or grooves are made with help of chisels at all the four edges and any excessive stone from the centre is then removed. These stones are specially used in plinths and corners of buildings (Figure 1.2).

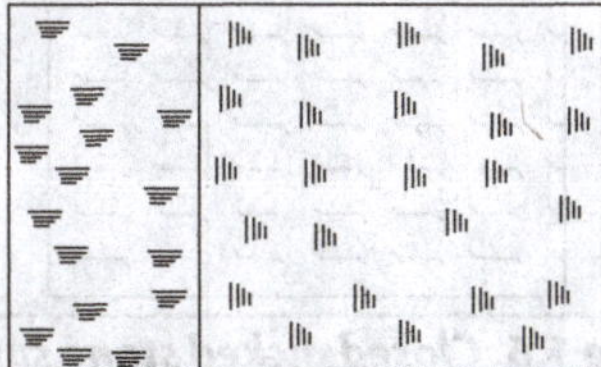

Figure 1.2 Chisel drafting

c. *Fine tooling:* This involves removing most of the projections and a fairly smooth surface is obtained (Figure 1.3).

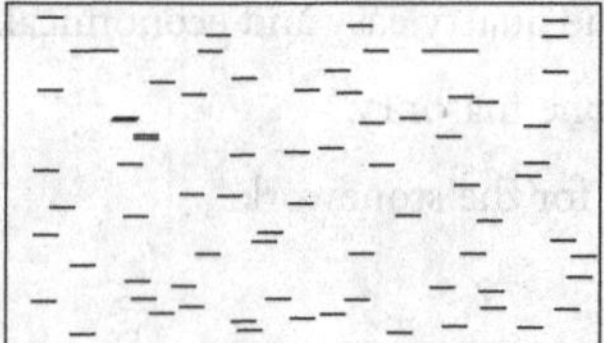

Figure 1.3 Fine tooling

d. *Rough tooling:* A rough-tooled surface has a series of bands, 4-5 cm wide, more or less parallel to tool marks all over the surface (Figure 1.4).

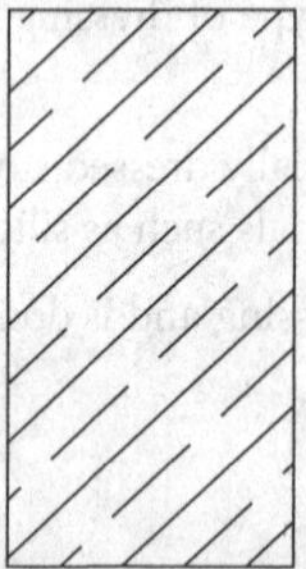

Figure 1.4 Rough tooling

e. *Punched dressing:* A rough-tooled surface is further dressed to show the series of parallel ridges. Chisel marks are left all over the face (Figure 1.5).

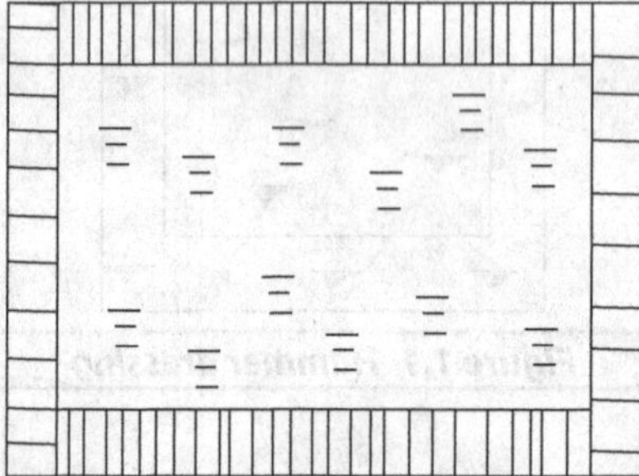

Figure 1.5 Punched dressing

f. *Closed picked dressing:* A punched stone is further dressed so as to obtain a finer surface (Figure 1.6).

Figure 1.6 Closed picked stone surface

1.4 PROPERTIES OF GOOD STONE

1. *Appearance:* The appearance of a stone in relation to the design is of great importance from an architectural point of view. Appearance depends upon the colour and ease with which the stone can be dressed, rubbed or polished. Deep colours, however, in sedimentary rocks are due to oxides of iron in the cementing material, which, upon exposure to the atmospheric influences, either fade away or disfigure and strain the surface on account of the rusting of iron. The stones which are to be used for face work should be attractive in appearance and should be of uniform colour and free from clay holes, spots of other colour, bands, etc. Stones with lighter shades are preferable, because even if they fade a little they will not show a striking difference and spoil the appearance.
2. *Durability:* It denotes the period in years for which a stone may stand practically unaltered after being used in construction. A variety of factors affect the durability of a stone. Of these, the mineral composition, texture and structure of rocks and their capacity to absorb moisture are very important.
3. *Texture:* Texture relates to the grains or particles composing the stone in the strata. A good building stone should have a compact fine crystalline structure, free from cavities, cracks or patches of soft or loose material.
4. *Crushing strength:* It is also called compressive strength of a stone and is defined as the load per unit area at which a given stone starts cracking or failing. For a good structural stone the crushing strength should be greater than 100 N/mm^2.
5. *Resistance to fire:* The fire resistance of a stone may be defined as its capacity to withstand very high temperature without disintegrating. For this requirement, the different materials constituting the composition of a stone should have different coefficients of expansion. In igneous rocks, like granite, free quartz is the most dangerous material as it undergoes a sudden expansion at less than 600°C and flies into splinters; even timber is able to withstand a much higher temperature (about 800°C). It then crumbles to powder and also increases in bulk. Sandstone with silicates as binding material are fire resistant. Clay stones have good fire resistance but are poor in strength and durability.
6. *Specific gravity:* For a good building stone, the specific gravity should be greater than 2.7 or so. Heavy stones are suitable for construction of abutments, dams, docks, harbours, etc. while lighter varieties are used in building construction.
7. *Hardness:* For use in structures subjected to very heavy loadings, such as for constructing bridges, piers and abutments and marine structures, and particularly where they are subjected to abrasion, hardness of the stone is a necessary requirement.
8. *Water absorption:* Moisture reduces the strength of the rocks and as such rocks that contain or absorb great amounts of moisture show lower strength values. All the stones are more or less porous, but for a good stone percentage absorption by weight after 24 hours should not exceed 0.60.
9. *Weathering:* A good building stone should possess better weathering qualities. It should be capable of withstanding adverse effects of various atmospheric and external agencies such as rain, frost, wind, etc.
10. *Facility of dressing:* For facility of dressing, a stone must be comparatively soft, yet durable, compact-grained and homogeneous. It must be free from veins and planes of cleavage. Such a stone is called freestone. The stones should be such that they can be easily carved, moulded, cut and dressed. It is an important consideration from the economic point of view.

1.5 COMMON BUILDING STONES IN INDIA

The commonly used building stones in India include granite, basalt and trap, limestone, marbles, gneiss, laterite, slate, etc.

1. *Granite:* Granite is essentially an igneous rock and is hard and durable. Most of these rocks possess excellent building properties, like high strength, very low abrasion value, good resistance to frost and other weathering agencies, and are available in different appealing colours. It is used for facing work, walls, bridge piers, columns, steps, etc. These rocks are mainly found in Kashmir, Rajasthan, Punjab, Uttar Pradesh, Bihar, Orissa, Kerala, Tamil Nadu and Gujarat.

2. *Limestone:* They are sedimentary rocks composed mainly of calcium carbonate. They show great variation in their properties and, hence, all types are not useful as building stones. They can be used as road metal for construction of floors, steps, walls, etc. Limestone is also used for the manufacture of cement and lime. The use of limestone as facing stones should be avoided in areas where the air is polluted with industrial gases or in coastal regions where air from the sea can attack them. India has extensive deposits of limestone in Maharashtra, Rajasthan, Delhi, Andhra Pradesh, Madhya Pradesh, Bihar and Bengal.

3. *Marble:* These are metamorphic rocks and have been formed from limestone under high temperatures. Marbles vary greatly in colour, structure and texture and most of them are suitable both as an ornamental stone and as a construction material. These stones can take brilliant polish. India has got fairly widespread deposits of marble and it is mainly exploited from Rajasthan, Maharashtra and Gujarat.

4. *Basalt and Trap:* These are also igneous rocks and are generally heavier and darker than granites and also stronger, but may contain cavities and pores within them. They are extensively used for rubble masonry, foundation work and road construction. They are mainly found in Maharashtra, Bihar, Bengal and Madhya Pradesh.

5. *Sandstone:* These are sedimentary rocks and consist mainly of quartz. Sandstones occur in many colours. The most suitable and durable type is that which is light coloured, having silica, cement and a homogeneous, compact texture. It can be used for steps, facing work, flooring, columns, etc. India has immense reserves of really good quality sandstones in Madhya Pradesh, Uttar Pradesh, Orissa and Bihar.

6. *Laterite:* It is a sedimentary rock composed mainly of oxides of iron and aluminium. Laterites are of dull red or brown colour and their important property is that they are quite soft and plastic when cut from the natural bed rock but become hard on exposure. It is used for rough stone masonry work and is sometimes used in place of bricks. They are mainly found in Maharashtra, Tamil Nadu, Kerala, Karnataka and Andhra Pradesh.

7. *Gneiss:* Gneiss is a metamorphic rock and closely resembles granite in its building properties. But sometimes it may be rich in mica and useless as a building stone. They can be used for street paving, rough stone masonry work, etc. They are mainly found in southern states like Tamil Nadu, Kerala, Karnataka, Andhra Pradesh and also in states like Bihar, Bengal and Orissa.

8. *Slate:* Slate is a metamorphic rock and splits into thin sheets having smooth surfaces along the natural bedding planes. In building construction their use is limited to roofing purpose for ordinary buildings or as paving or insulating materials. Slate occurs in Gujarat, Rajasthan, Andhra Pradesh and Bihar.

9. *Kankar:* Kankar is a sedimentary rock and is a form of impure limestone. It is used as road metal for the manufacture of hydraulic lime, etc. It occurs among the different parts of north and central India.

10. *Murum:* Murum is a metamorphic rock. It is a form of decomposed laterite and is deep brown or red in colour. It is a soft rock and can be used for fancy paths and garden walls. It occurs mainly in Maharashtra, Tamil Nadu, Orissa, Bihar and Madhya Pradesh.

REVIEW QUESTIONS

1. What are the main uses of stone in building construction?
2. How are rocks classified? Briefly discuss the different classification of rocks.
3. Give short notes on
   a. Chemical classification of rocks
   b. Different stages of dressing of stones
   c. Common building stones in India
4. What are the purposes of dressing of stones?
5. What are the different types of dressing of stones?
6. Briefly discuss the properties of a good stone.

chapter 2

Sand

Sand is an important building material. It abundantly occurs in nature and is formed by the decomposition of rocks. Sand particles consist of small grains of silica (SiO_2). It forms a major ingredient in concrete, lime mortar, cement mortar, etc.

2.1 NATURAL SOURCES OF SAND

Sand is formed by the weathering of rocks. Based on the natural sources from which sand is obtained, it is classified as follows:

1. Pit sand
2. River sand
3. Sea sand

2.1.1 Pit sand

This sand is obtained by forming pits in soils. It is excavated from a depth of about 1-2 m from the ground level. This sand is found as deposits in soil and it consists of sharp angular grains, which are free from salts. It serves as an excellent material for mortar or concrete work. Pit sand must be made free from clay and other organic materials before it can be used in mortar. Also, the coating of oxide of iron over the sand grains should be removed.

2.1.2 River sand

This sand is widely used for all purposes. It is obtained from the banks or beds of rivers and it consists of fine rounded grains. The presence of fine rounded grains is due to mutual attrition under the action of water current. The river sand is available in clean conditions. The river sand is almost white in colour.

2.1.3 Sea sand

Sea sand is obtained from the sea shores. It consists of fine rounded grains like the river sand. Sea sand is light brown in colour. Since the sea sand contains salts, it attracts moisture from the atmosphere. Such absorption causes dampness, efflorescence and disintegration of work. Sea sand increases the setting time of cement. Hence, it is the general rule to avoid use of sea sand for engineering purposes even though it is available in plenty. However, after removing the salts by washing, it can be used as a local material.

2.2 CLASSIFICATION OF SAND

Based on the grain size distribution, sand is classified as fine, coarse and gravelly.

1. *Fine sand:* The sand passing through a sieve with clear openings of 1.5875 mm is known as fine sand. Fine sand is mainly used for plastering.
2. *Coarse sand:* The sand passing through a sieve with clear openings of 3.175 mm is known as coarse sand. It is generally used for masonry work.
3. *Gravelly sand:* The sand passing through a sieve with clear openings of 7.62 mm is known as gravelly sand. It is generally used for concrete work.

2.3 BULKING OF SAND

The increase in the volume of sand due to the presence of moisture is known as bulking of sand. This is due to the fact that moisture forms a film of water around the sand particles and this results in an increase in the volume of sand. The extent of bulking depends on the grading of sand. The finer the material the more will be the increase in volume for the given moisture content. Bulking of sand can be expressed in a graphical way as shown in Figure 2.1.

For a moisture content of 5–8 per cent, the increase in volume may be about 20–40 per cent depending upon the gradation of sand. When the moisture content is further increased, the sand particles pack near each other and the amount of bulking is decreased. Hence, dry sand and the sand completely flooded with water have practically the same volume.

The volumetric proportioning of sand depends upon the extent of bulking. It is more with fine sands than with coarse sands. If proper allowances are not made for bulking of sand, the cost of concrete and mortar increases and it results in mixes with inadequate sand. This makes the mix harsh and difficult for working and placing.

2.4 PROPERTIES OF GOOD SAND

Good sand should possess the following properties:

1. It should be clean and coarse.
2. It should be free from any organic or vegetable matter; usually 3-4 per cent clay is permitted.

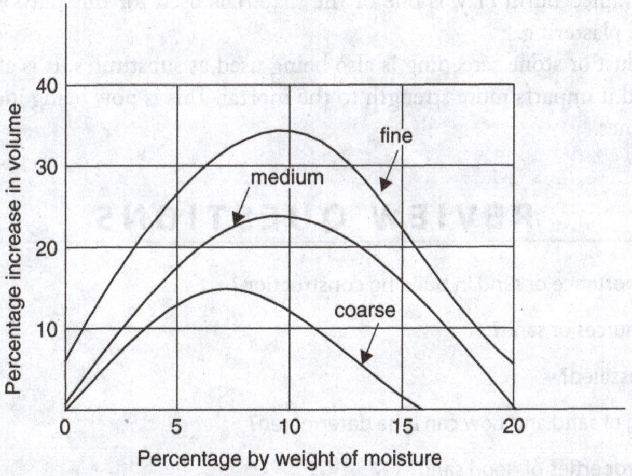

Figure 2.1 Graph showing the percentage increase in volume to the percentage by weight of moisture

3. It should be chemically inert.
4. It should contain sharp, angular, coarse and durable grains.
5. It should not contain salts which attract moisture from the atmosphere.
6. It should be well graded, i.e., it should contain particles of various sizes in suitable proportions.
7. It should be strong and durable.
8. It should be clean and free from coatings of clay and silt.

2.4.1 Functions of sand in mortar

Sand is used in mortar and concrete for the following functions:

a. *Strength:* It helps in the adjustment of the strength of mortar or concrete by variation of its proportion with cement or lime. It also increases the resistance of mortar against crushing.
b. *Bulk:* It acts as an adulterant. Hence, the bulk or volume of mortar is increased which results in reduction of cost.
c. *Setting:* In the case of fat lime, carbon dioxide (CO_2) is absorbed through the voids of sand and setting of fat lime occurs effectively.
d. *Shrinkage:* It prevents excessive shrinkage of mortar in the course of drying and, hence, the cracking of mortar during setting is avoided.
e. *Surface area:* It subdivides the paste of the binding material into a thin film and, thus, more surface area is offered for its spreading and adhering.

2.4.2 Substitutes for sand

Sand has now become a scarce and costly material and extraction of river sand is now said to affect the ecological balance. The use of substitutes for sand has gained great importance.

Surkhi, or finely grained burnt clay, is one of the materials used for this purpose. It can be used in all mortars, except that for plastering.

Processed quarry dust or stone screening is also being used as substitutes. It is abundantly available and if it is properly screened it imparts more strength to the mortar. This is now being industrially manufactured under various trade names.

REVIEW QUESTIONS

1. What is the importance of sand in building construction?
2. What are the sources of sand?
3. How is sand classified?
4. What is bulking of sand and how can it be determined?
5. What are the properties of good sand?
6. Describe the functions of sand in mortar.
7. What are the substitutes of sand used due to the scarcity of sand?

chapter 3

Lime

Lime is one of the most important and largely used building materials. In fact, it used to be the main cementing material before the advent of Portland cement. Egyptians and Romans made remarkable application of this material for various constructional purposes. Even in India, various engineering structures such as big palaces, bridges, temples, forts and monuments were constructed with lime as a cementing material and some of these structures still exist in good condition. This is attributed to some of the unique properties of lime, such as its better workability, early stiffening, good strength and resistance to moisture and excellent adherence to masonry units. Although it has been largely replaced by cement in India and elsewhere, the material still stands comparable to cement in most important properties.

3.1 SOURCES OF LIME

Lime is not usually available in nature in free state. It is chiefly prepared by burning limestone. Depending on the percentage of calcium carbonate ($CaCO_3$) in limestone, lime is classified into A, B and C types, which are used for masonry work, mortar and plaster and white washing, respectively. The type C variety is also known as pure lime or fat lime. Class A variety is only available in slaked form, while class B and C are available in slaked as well as unslaked forms. Lime is also obtained by burning kankar, shells of sea animals and boulders of limestone from beds of old rivers.

The lime that is obtained from the calcination of pure limestone is known as quicklime. It mainly consists of oxides of calcium and it is not crystalline. It shows great affinity to moisture.

3.2 CLASSIFICATION OF LIME

Lime has been conventionally classified into the following three types, namely

1. Fat lime
2. Hydraulic lime
3. Poor lime

3.2.1 Fat lime

This lime is also known as high calcium lime, pure lime, rich lime or white lime. Fat lime is obtained from pure limestone, shell and coral. When it is left in air, it absorbs carbon dioxide (CO_2) from air and gets transformed into calcium carbonate ($CaCO_3$). It is popularly known as fat lime as it slakes vigorously and its volume gets increased to about 2-2½ times the volume of that of quick lime.

The following are the properties of fat lime:

a. It hardens very slowly.

b. It has a high degree of plasticity.

c. Its colour is perfect white.

d. It sets slowly in the presence of air.

e. It slakes vigorously.

The following are the uses of fat lime:

a. It is used in white washing and plastering of walls.

b. With sand, it forms lime mortar, which sets in thin joints. Such mortar can be used for thin joints of brickwork and stonework.

3.2.2 Hydraulic lime

Hydraulic lime is different in composition from quick lime in that it contains a definite amount of clay, which gives it the hydraulic property, i.e., the capacity to set and harden even under water. Quick lime does not set under water. Hydraulic lime contains CaO between 70 and 80 per cent and clay about 15–30 per cent.

Hydraulic lime is generally manufactured from a limestone that is rich in clay, or by adding clay materials to the limestone during burning of limestone. This lime is further divided into feebly hydraulic, moderately hydraulic and eminently hydraulic lime, depending upon its hydraulicity. The comparison between these types of hydraulic limes is shown in Table 3.1.

3.2.3 Poor lime

This lime is also known as impure lime or lean lime. The following are the properties of poor lime.

a. This lime contains more than 30 per cent clay.

b. It slakes very slowly.

c. It forms a thin paste with water.

d. It does not dissolve in water though it is frequently changed.

e. It hardens very slowly.

f. It has poor binding properties.

Table 3.1 Comparison Between the Types of Hydraulic Lime

Number	Item	Feebly hydraulic lime	Moderately hydraulic lime	Eminently hydraulic lime
1.	Clay content	5%–10%	11%–20%	21%–30%
2.	Slaking action	Slakes after few minutes	Slakes after 1 or 2 hours	Slakes with difficulty
3.	Setting action	Sets in water in 3 weeks or so	Sets in water in 1 week or so	Sets in water in a day or so
4.	Hydraulicity	Feeble	Moderate	Eminent
5.	Uses	Mortar produced by this lime is reasonably strong, and hence it can be used for ordinary masonry work	Mortar produced by this lime is strong, and hence it can be used for superior type of masonry work	Mortar produced by this lime is similar to cement and hence it can be used for damp places

g. Its colour is muddy white.

h. It can be used for inferior type of work where there is scarcity of good lime.

3.3 CALCINATION OF LIME IN CLAMPS AND KILNS

3.3.1 Calcination

Lime is manufactured by the burning of limestone to bright red in suitable kilns or clamps. Theoretically, limestone dissociates when heated at 880°C into carbon dioxide and calcium oxide, which is also reversible.

$$CaCO_3 \leftrightarrow CaO + CaO_2$$

The burning or calcination of limestone can be carried out in one of the following ways:

a. Clamps

b. Kilns

  i. Intermittent kilns

  ii. Continuous kilns

3.3.1.1 *Clamp burning*

It is a very common and quick method of obtaining small supplies of ordinary type of quick lime. No constructions have to be made. Simply the ground is levelled and cleaned and the limestone and fuel are stocked in alternative layers, if the fuel is wood. But if the fuel is coal or charcoal, the limestone and fuel are mixed together and placed in a heap form. Any type of burning material locally available is used. The whole heap is then covered with mud plaster and an attempt is made to preserve as much heat as possible. Small holes are left at the top of the plaster and also at the bottom. When the blue flame at the top disappears, it indicates the completion of the burning of lime. The clamp is then allowed to cool and the pieces of quick lime are hand-picked subsequently. Even though burning is quick and cheap for ordinary type of lime, it is not suitable for large supplies because of the following reasons:

i. It proves to be uneconomical to manufacture lime on a large scale.

ii. The burning is not complete which results in poor quality lime containing unburnt limestone.

iii. Wastage is considerable, both in terms of heat generated and material produced.

iv. The quantity of fuel required is more and hence can only be practised where fuel is abundant (Figure 3.1).

3.3.1.2 *Kiln burning*

Most of the commercial lime is made by burning limestones in permanent structures called kilns. Kilns used in the manufacture of lime are of a number of variety and designs. A kiln may be of intermittent or continuous type. The kiln may be mixed feed type where the fuel and limestones are mixed up during burning. In separate feed type, the fuel is burnt separately and does not come into contact with the limestones.

Intermittent kilns

This is also known as batch type kiln. They are permanent structures of rectangular, oval or cylindrical shape. It may be made of bricks or stones. The interior of this kiln is lined with refractory bricks and does not break

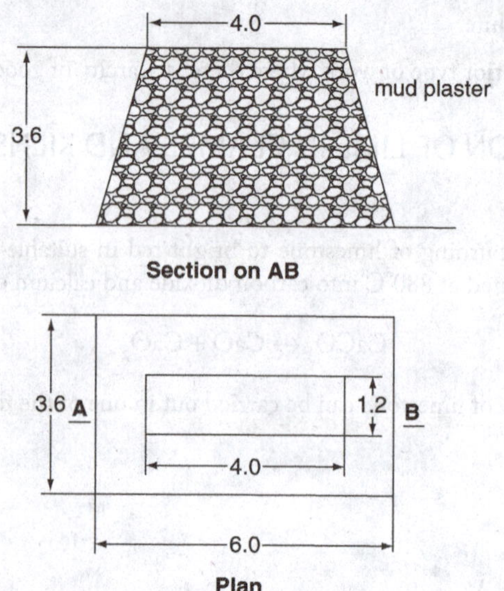

Figure 3.1 Clamp

even at high temperatures. The kiln is provided with openings or flues for supply of air during burning. The roof may be covered or partly covered. An escape hole for gases is always provided at the top.

In mixed feed type, the kiln is first loaded with a calculated amount of fuel and limestone, being deposited in alternative layers. The top of the kiln is covered with unburnt material. The kiln is ignited from the bottom for a required number of days till calcination is complete. It is allowed to cool and then unloaded. The next batch of fuel and limestone is charged to repeat the process (Figure 3.2).

In separate feed type, the fuel is not allowed to come in contact with the limestones. Bigger pieces are stocked in lower regions and smaller pieces above them, leaving open spaces for circulation of hot gases. Fuel is burnt in arch-type gates from where the hot gases rise and circulate between the limestones. When the limestones are sufficiently burnt, the kiln is cooled and unloaded. In this process the burning is more complete

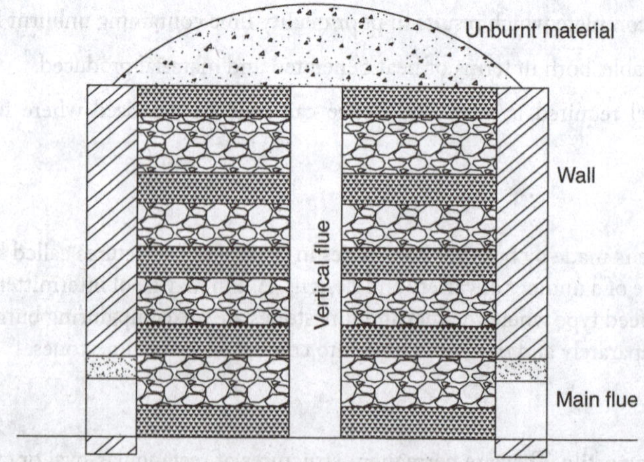

Figure 3.2 Intermittent flame kiln

and mixing of ash and lime is avoided. There is a considerable wastage of lime in intermittent kilns as every operation includes loading, burning, cooling and unloading.

Continuous kilns

These kilns are of such designs that from one end they are charged with raw materials and from the other end finished materials are taken out. Consequently, the kiln is not stopped for emptying or cooling operations. Naturally, the rate of production of lime is much higher.

In continuous flame kiln or mixed feed continuous kiln, the mixture of limestone and fuel is fed from the top. The widening of the middle portion is done so as to accommodate the hot gases of combustion. The bottom is covered by grating with holes. After burning, the lime is collected at the bottom and removed through the access shaft. The kiln is partly under the ground and partly above the ground. The inside surface of the kiln is covered with fire brick lining. The loading platform is provided at the top. As the level of material falls, the required amount of the mixture of fuel and limestone is fed from the top.

In continuous flare kiln or separate feed continuous kiln, the fuel is not allowed to come in contact with the limestone. The kiln consists of two sections. The upper section serves for the storage of limestones. The lower portion is provided with fire brick lining. Initially, a small quantity of fuel is mixed with limestone and ignited. The fuel is then fed through shafts around the lower and upper section of the kiln. The feeding of limestone is from the top. The removal of the calcinated material is done through a grating placed at the bottom of the kiln from where it can be removed (Figure 3.3).

There is a considerable saving in time and fuel, but the initial cost is high. Hence, this method can be adopted to manufacture lime on a large scale.

3.4 SLAKING OF LIME

The process of mixing water in quick lime is known as slaking of lime. It is an important operation in the preparation of lime at site for use in building construction. Improper slaking results in serious defects in mortars and plasters.

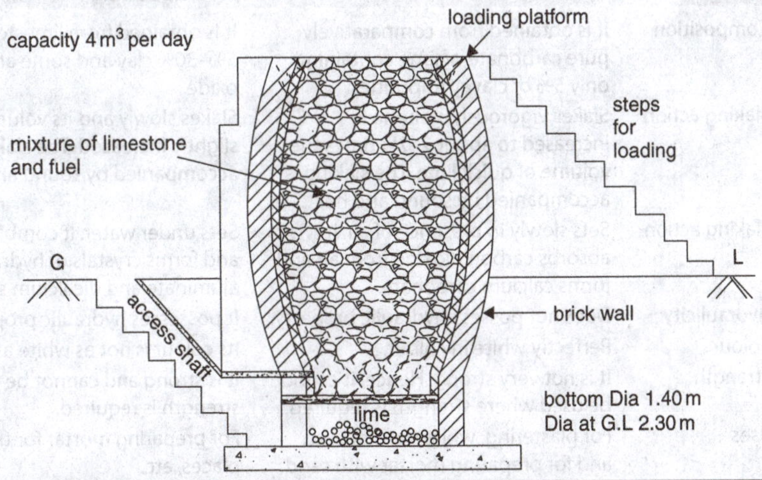

Figure 3.3 Continuous flare kiln

Two methods are commonly used for slaking of lime.

3.4.1 Tank slaking

In this method, two brick lined tanks are required. The first tank is about 45 cm deep and is made at a higher level, whereas the second tank is 60–75 cm deep and is made at a lower level, adjoining the first tank. Water is filled in the upper tank to about three-fourth its depth and quick lime is gradually added to it in small amounts. Water is constantly stirred during the process of addition of lime. This results in the formation of 'milk of lime', which is allowed to fall through a sieve into the lower tank.

After the lime-milk has been allowed to stand for the requisite time (2-3 days), it matures itself and forms lime putty, which is taken for use.

3.4.2 Platform slaking

The method provides dry slaking of lime which can be used as it is or may be converted to putty. A watertight masonry platform is built, over which the lime is spread in a 15 cm layer. Water is then sprayed over it using a hose pipe till lime disintegrates into a fine powder. During the water sprinkling process, the lime heap is turned over and over again. It is then left for 24 hours during which further slaking occurs.

The rate of hydration or slaking is greatly a function of the composition, physical state and degree of burning of lime. It is to be noted that over-burnt or under-burnt limestones do not slake easily. Hence, such undesirable pieces should be removed before slaking. It is also necessary to convert all lumps into powder or pulp form. It is observed that one part of fat quick lime is converted into about 1½ parts in paste form and about 2 parts in powder form.

The quantity of water required for hydrating 100 kg of lime is 32 litres, but practically even 100 litres of water may have to be added. This is generally determined by experience and depends on other factors like composition of quick lime, method of slaking and the form in which hydrated lime is required, i.e., either as putty or powder form.

3.5 COMPARISON BETWEEN FAT LIME AND HYDRAULIC LIME

Number	Item	Fat lime	Hydraulic lime
1	Composition	It is obtained from comparatively pure carbonate of lime containing only 5% of clayey impurities.	It is obtained from limestones containing 5%–30% clay and some amount of ferrous oxide.
2	Slaking action	Slakes vigorously, volume is increased to about 2–2½ times the volume of quick lime. The slaking is accompanied by sound and heat.	Slakes slowly and its volume is only slightly increased. The slaking is not accompanied by sound and heat.
3	Slaking action	Sets slowly in presence of air. It absorbs carbon dioxide from air and forms calcium carbonate.	Sets under water. It combines with water and forms crystals of hydrated tricalcium aluminate and dicalcium silicate.
4	Hydraulicity	Does not possess hydraulic property.	It possesses hydraulic property.
5	Colour	Perfectly white in colour.	Its colour is not as white as fat lime.
6	Strength	It is not very strong. Hence, it cannot be used where strength is required.	It is strong and cannot be used where strength is required.
7	Uses	For plastering, whitewashing, etc. and for preparing mortar with sand or surkhi.	For preparing mortar for thick walls, damp places, etc.

3.6 USES OF LIME

Lime can be used for the following purposes:

1. *Lime mortar:* Lime mortar has been extensively used in construction work from times immemorial. Lime mortar is used as a building medium in brick, stone and other masonry works as well as for plastering and pointing. The composition of the lime mortar for building work depends on:

   a. The type of masonry
   b. Situation/location of work
   c. Load which the masonry will have to take
   d. Condition of exposure to weather or soil conditions
   e. In case of hydraulic structures, weather conditions under water.

2. *Plastering:* Plastering serves the following functions:

   a. To smoothen the surface of masonry
   b. To protect the masonry surface from weathering
   c. To cover unevenness of masonry
   d. To prepare surface for decorative treatment.

3. *Whitewashing:* Whitewashing is applied on internal and external plastered surface as a decorative feature. Apart from decorative effect, the whitewashed outer surface reflects away the sun's rays and reduces the heating effect.
4. *Lime concrete:* In situations where quick setting and high strength are not required, lime concrete serves as an economical substitute. Lime concrete can be used for foundation, terraced roofing, flooring, ditches for sullage water, etc.
5. *Lime sand bricks:* It is a pearl gray brick like dry pressed burnt clay brick. This can be used for low-cost constructions and as a refractory material for lining open hearth furnaces.
6. It is used as a chemical raw material in the purification of water and for sewage treatment.
7. It is used for soil stabilization and for improving soil for agricultural purposes.
8. It is used as a flux in the metallurgical industry.
9. It is used for the production of glass.

REVIEW QUESTIONS

1. Explain the importance of lime as a building material.
2. What are the different sources of lime?
3. Briefly discuss the classifications of lime.
4. What is fat lime? What are the properties of fat lime?

5. What is poor lime? What are the properties of poor lime?
6. What is clacination of lime? How is clacination of lime carried out?
7. Explain briefly
    a. Clamp burning of lime
    b. Kiln burning of lime
    c. Hydraulic lime
    d. Slaking of lime
    e. Uses of lime
8. What is slaking of lime? What are the different methods used for slaking of lime?

chapter 4

Cement

Natural cement is brown in colour. It sets very quickly after the addition of water and is not as strong as artificial cement, and hence it has limited use.

It was in the eighteenth century that the most important advances in the development of cement were made, which finally led to the invention of Portland cement. In 1756, John Smeaton showed that the hydraulic lime which can resist the action of water can be obtained not only from hard lime but also from a limestone which contains a substantial proportion of clay.

In 1796, Joseph Parker found that the modules of argillaceous limestone made excellent hydraulic cement when burnt in the usual manner. Later, several experiments with several mixtures of limestone and argillaceous were carried out so that the properties of the product could be kept under more uniform and proper control by using varying lime and clay proportions. In 1824, Joseph Aspidin of Leeds in England introduced Portland cement.

4.1 PROPERTIES OF CEMENT

The properties of cement are:

1. It gives strength to the masonry.
2. It acts as an excellent binding material.
3. It offers good resistance to moisture.
4. It possesses good plasticity.
5. It stiffens or hardens early.
6. It is easily workable.

4.2 INGREDIENTS OF CEMENT

1. *Lime (CaO):* The chief constituent of cement is lime. Its proportion varies from 60 to 67 per cent. The lime in excess makes the cement unsound and causes the cement to expand and disintegrate and also retards the setting property. On the other hand, if lime is in deficiency, it reduces the strength of cement.
2. *Silica (SiO_2):* It forms 17 to 25 per cent of Ordinary Portland Cement. It imparts strength to the cement due to the formation of dicalcium and tricalcium silicates. Excess of silica increases the strength of cement, but at the same time the setting time is prolonged.
3. *Alumina (Al_2O_3):* It acts as a flux and lowers the clinkering temperature. It imparts quick setting property to cement. If in excess, it weakens the strength of cement.

4. *Calcium sulphate ($CaSO_4$):* This ingredient is in the form of gypsum. It is generally added in very small amounts (2 per cent of wt.) to cement towards the last stage of manufacture with a view of retarding the setting time of cement.
5. *Iron oxide (Fe_2O_3):* This is responsible for imparting the characteristic grey colour to cement. Its percentage varies from 0.5 to 6 per cent.
6. *Magnesia (MgO):* Magnesia varies from 0.1 to 45 per cent. Excess of magnesia reduces the soundness of cement. It imparts hardness and colour to the cement.
7. *Sulphur:* It varies from 1 to 2.5 per cent. If it is in excess, it makes the cement unsound.
8. *Alkalies:* Most of the alkalies present in raw materials are carried away by the flue gases during heating. If they are in excess in cement, they result in alkali-aggregate reaction, efflorescence and staining when used in masonry work.

4.2.1 Harmful constituents of cement

The presence of alkali oxides like K_2O and Na_2O and magnesium oxides like MgO adversely affects the quality of cement. If the amount of alkali oxides exceeds 1 per cent, it leads to the failure of concrete. If the content of magnesium oxide exceeds 5 per cent, it causes cracks after mortar and the concrete hardens. Table 4.1 shows the admissible average (in %) and limits (in %) of ingredients in ordinary cement.

Table 4.1 Admissible Average (in %) and Limits (in %) of Ingredients in Ordinary Cement

Ingredient	Limits (%)	Average (%)
1. Lime	60–66	63.5
2. Silica	18–25	22.5
3. Alumina	3–8	6
4. Iron oxide	0.5–5	2.5
5. Magnesia	1–5	1
6. Sodium and potassium oxides	0.5–5	1
7. Sulphuric anhydride	0.5–5	1

4.3 SETTING TIME OF CEMENT

When water is added to cement, the ingredients of cement react chemically with water and form a complicated chemical compound. The mixing of cement with water results in a sticky cement paste and it gradually goes on thickening in course of time. It is found that ordinary cement achieves 70 per cent of its final strength in 20 days and 90 per cent in 1 year or so.

The time of setting is greatly influenced by the following factors:

1. The temperature at which the cement paste is allowed to set.
2. The percentage of water mixed to cement in making the paste.
3. The humidity at which the setting is allowed.

Setting time is distinguished into initial setting time and final setting time on the basis of the time taken by the test specimen to set to a specified minimum depth.

A Vicat needle apparatus is used for the determination of setting time (Figure 4.1).

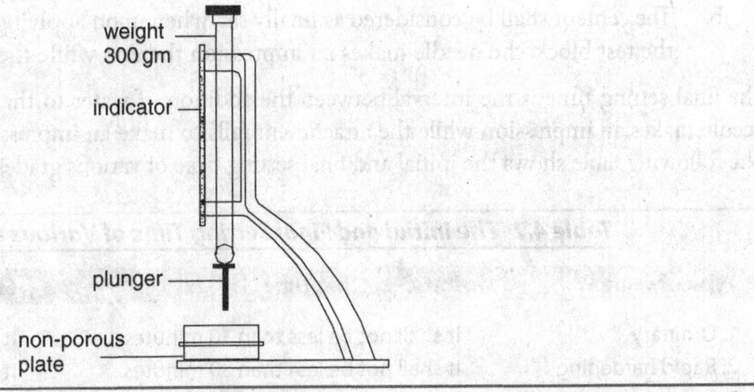

Figure 4.1 Vicat apparatus

Apparatus

1. It consists of a frame with a movable rod fitted with a cap.
2. A needle of 1 mm square cross section is attached to the lower end of the rod for the determination of initial setting time. The total weight of the rod along with the needle is 300 g.
3. Another needle like the above mentioned but with a hollow metallic attachment with a circular cutting edge of 5 mm diameter and having a 0.5 mm projection at the end is used to determine the final setting time.
4. A standard Vicat mould in which the specimen is allowed to set.

4.3.1 Initial setting time – procedure

a. Take 300 g by weight of cement and mix with 0.85 times the water required to give a paste of standard consistency.

b. Start the stop watch at the instant water is added to the cement.

c. Fill the Vicat mould with the cement paste and smooth the surface.

d. Place the square needle of cross section 1 mm to the moving rod of the Vicat apparatus.

e. Lower the needle gently bringing it in contact with the surface and quickly release allowing it to penetrate the paste.

f. In the beginning the needle will completely pierce the test block. Repeat the procedure in a fresh place until the needle, when brought in contact with the test block and released, fails to pierce the block for 5 mm measured from the bottom.

The initial setting time is the interval between the addition of water to the cement and the stage when the needle fails to pierce the test block for 5 mm measured from the bottom.

4.3:2 Final setting time – procedure

a. Replace the needle for initial setting time by the needle with an annular attachment for the final setting time.

b. The cement shall be considered as finally set, when upon applying the needle gently to the surface of the test block, the needle makes an impression thereon while the attachment fails to do so.

The final setting time is the interval between the addition of water to the cement and the time at which the needle makes an impression while the attachment fails to make an impression on the surface of the test block. The following table shows the initial and final setting time of various grades of cements.

Table 4.2 The Initial and Final Setting Time of Various Grades of Cements

Type of cement	Initial setting time	Final setting time
1. Ordinary	It shall not be less than 30 minutes.	It shall not be more than 10 hours.
2. Rapid hardening	It shall not be less than 30 minutes.	It shall not be more than 10 hours.
3. Low heat	It shall not be less than 60 minutes.	It shall not be more than 10 hours.

4.4 MANUFACTURE OF CEMENT

4.4.1 Wet process

In the earlier part of the century, from 1913 to 1960, the wet process was used for the manufacture of cement.

4.4.1.1 *Mixing of raw materials*

The calcareous materials such as limestones are crushed and stored in silos or storage tanks. The argillaceous materials, such as clay, are thoroughly mixed with water in a container known as wash mill and they are stored in basins. Now in correct proportions, the limestones from storage tanks and wet clay from basins are allowed to fall in a channel. This channel leads the material to grinding mills where they are brought to form a slurry. The grinding is carried out in either ball mill or tube mill or both. The slurry is lead to correcting basins where it is constantly stirred and at this stage the chemical compositions are adjusted as necessary. This corrected slurry is then stored in a different storage tank from where it is fed to the rotary kiln for burning.

4.4.1.2 *Burning*

The burning is carried out in the rotary kiln. The rotary kiln is formed of steel tubes whose diameter varies from 250 to 300 cm. The length varies from 90 to 120 m. It is laid at a gradient of 1 in 25 to 1 in 30. The kiln is supported at intervals by columns of masonry. A refractory lining is provided inside the kiln. It is arranged in such a way that the kiln rotates at 1–3 revolutions per minute about its longitudinal axis. The corrected slurry is charged into the rotary kiln for the wet process. Coal in finely pulverized form, fuel oil and gas are the common fuels for burning these kilns. The portion of the kiln near its upper end is known as dry zone and in this zone the water of the slurry is evaporated. As the slurry descends to the next zone, there is a rise in temperature from where the carbon dioxide from the slurry is evaporated. Small lumps known as nodules are formed at this stage. These nodules gradually pass through zones of rising temperature and ultimately reach the burning zone where temperature is around 1,500°C. In the burning zone, the calcined product is formed and nodules are converted into small, hard, dark, greenish blue balls which are known as clinkers. The size of the clinkers varies from 3 to 20 mm. Rotary kilns of small size are provided to cool down the clinkers and the cooled clinkers having temperature around 95°C are collected in containers of suitable sizes.

4.4.1.3 *Grinding*

The clinkers obtained from the rotary kiln are finely ground in ball mills and tube mills. During grinding, a small quantity, around 3-4 per cent, of gypsum is added. Gypsum controls the initial setting time of cement. If gypsum is not added, the cement would set as soon as water is added. After grinding, the product is stored in storage tanks and finally they are packed in bags of different types to ensure a 50 kg net weight of cement bag with ±200 g. Each bag contains 50 kg or about 0.035 m^3 of cement. The bags are automatically discharged from the packer to the conveyor belt to different loading areas and are carefully stored in the right place (Figure 4.2).

4.4.2 Dry process

Nowadays the dry process of manufacture of cement is most often adopted and this improves the quality of cement produced, with less consumption of power. In this process, the raw materials which are ground to about 25 mm size in crushers are dried by passing dry air over it. They are then pulverized to a very fine powder in ball mills and tube mills. This is done separately for each raw material and then they are mixed in the correct proportion and made ready for the feed of the rotary kiln.

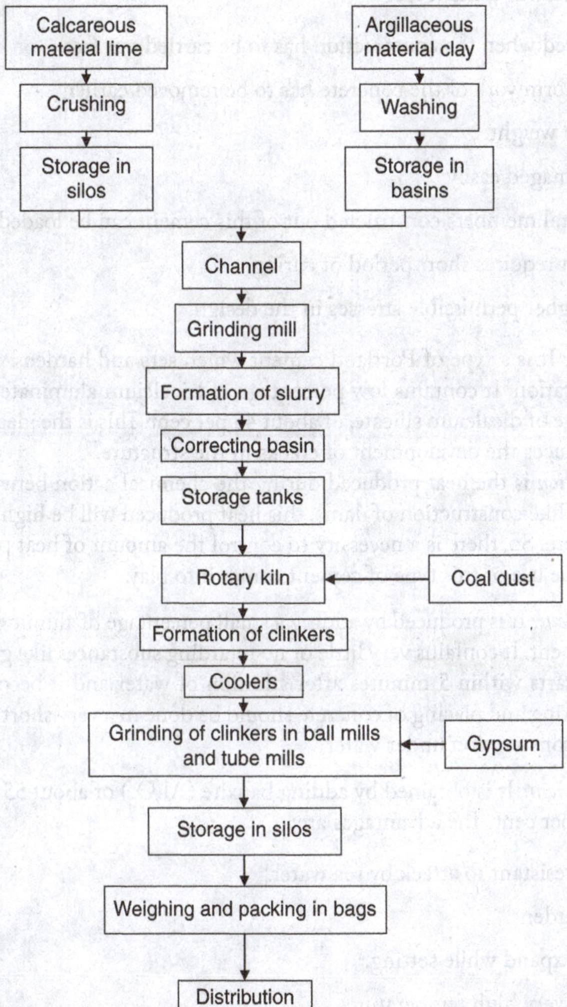

Figure 4.2 Schematic diagram of different processes involved in the manufacturing of cement

4.5 DIFFERENT TYPES OF CEMENT AND USES

1. *Ordinary Portland Cement:* It derives its name from the name of a stone (Portland) which resembles its colour. It is the most commonly used building material in mortar for masonry work, in mortar for plastering and pointing and as a binding medium in cement concrete, reinforced cement concrete and prestressed cement concrete construction.

2. *Rapid hardening cement:* The rapid hardening property is imparted to the cement primarily by burning at a higher temperature and secondly by finer grinding of the particles. The initial and final setting time of the cement is the same as ordinary cement, but it attains high strength in the early stages. It is useful in emergency situations as it develops the same strength in 4 days which ordinary cement acquires in 28 days. It is comparatively costlier than Ordinary Portland Cement. The uses and advantages of this cement are:

    a. It can be used when the construction has to be carried out fast.

    b. When the formwork of the concrete has to be removed earlier.

    c. It is light in weight.

    d. It is not damaged easily.

    e. The structural members constructed out of this cement can be loaded earlier.

    f. This cement requires short period of curing.

    g. It allows higher permissible stresses in the design.

3. *Low heat cement:* It is a type of Portland cement which sets and hardens with the evolution of very low heat of hydration. It contains low percentage of tricalcium aluminate, of about 5 per cent, and higher percentage of dicalcium silicate, of about 45 per cent. This is the ideal cement for construction of dams as it reduces the development of cracks in the structure.

    Heat of hydration is the heat produced during the chemical action between cement and water. In mass concreting like construction of dams, this heat produced will be high and will affect the stability of the structure. So, there is a necessity to control the amount of heat produced and it is in these situations that the use of this type of cement comes into play.

4. *Quick setting cement:* It is produced by adding a small percentage of aluminium sulphate and by finely grinding the cement. It contains very little or no retarding substances like gypsum. The setting action of the cement starts within 5 minutes after addition of water and it becomes hard in less than 30 minutes. The mixing and placing of concrete should be done in a very short time. This type of cement can be used for construction under water.

5. *High alumina cement:* It is obtained by adding bauxite (Al_2O_3) of about 55 per cent and lime (CaO) of about 35–45 per cent. The advantages are:

    a. It is highly resistant to attack by sea water.

    b. It rapidly hardens.

    c. It does not expand while setting.

    d. It can stand very high temperatures.

    e. It resists the action of frost.

The disadvantages are:

a. It cannot be used for massive concrete work.

b. It is much costlier.

c. Extreme care should be taken to see that it does not come in contact with ordinary cement or lime as it reduces the strength.

6. *Coloured cement:* This cement will produce a surface of desired colour and is manufactured by the addition of a small proportion of some colouring material, generally a mineral pigment to the clinker. The amount of colouring material may vary from 5 to 10 per cent. Chromium oxide gives green colour and cobalt imparts blue colour. Iron oxides in different proportions give brown, red and yellow colour and manganese dioxide produces black and brown colour.

7. *Expanding cement:* This cement is used to neutralize the effect of shrinkage of ordinary concrete. It is produced by adding an expanding medium like sulpho-aluminate and a stabilizing agent to the ordinary cement. It is used for the construction of water-retaining structures and also for repairing damaged concrete surfaces.

8. *Hydrophobic cement:* It contains admixtures which decrease the wetting ability of cement. The admixtures usually used are acidol, naphtenesoap, etc. These substances form a thin film around the cement grains. When water is added to this cement, the absorption films are torn off the surface and they do not in any way prevent the normal hardening of cement. However, in the initial stage the gain of strength is less as the hydrophobic films of cement grains prevent the interaction with water.

9. *Air entraining cement:* Air content of 2–6 per cent is introduced in the cement by grinding air entraining agents with the cement clinker during the manufacture of cement. The addition of air entraining agents introduces large amount of air which results in the formation of voids and increases the workability of concrete. The weight as well as the strength of the concrete is reduced.

10. *White cement:* White cement is manufactured from china clay and white chalk in place of limestone and clay. It is used as a decorative feature for high-quality plasterwork. The white colouring effect is due to the absence of iron oxide. The cement is about four times costlier than Ordinary Portland Cement. It has quick drying properties, high strength and superior aesthetic values. It is used in swimming pools where it replaces the use of glazed tiles with coloured shades, for moulding sculptures and statues, for painting garden furniture and for fixing marbles and glazed tiles.

11. *Blast furnace slag cement:* The iron and steel industry produces large quantities of blast furnace slag as a by-product. The slag is a waste product in the manufacturing of pig iron and it contains the basic elements of cement, namely alumina, lime and silica. The clinkers of cement are ground with 60–65 per cent of the slag. This cement has a slow rate of hardening and less heat of hydration. It is not affected by sea water and, hence, is used for marine structures. Its strength in the early days is less and, hence, requires longer curing period.

4.6 DIFFERENT GRADES OF CEMENT

Prior to 1987, there was only one grade of Ordinary Portland Cement which was governed by IS 269-1976. After 1987 higher-grade cements were introduced to India. The Ordinary Portland Cement was classified into three grades, namely 33 grade, 43 grade and 53 grade depending upon the strength of the cement at 28 days when tested as per IS 4031-1988. If the 28-day strength is not less than 33 N/mm^2, it is called 33 grade cement. If the 28-day strength is not less than 43 N/mm^2, it is called 43 grade cement. If the 28-day strength is not less than 53 N/mm², it is called 53 grade cement. But the actual strength obtained by the cement at the factory is much higher than the BIS specification. Table 4.3 shows compressive strength of different grades of cement.

Table 4.3 Compressive Strength of Different Grades of Cement

Sl. No.	Type of cement	Compressive strength 1 day min. MPa	3 days min. MPa	7 days min. MPa	28 days min. MPa
1	33 grade OPC (IS 269-1989)	N.S.	16	22	33
2	43 grade OPC (IS 269-1989)	N.S.	23	33	43
3	53 grade OPC (IS 269-1989)	N.S.	27	37	53

N.S. – Not specified

The compressive strength of Ordinary Portland Cement increases with time. For example, 33 grade OPC (IS 269-1989) acquires a compressive strength of 16 N/mm² at 3 days, 22 N/mm² at 7 days and 33 N/mm² at 28 days.

4.7 STORAGE OF CEMENT

Cement absorbs moisture from nature and gets hardened. So suitable precautions should be taken in storing cement.

An absorption of 5 per cent moisture means the cement becoming useless and so the cement is to be stored in a moisture-free atmosphere. It is advisable not to store cement in jute bags for a period of more than 3 months. The cement bags should be stored in piles of one above the other, at a minimum distance of 300 mm from the exterior walls. Between the piles, a passage of 900 mm width should be kept. The top and bottom of the piles should be covered and waterproofed for long storage.

Storage for longer periods makes the cement weaker, even under favourable conditions.

REVIEW QUESTIONS

1. What are the properties of cement?
2. Briefly discuss the ingredients of cement.
3. What is the setting time of cement and what are the factors affecting it?
4. How is the setting time of cement determined? Explain briefly.
5. What is the difference between initial and final setting time of cement?
6. How is cement manufactured in the wet process?
7. Draw a schematic diagram showing the different processes involved in the manufacture of cement.
8. What are the different types of cement?
9. Write short notes on
   a. Low heat cement
   b. Quick setting cement
   c. Rapid hardening cement
10. What are the different grades of cement and how is it stored?

chapter 5

Bricks

Manufacture of bricks is mostly a village industry. Bricks have been produced since the dawn of civilization in the sun dried form. The Great Wall of China was made of both burnt and sun dried bricks. Bricks have been used all over the world in every class and kind of building. In places where plenty of clay is available, brickwork is cheaper. The cost of construction work is less with bricks. Bricks resist fire and, hence, they do not easily disintegrate. The atmospheric effects are resisted by bricks of good quality.

5.1 COMPOSITION OF GOOD BRICK EARTH

The constituents of good brick earth are:

1. *Alumina:* A good brick earth should contain 20–30 per cent of alumina. It imparts the property of plasticity to the earth. An excess of alumina causes shrinkage and warping of bricks during drying and burning and it becomes too hard when burnt.
2. *Silica:* Silica forms 50–60 per cent of good brick earth. It is seen either in the free or combined state. In the free state, it is mechanically mixed with clay and in the combined form it exists in a chemical composition with alumina. The cracking, shrinking and warping of raw bricks are being prevented by the presence of silica. The durability of bricks depends upon the proportion of silica. An excess of silica destroys the cohesion between particles and the bricks become brittle.
3. *Lime:* A good brick earth should contain lime not exceeding 5 per cent. It should be present in a very finely powdered state in order to prevent the flaking of bricks. Lime prevents the shrinkage of bricks. An excess of lime causes the bricks to melt and, hence, to lose its shape.
4. *Oxides of iron:* A small quantity of the oxide of iron to the extent of 5-6 per cent is desirable in good brick earth. It imparts red colour to the bricks. But excess of lime makes the colour dark blue or blackish. On the other hand, if the quantity of lime is less, the bricks will be yellowish in colour. It also helps to fuse sand and, thereby, increases the hardness of bricks.
5. *Magnesia:* Presence of magnesia in small quantity imparts a yellowish tint to the bricks and decreases the shrinkage. But if in excess, it causes the decay of bricks.

The ingredients which are undesirable in the brick earth include excess of lime, the presence of iron pyrites, pebbles, alkalies, vegetation and organic matter.

5.2 MANUFACTURE OF BRICKS

The manufacture of bricks is carried out in a number of stages. It includes the following:

1. Selection and preparation of clay
2. Shaping and moulding of units

3. Drying
4. Burning

5.2.1 Selection and preparation of clay

As a practise, suitable deposits of clay are first located and thoroughly tested for the quality for brick making. Clay for bricks is prepared in the following order.

a. *Unsoiling:* The top layer of the soil is taken out. This is because the clay in the top layer is full of impurities and, hence, it is to be rejected for the purpose of preparing bricks.
b. *Digging:* The clay which is dug out is spread on a level ground, just little deeper than the general ground level. The height of the heap of clay is about 60–120 cm.
c. *Cleaning:* The clay should be made free from stones, pebbles and vegetable matter. If these particles are in excess, the clay is to be washed and screened, which is considered to be uneconomical.
d. *Weathering:* The softening of clay is done by exposing it to the atmosphere. The period of exposure varies from weeks to full seasons.
e. *Blending:* The clay is made loose and any ingredient to be added is spread out at its top. Blending indicates intimate mixing. A small portion of clay is taken every time for mixing.
f. *Tempering:* In this stage, the clay is brought to a proper degree of hardness and it is made fit for moulding. Water in the required quantity is added and the whole mass is mixed so as to form a mass of uniform character. A large-scale tempering is usually done in a pug mill. The process of grinding clay with water and making it plastic is known as pugging.

A pug mill consists of a conical iron tub with a cover at the top. It is fixed on a timber base which is made by fixing two wooden planks at right angles. The diameter of the pug mill at the bottom is about 80 cm and at the top is about 1 m. A vertical shaft with horizontal arms is provided at the centre of the iron tub. The small wedge-shaped knives of steel are fixed on the horizontal arms. Openings are provided at the top and bottom for charging clay and water and removing the mix respectively. The height of the pug mill is about 2 m.

5.2.2 Moulding

Moulding is the process of making rectangular-shaped brick units from properly tempered clay. The two types of moulding are

a. Hand moulding
b. Machine moulding.

5.2.2.1 Hand moulding

This is presently the most common method for brick manufacture. This is adopted where manpower is cheap and readily available. The moulds used for hand moulding are rectangular boxes made from well-seasoned wood or steel open at the top and bottom. Hand moulding is of two types:

i. Ground moulding
ii. Table moulding

Ground moulding

In this method, the ground is first levelled and fine sand is sprinkled over it. The mould is dipped in water and placed over the ground. The clay is pressed in the mould in such a way that it fills all the corners of the mould. Any surplus earth from the top of the mould is removed using a cutting wire or a metal with a sharp edge, which has to be dipped in water every time it is used. The mould is then lifted up and the brick is left on the ground. The mould is dipped in water and placed close by and another brick is moulded in the same way. If the mould is dipped in water every time, such preparation of bricks is known as slope moulded bricks. If fine sand or ash is sprinkled on the inside surface of the mould instead of dipping the mould in water, such bricks are called sand moulded bricks.

In pallet moulding, bricks of higher quality and with frogs are produced. The frogs are made using a pair of pallet boards and a wooden block. A frog is a mark of depth about 10–20 mm made on raw bricks during moulding. The frog is provided for mainly two purposes:

i. It serves as a key of mortar when other bricks are placed over it.
ii. It indicates the trade name of the manufacturer.

Table moulding

The process of moulding operations are carried out on a specially designed moulding table. The clay, the mould, water pots, stock board, etc. are placed on this table. The bricks are moulded similar to the ground moulding on the table. The cost of brick moulding increases when table moulding is adopted.

5.2.2.2 *Machine moulding*

The moulding can also be achieved by using machines. It is quite economical when bricks are produced in huge amounts. The machine moulding is broadly classified into two categories:

i. Plastic clay machines
ii. Dry clay machines

Plastic clay machines

The machines contain rectangular openings of size equal to the length and width of a brick. The pugged clay is placed in the machine and as it comes out through the openings it is cut into strips by wires fixed in frames. Hence, it is known as wire cut bricks.

Dry clay machines

In this machine, the strong clay is first converted to powder form. A small quantity of water is added to the stiff plastic paste. Such paste is placed in the mould and pressed by machines to form well-shaped hard bricks. These bricks are known as pressed bricks.

The machine moulded bricks have regular shape, sharp edges and corners; they are heavier and stronger than hand moulded bricks.

5.2.3 Drying

The drying of bricks is necessary, firstly to make them strong enough for rough handling during subsequent stages and secondly to save fuel during burning. For drying the bricks are laid longitudinally in stocks of bricks with width equal to two bricks. Drying of bricks is achieved by either natural or artificial methods.

The important facts to be remembered while drying of bricks are as follows:

a. The bricks are generally dried by natural process. But when bricks are to be rapidly dried, artificial drying may be adopted. In artificial drying, bricks are made to pass through driers in the form of tunnels or hot channels or floors. The tunnel driers are more economical than hot floor driers.
b. The brick in stocks should be arranged in such a way that sufficient air space is left between them for circulation of air.
c. Special drying yards should be prepared and accumulation of rainwater should be prevented.
d. The period of drying depends upon the prevailing weather conditions.

5.2.4 Burning

Burning of dried bricks is essential to develop the desired engineering properties, like hardness, durability and resistance to decay. Three chemical changes are known to take place in the brick earth during burning, namely dehydration, oxidation and vitrification.

Dehydration is completed within 425–750°C temperature range and it results in expulsion of most of the water from the bricks.

During oxidation, carbon and sulphur are eliminated as oxides, whereas the fluxes are also oxidized. Oxidation starts at the range of dehydration temperatures and is completed at about 900°C.

Vitrification is the extreme reaction and occurs when heating is carried out beyond 900°C. This is commonly not required in building bricks although in other clay products like sewer pipes it is necessary.

Burning of bricks is either done in clamps or kilns. Clamps are temporary structures while kilns are permanent structures. Clamps are adopted to manufacture bricks on a small scale while kilns are adopted to manufacture bricks on a large scale.

5.2.4.1 *Clamps*

The shape of the clamp is generally trapezoidal. The brick wall is constructed on the short end and a layer of fuel is placed on the prepared floor. The fuels generally used are cow dung, litter, husk of rice, wood, coal, etc. The thickness of fuel layer varies from 70 to 80 cm. The layer consisting of 4 or 5 courses of raw brick is then put up. Sufficient space for circulation of air is provided. Alternate layers of fuel and bricks are placed over this. The total height of the clamp is around 3-4 m. When nearly one-third of the height is reached, the lower portion of the clamp is ignited so as to burn the bricks in the lower part when the construction of bricks in the upper part is in progress. After construction is complete, it is completely plastered with mud in order to prevent the escape of heat. The clamp is allowed to burn for 1 or 2 months and cooling is also done for 2 months and later the burnt bricks are taken out.

5.2.4.2 *Kilns*

The kilns used for the manufacture of bricks are of two types:

i. Intermittent kilns
ii. Continuous kilns

Intermittent kilns

These kilns may be underground or overground in model. They are classified in two ways:

Intermittent up-draught kilns These kilns are in the form of rectangular structures with thick outside walls. Doors are provided at each end for loading and unloading of kilns. The flues are channels or passages which are provided to carry flames or hot gases through the body of the kiln. A roof is provided to protect the raw bricks from rain.

The quality of the bricks is not uniform; the bricks at the bottom are overburnt and at the top are under-burnt. The supply of bricks is not continuous and there is a considerable wastage of fuel in the kiln.

Intermittent down-draught kilns These kilns are rectangular or circular in shape. They are provided with permanent walls and a closed tight roof. The floor of the kiln has openings which are connected to a common chimney stacked through flues. They are so arranged that in this kiln the hot gases are carried through the vertical flues upto the level of roof and then released. As a result the bricks are evenly burnt and the performance is much better than intermittent up-draught kilns. Here, there is close control of heat and the bricks obtained are evenly burnt.

Continuous kilns

These kilns are continuous in operation where loading, firing, cooling and unloading are simultaneously carried out.

Bull's trench kiln This is one of the continuous type kilns. These kilns are rectangular, circular or oval shaped in plan. These kilns are constructed in a trench excavated on the ground. It may be fully underground or partially projecting above the ground. The outer and inner walls are to be constructed in bricks. Openings are provided on the outer walls to act as flue holes. Iron plates are used to divide the kiln into suitable sections. The fuel is placed in flues and is ignited after covering the top surface with earth to prevent the escape of heat. Usually, two movable iron chimneys are employed to form draught. The chimneys are placed in advance of the firing sections so that the warm gases leaving the chimney warm up the bricks in the next section. As the section has burnt, the flue holes are closed and allowed to cool down. Later the fire is advanced to the next section.

Hoffmann kiln This kiln is constructed under ground, is circular in plan and consists of a number of chambers. A permanent roof is provided so that the kiln can function even in the rainy season. The chamber in Hoffmann's kiln is provided with a main door for loading and unloading bricks. Communicating doors should act as flues in the open condition. A radial flue connected with a central chimney and fuel holes are also provided. The advantages are that the bricks are uniformly, equally and evenly burnt and that there is no air pollution in the locality. Also, there is saving in fuel and a high percentage of good bricks are produced.

Tunnel kiln This type of kiln is in the form of a tunnel, which is oval, circular or straight in plan. It contains a stationary source of fire. The raw bricks are placed in trolleys which are then moved from one end to another end of the tunnel. The raw bricks get dried and preheated as they approach the zone of fire and in the zone of fire the bricks are burnt and pushed forward for cooling. Later, after cooling, they are unloaded.

Table 5.1 shows the comparison between clamp and kiln burning.

Table 5.1 Comparison Between Clamp and Kiln Burning

Number	Item	Clamp burning	Kiln burning
1.	Structure	Temporary	Permanent
2.	Initial cost	Very low as no structures are to be built	Very high as permanent structures are to be built
3.	Cost of fuel	Low as grass, cow dung etc. is being used	High as coal dust is being used
4.	Quality of bricks	Percentage of good bricks is less, around 60%	Percentage of good bricks is more, around 90%
5.	Supervision	Not necessary throughout the process	Skilled supervision required
6.	Wastage of heat	More	Less
7.	Capacity	About 20,000–1,00,000 bricks at a time	25,000 bricks per day
8.	Suitability	For small scale	For large scale
9.	Time for burning and cooling	2–6 months	24 hours for burning and 12 days for cooling

5.3 SIZE AND WEIGHT OF BRICKS

Bricks are prepared in various sizes. If the bricks are too large, it is difficult to burn them and handle them. But, on the other hand, if the bricks are small, more quantity of mortar is required while placing.

In India, the standard size recommended by the Bureau of Indian Standards (BIS) is 19 cm × 9 cm × 9 cm and the size of brick including the mortar thickness is 20 cm × 10 cm × 10 cm, which is known as the nominal size of the brick.

The test carried out for inspecting the size is that 20 bricks of standard size (19 cm × 9 cm × 9 cm) are stacked length wise, along the width and along the height. For good quality bricks the results should be within the following permissible limits:

Length: 368–392 cm

Width: 174–186 cm

Height: 174–186 cm

The weight of 1 m^3 of brick earth is 18 kN. Hence, the average weight of a brick will be around 30–35 N.

5.4 QUALITIES OF A GOOD BRICK

A good brick should possess the following properties:

1. The brick should be uniform in shape and should be of standard size.
2. The brick when broken should show a uniform compact and homogeneous structure free from voids.
3. The brick should not absorb water more than 20 per cent for first-class bricks and 22 per cent for second-class bricks when soaked in cold water for a period of 24 hours.
4. The brick should be hard enough. No impression should be left when scratched.

5. The brick should not break into pieces when dropped from a height of 1 m.
6. The brick when soaked in water for 24 hours should not show deposits of white salts when allowed to dry in shade.
7. The brick should have low thermal conductivity and should be sound proof.
8. The crushing strength of brick should not be below 5.5 N/mm^2.
9. The brick should be table moulded, well burnt and free from cracks with sharp and square edges.
10. The colour should be uniform and bright.
11. The bricks should give a good metallic sound when struck with each other.

Concrete blocks, hollow blocks and bricks made of various materials like fly ash are being used successfully as substitutes.

5.5 FALG BRICKS

The rapid increase in the capacity of thermal power generation in India has resulted in the production of a huge quantity of fly ash, which is approximately 50 million tons per year. The prevailing disposal methods are not free from environmental pollution and ecological imbalance. Large stretches of scarce land, which can be used for shelter, agriculture or some other productive purposes, are being wasted for disposal of fly ash. Fly ash, lime and gypsum (FALG) can be used to make bricks and hollow blocks of adequate strength, an economical alternative to conventional burnt clay bricks will be available. Lime and gypsum are either available from mineral sources or can be procured from industrial wastes.

Fly ash bricks are made of fly ash, lime, gypsum and sand. Fly ash, lime sand and gypsum are manually fed into a pan mixer where water is added in the required proportion for intimate mixing. The proportion of the raw material is generally in the ratio of 60–80 per cent of fly ash, 10–20 per cent lime, 10 per cent Gypsum and 10 per cent sand, depending upon the quality of raw materials. The mixture is slow-setting pozzalona cement mix. After mixing, the mixture is shifted to the hydraulic/mechanical presses. A specially designed machine which gives a very high pressure load at slow rate (in the order of 280–350 kg/inch) is used to mould the bricks. Holding the pressure at specific times gives more strength to the finished product. The moulded bricks are then transferred to hydraulic-operated wooden pellets manually and stored in covered space for 3 days (minimum) for setting.

Then the bricks are taken to the yard for water curing for 15–20 days. Then it is sorted and tested before despatch. These can be extensively used in all building constructional activities similar to that of common burnt clay bricks. The fly ash bricks are comparatively lighter in weight and stronger than common clay bricks. Fly ash bricks are used in multi-storeyed apartment houses for non-load bearing purposes and in making curtain and partition walls of these houses. Use of fly ash bricks in this type of construction is meant mainly to achieve economy and make profits. The domestic buildings of low- or middle-income groups mostly have single or two-storied dwelling units. Therefore the cost effectiveness along with the strength and durability of fly ash bricks are very important for them.

5.6 FLY ASH BRICKS

Fly ash is one of the residues generated in the combustion of coal. It is generally captured from the chimneys of coal-fired power plants and is one of two types of ash that jointly are known as coal ash; the other, bottom ash, is removed from the bottom of coal furnaces. Depending upon the source and makeup of the coal being burnt, the components of fly ash vary considerably, but all fly ash includes substantial amounts of silicon dioxide (SiO_2) (both amorphous and crystalline) and calcium oxide (CaO).

Fly ash has been used for over 50 years to make concrete building blocks. They are widely used for the inner skin of cavity walls. They are naturally more thermally insulating than blocks made with other aggregates.

Fly ash bricks have been used in house construction since the 1970s. There is, however, a problem with the bricks in that they tend to fail. This happens when the bricks come into contact with moisture and a chemical reaction occurs causing the bricks to expand.

REVIEW QUESTIONS

1. What are the constituents of good bricks?
2. What are the different stages involved in the manufacture of bricks?
3. Write short notes on
   a. Moulding of bricks
   b. Drying of bricks
   c. Different types of kilns used for the manufacture of bricks
   d. Size of bricks
   e. Fly ash bricks
4. What are the qualities of a good brick?
5. Briefly discuss FALG bricks.

chapter 6

Tiles

Tiles can be defined as thin slabs or bricks, which are burnt in kilns. The tiles are thinner than bricks. Tiles are classified into two types: common tiles, which are available in different shapes and sizes and used for paving, flooring and roofing, and encaustic tiles, which are used for decorative purposes in floors, walls, ceilings and roofs.

6.1 DIFFERENT TYPES OF TILES

The different types of tiles are

1. Drain tiles
2. Flooring tiles
3. Roofing tiles

6.1.1 Drain tiles

Drain tiles are prepared in such a way that they retain their porous texture after burning. Hence, they are suitable to be laid in waterlogged areas. They allow water to pass. They are also used to convey irrigation water. These drains may be circular, semicircular or segmental.

6.1.2 Flooring tiles

The flooring tiles should be hard enough to resist wear and tear. They are thin tiles of thickness 12–50 mm and can also be adopted for ceilings. Colouring substances can be added to the clay during preparation to impart colour to floor tiles. Low-strength floor tiles can be used for fixing on the surface of walls. They are easier to lay as they are small in size and much lighter than mosaic and marbles. They do not require polishing. They are scratch, stain and damp proof in nature.

6.1.2.1 Wood tiles

Wood flooring requires a protective coating such as varnish or wax. This type is suitable for gymnasium, skating rinks and air-conditioned rooms.

6.1.2.2 Cork tiles

Different colours and designs are available. These are warm, quiet and resilient but not durable.

6.1.2.3 Cement concrete tiles

Plain concrete tiles, plain coloured tiles and terrazo tiles are the three different types of tiles coming in this category. These are easy to clean and shine well if the quality is good. Their cost is comparatively reasonable.

6.1.2.4 Magnesium flooring tiles

This flooring is used as a substitute for asphalt flooring.

6.1.2.5 Ceramic tiles

These are non-slippery and used in wet areas like bathrooms, kitchens, etc. They are available in a wide range of colours and textures. They are used in living rooms also.

6.1.3 Roofing tiles

They act as a covering to the roof. The important varieties of roofing tiles are as follows.

6.1.3.1 Allahabad tiles

They are tiles made from clay. The moulding is done under pressure in machines. Interlocking facility is attained by the projection provided in the tiles. These tiles can also be adopted for the hip, ridge and valley portions of the roof.

6.1.3.2 Corrugated tiles

Corrugations are provided for the side lap when they are placed in position. These tiles give a pleasing appearance. The placing of tiles gives an appearance of corrugated iron sheets.

6.1.3.3 Mangalore tiles

These tiles are provided with suitable projections so that they interlock with each other. These tiles are of flat pattern but special Mangalore pattern tiles are available for the hip, valley and ridge portions. The life of these tiles is estimated as about 25 years. These are red in colour and it is found that 15 Mangalore tiles are required for covering 1 m^2 of roof area.

These are the most commonly used roofing tiles in Kerala. The length of the tiles varies from 32 to 35 cm and width from 21 to 23 cm. Maximum water absorption percentage is 20 and minimum average breaking load is 1.00 kN. More than 40 per cent of the tiles get broken due to wrapping. The number of breakages can be reduced by adding a small amount of aluminium chloride to the clay.

6.1.3.4 Guna tiles

They are hollow, tapered, burnt tiles. They are conical in shape with a base of 100 mm diameter at the broader end and 75 mm at the narrower end. These may be made of suitable shapes, like parabolic, elliptical, etc.

6.1.3.5 Pan tiles

They are short and heavy. They are first moulded as flat sections and later given the required curvature by moulding in suitable forms. These tiles have a length between 30 and 40 cm and width between 20 and 30 cm.

6.1.4 Encaustic tiles

They are manufactured from ordinary clays with colouring materials and finer clays. The encaustic tiles consist of three layers: body which is made of coarser clay, face which comprises of 6 mm coat of finer clay and colouring material and back which is a thin coat of clay to prevent the tile from warping.

6.2 CHARACTERISTICS OF A GOOD TILE

1. It should possess uniform colour.
2. It should give an even and compact structure when seen on its broken surface.
3. It should be sound, hard and durable.
4. It should be regular in shape and size.
5. It should fit in properly when placed in the proper position.
6. It should be free from cracks, bends and warps.

6.3 PORCELAIN GLAZED TILES

The purpose for which glazing is done are:

1. To improve the appearance.
2. To produce decorative effects.
3. To provide smooth surfacing.
4. To protect the surface from the action of atmospheric agencies.
5. To make the particles durable.

The term porcelain is used to indicate fine earthenware which is white, thin and transparent. Clay of sufficient purity and possessing high degree of tenacity and plasticity is used for preparing porcelain. It is hard, brittle and non-porous. It is prepared from clay, feldspar, quartz and minerals. Porcelains are of two types – low-voltage porcelain and high-voltage porcelain. Low-voltage porcelain is prepared from dry process and high-voltage porcelain is prepared from wet process.

Porcelain enamelling is a common process for making the surface smoother, chemical proof and resistant to deterioration. The process involves fusing thoroughly the mixture of raw materials in a furnace and then grinding the mixture to a fine powdered stage. The material to be enamelled is then given a coating of this powder either directly by dusting the powder over the surface or by mixing the mixture with water and spraying on the surface. They are adopted for various uses like sanitaryware, electrical insulators, storage vessels, etc.

REVIEW QUESTIONS

1. Write short notes on
   a. Different types of tiles used for building construction
   b. Flooring tiles
   c. Mangalore pattern roofing tiles
2. What are the different types of tiles used for building construction?
3. What are the characteristics of a good tile?
4. What are the purposes of glazing of tiles?
5. How is glazing done for the tiles?
6. What is porcelain enamelling?

chapter 7

Timber

Timber denotes wood which is suitable for building or carpentry and for various engineering and other purposes. The word timber is derived from an old English word 'timbrian' which means to build. Timber or wood as a building material possesses a number of valuable properties, such as low heat conductivity, amenability to mechanical working, low bulk density and relatively high strength.

Timber has been a very important structural member from time immemorial. It has been extensively used as beams, columns and plates in construction in a variety of situations, such as foundation, flooring, stairs and roofing. Even today its use as a building material is quite popular, though it has to face tough competition from structural steel and reinforced concrete.

7.1 CLASSIFICATIONS OF TREES

Trees are classified into two types, namely,

1. Endogenous
2. Exogenous

7.1.1 Endogenous

In the endogenous trees, a plant grows by the addition of new cells only at the tip or end, i.e., the trees grow inwards and fibrous mass is seen in their longitudinal section. Such trees show very little branches. The timber from these trees has very limited engineering applications. The examples of endogenous trees are bamboo, cane, palm, etc.

7.1.2 Exogenous

These trees increase in bulk by growing outwards and distinct consecutive rings are formed in the horizontal section of such trees. These rings are known as annual rings, because one such ring is added every year. Such trees grow bigger in diameter as well. The timber, which is mostly used for engineering purposes, belongs to this category.

The exogenous trees are further classified as

a. Conifers

b. Deciduous

Conifers: The conifers are known as evergreen trees. These trees bear cone-shaped fruits. These trees yield soft woods, which are generally light in colour, resinous, light in weight and weak.

Deciduous: They are also known as broad leaf trees. The leaves of these trees fall in autumn and new ones appear in spring. Timber for engineering work is mostly derived from deciduous trees.

Trees can also be classified based on hardness into the following categories.

a. Hard wood

b. Soft wood

The soft wood forms a group of evergreen trees while the hard wood forms a group of broad leaf trees. Examples of soft woods are deodar, pine and other conifers. Hard woods include sal, mahogany, teak, oak, etc.

7.2 STRUCTURE OF A TREE

From the visibility aspect, the structure can be divided into two categories:

1. Macrostructure
2. Microstructure

7.2.1 Macrostructure

The structure of wood visible to the eye or at a small magnification is called a macrostructure. The following are the different components (Figure 7.1).

a. Pith: The pith or medulla is the innermost part, seen only in old and immature trees. Wood of this zone is of black, brown or grey appearance. It is normally found in the first year of growth of the tree.

b. Heartwood: The inner annual rings surrounding the pith constitute the heartwood. It indicates the dead portion of the tree. It does not take active part in the growth of a tree. However, it imparts rigidity to the tree and, hence, forms durable timber for engineering purposes.

c. Sap wood: The sap wood comprises of new and lighter cells that line nearer to the skin of the tree. It indicates recent growth and contains sap. It takes active part in the growth of a tree and the sap moves in an upward direction through it.

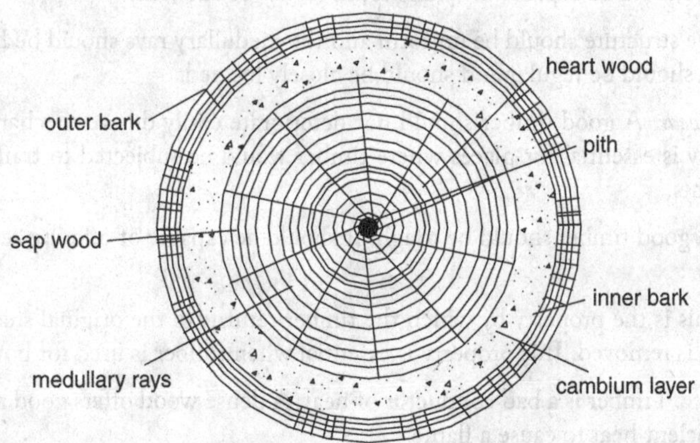

Figure 7.1 Cross section of an exogenous tree

d. Cambium: The thin layer of wood between the sap wood and the inner bark is known as cambium layer. If the bark is removed for any reason, the cambium layer gets exposed and results in the death of the tree.

e. Bark: It is the outermost zone and makes the skin of the tree. The function of the bark is to protect the inner tissue from heat, rain and injury. Sometimes a second thin membrane is also present inside the bark and it is called the inner bark.

f. Medullary rays: The thin radial fibres extending from the pith to the cambium is known as medullary rays. The function of it is to hold together the annual rings of hard wood and sap wood.

7.2.2 Microstructure

The structure of the wood apparent only at great magnification is called microstructure. A living cell consists of four parts, namely membrane, protoplasm, sap and core. The cells according to their function are classified into conductive cells, mechanical and storage cells.

The conductive cells serve mainly to transmit nutrients from the roots to the branches and leaves. The mechanical cells impart strength to the wood and the storage cells serve to store and transmit nutrients to the living cells.

7.3 QUALITIES OF GOOD TIMBER

The following are the qualities of a good timber:

1. *Appearance:* A freshly cut surface of timber should exhibit a hard and shining appearance.
2. *Colour:* The colour of the timber should be preferably dark. A light colour indicates low strength.
3. *Hardness:* A good timber should be hard, i.e., it should offer resistance when it is being penetrated by another body. The chemical present in heartwood and the density of wood imparts hardness to timber.
4. *Durability:* A good timber should be durable. It should be capable of resisting the action of fungi, insects, chemicals, physical agencies and mechanical agencies.
5. *Strength:* A good timber should be strong for working as a structural member such as joist, beams and rafter. It should be capable of taking loads slowly or suddenly.
6. *Structure:* The structure should be uniform and the medullary rays should be hard and compact. The annual rings should be regular and should be closely located.
7. *Mechanical wear:* A good timber should not deteriorate easily due to mechanical wear or abrasion. This property is essential for places where timber would be subjected to traffic, like wooden floors and pavements.
8. *Toughness:* A good timber should be tough. It should be capable of offering resistance to shocks due to vibrations.
9. *Elasticity:* This is the property by which the timber returns to the original shape when load causing deformation is removed. This property is essential when timber is used for bows, carriage shaft, etc.
10. Fire resistance: Timber is a bad conductor of heat. A dense wood offers good resistance to fire and it requires sufficient heat to cause a flame.
11. *Defects:* A good timber should be free from serious defects such as dead knots, flaws and shakes.

12. *Fibres:* Timber should have straight fibres.
13. *Shape:* A good timber should be capable of retaining the shape during conversion or seasoning.
14. *Smell:* A good timber should have a sweet smell.
15. *Sound:* A good timber should give a clear ringing sound when struck.
16. *Weight:* A timber with heavy weight is considered to be sound and strong.
17. *Working condition:* Timber should be easily workable. It should not clog the teeth of saw and should be capable of being easily planed or made smooth.

7.4 SEASONING OF TIMBER

Wood from freshly felled trees cannot be used in construction because it contains more moisture and is undesirable in many accounts. The water is to be removed before the timber can be used for any engineering purpose. This process of drying out the timber is known as seasoning of timber and the moisture should be extracted during seasoning under controlled conditions at a uniform rate from all parts of the timber. The remaining moisture, which cannot be extracted, should be uniformly distributed throughout the mass. The major objectives of seasoning are as follows:

1. To reduce the weight of the timber.
2. To increase the strength, stiffness and durability of the timber.
3. To make the timber easily workable.
4. To reduce the tendency of timber to crack, shrink and warp.
5. To allow the timber to burn readily if used as a fuel.
6. To make the timber fit for receiving treatment of paints, varnishes, etc.
7. To make the timber safe from the attack of fungi and insects.

7.4.1 Methods of seasoning

The two main methods of seasoning of wood are

a. Natural seasoning
b. Artificial seasoning

7.4.1.1 Natural seasoning

In this method, the seasoning of timber is carried out by natural air and, hence, it is referred to as air seasoning. Stacking of timber is done very carefully as to allow free circulation of air between the individual pieces. The length of stack is equal to the length of timber pieces. The stack is to be protected from fast blowing wind, rain and extreme heat. Hence, a roof of suitable material should preferably cover the stack. Loss of moisture from wood is by the simple process of evaporation.

This method of seasoning is cheap and simple and does not require skilled supervision, but the drying of different surfaces may not be even and uniform. The process is very slow and may take 1–4 years before the timber is properly seasoned.

7.4.1.2 *Artificial seasoning*

In artificial seasoning, the drying of different surfaces is even and uniform and it reduces the period of seasoning. Table 7.1 shows the comparison between natural and kiln seasoning. The various methods of artificial seasoning are as follows.

Table 7.1 Comparison Between Natural and Kiln Seasoning

Number	Item	Natural seasoning	Kiln seasoning
1	Moisture content	Difficult to reduce moisture content below 15%–18%	Moisture content can be reduced to any desired level.
2	Nature	Simple and economical	Expensive and technical
3	Quality of timber	More liable to attack of insects and fungi	Less liable to attack of fungi and insects
4	Space	Requires more space for stacking	Requires less space for stacking
5	Speed	Slow process	Quick process
6	Strength	Gives stronger timber	Gives a little weaker timber

i. *Boiling:* In this method, the timber is immersed in water and the water is then boiled. This is a very quick method, but it affects the elasticity and strength of wood and also this method proves to be costly.

ii. *Chemical seasoning:* In this method, the timber is immersed in a solution of suitable salt. It is then taken out and seasoned in the ordinary way. Here the chances of formation of external cracks are reduced.

iii. *Electrical seasoning:* In this method, high-frequency alternating current is used. This is the most rapid method of seasoning, but it is uneconomical as the initial and maintenance cost is very high.

iv. *Kiln seasoning:* In this method, the drying is carried out inside an airtight chamber or oven. Depending upon the mode of construction and operation, the kilns are of two types, namely, stationary kilns and progressive kilns.

   In a stationary kiln, the process of seasoning is carried out in a single compartment only. This kiln is adopted for seasoning timber, which requires a close control of humidity and temperature. It gives better results. In a progressive kiln, the carriage with timber sections travels from one end of the kiln to the other and the hot air is applied from the discharging end. It is used for seasoning timber on a large scale.

v. *Water seasoning:* This is a quick method and removes organic materials contained in the sap of timber. It, however, weakens the timber and makes it brittle.

7.5 COMMON TIMBERS USED FOR BUILDING WORK

1. *Teak:* Teak forms one of the most valuable timber types of the world. It is durable, fire resistant and moderately hard. It takes up good polish. It is not attacked by white ants and dry rot. Its colour is deep yellow to dark brown. It is suitable for ship building, making fine furniture and in building construction.

2. *Sal:* Sal is also one among the valuable trees. It resembles teak in several aspects. It is hard, fibrous and close grained. It does not take up a good polish. It is durable under ground and under water. It requires careful and slow seasoning. It is used for railway sleepers, bridges, structural work, etc.

3. *Rosewood or black wood:* It is dark in colour. It is strong, tough and close grained. It is handsome and takes up a good polish. It maintains its shape well and is available in large sizes. It is used for furniture of superior quality, cabinet work, ornamental carvings, etc.
4. *Babul:* It is one of the common trees in India, growing in all parts of the country. It is strong, hard, very durable and tough. Its colour is whitish red. It takes up a good polish. Its structural uses include use as beams and rafters in ordinary type buildings and also for fabrication of door and window frames and lintels. Babul is chiefly used for making cartwheels and bodies, tools and handles for agricultural instruments like ploughs.
5. *Jack:* Its colour is yellow when freshly cut and it darkens with age. It is compact, even grained and moderately strong. It is easy to work with and takes a good finish. It maintains its shape well. It is used for making plain furniture, door panels, boat construction, cabinets, musical instruments, etc.
6. *Mango:* The mango tree is well known for its fruits. Its colour is deep grey and is easy to work with. It is moderately strong. It is practically found all over India. It is used for making cheap furniture, toys, packing boxes, ship building, cabinets, etc.
7. *Deodar:* It is the most important tree providing soft wood. Its colour is yellowish brown. It is strongly scented, oily, very durable and polishes well. It is moderately strong. It possesses distinct annual rings. It is used for making cheap and rough furniture, railway carriages, railway sleepers, packing boxes, etc.
8. *Bamboo:* It is an endogenous tree. It is flexible, strong and durable. It is found in most parts of the country. It is used for scaffolding, thatched roofs, rafters, temporary bridges, fancy goods, etc.
9. *Mahogany:* Its colour is shining reddish brown. It takes a good polish. It is easy to work and is durable under water. It is used for making furniture, patterns, cabinets, ornamental paneling, fancy goods, etc.
10. *Palms:* It contains ripe wood in the outer crust. The colour of this ripened wood is dark brown. It is strong and durable and is found practically all over India. It is used for making furniture, roof covering, rafters, joists, etc.

7.6 TYPES OF PLYWOOD AND THEIR USES

Plywood is a specially processed type of wood, which consists of essentially an odd number of plies (or thin sheets called veneers) glued together in such a manner that the grain of adjacent layers is perpendicular to each other. The placing of plies normal to each other increases the longitudinal and transverse strength of plywood. The layers are held in position by the application of suitable adhesives.

In the simplest form of plywood, the board consists of three layers or three plies arranged in such a manner that

1. One ply lies at the centre and is referred to as the central ply.
2. One ply each lies on the upper and lower surface of the central ply. One of these is referred as 'face ply' and the other as 'back ply'.

Plywood should always contain an odd number of panels and this is essential to obtain a balanced product. The most important properties that make plywood so advantageous over the ordinary wood are as follows:

1. Its better resistance to shrinkage, swelling and warping, which is achieved by making these defects uniform in both the directions.
2. Its almost perfect freedom from splitting.
3. Its increased strength as compared to the same thickness of ordinary wood.

Plywood is used for various purposes such as ceilings, doors, furniture, partitions, panelling walls, packing cases, formwork for concrete, etc. Plywood, however, is not suitable in situations subjected to direct shocks or impacts.

Plywood is available in different commercial forms, such as batten board, laminboard, metal-faced plywood, multi-ply, three-ply and veneered plywood.

The batten board is a solid block with core of sawn thin wood. The thickness of the core is about 20–25mm and the total thickness is about 50 mm. These boards are light and strong. They do not crack or split easily. They are widely used for making partition walls, packing cases, furniture pieces, ceilings, shutters of doors and windows etc.

The laminboard is similar to batten board except that the core is made of multi-ply veneers. The external plies are of thick veneers and they are firmly glued with core to form a solid block. These boards have the same uses as that of batten boards.

In metal-faced plywood, the core is covered by a thin sheet of aluminium, copper, bronze, steel etc. This plywood is rigid.

Plywood prepared from more than three plies are designated as multi-ply. The number of veneers is odd. The thickness may vary from 6 to 25 mm or more.

Plywood made from three plies only is known as three ply. The thickness may vary from 6 to 25 mm or more.

7.7 VENEERS

A veneer is defined as an essentially thin sheet of wood. The thickness of veneers varies from 0.4 to 6 mm or more. They are obtained by rotating a log of wood against a sharp knife of rotary cutter or saw. The veneers after being removed are dried in kilns to remove the moisture. Another essential character of the veneers is that its thickness is uniform throughout the length and breadth. The veneers form the basic units of plywood manufacture. Important points regarding the veneers are:

1. The edges of veneers are joined and sheets of decorative designs are prepared.
2. The Indian timbers which are suitable for veneers are mahogany, oak, rosewood, teak etc.
3. The process of preparing a sheet of veneers is known as veneering.
4. The veneers are used to produce plywoods, batten boards and laminboard.
5. The veneers may be fixed on corners or bent portions. It creates an impression that the whole piece is made of expensive timber.
6. The veneers may be glued with suitable adhesives on the surface of inferior wood. The appearance of inferior wood is then considerably improved.

The veneers form the starting point of the manufacture of plywoods. Besides, they find extensive application in the manufacture of many other articles of utility. Their other uses are in cabinet making and face decorations.

7.8 PARTICLE BOARD

There exist today a wide variety of wood polymer composite boards that are commonly called particle boards. Basically, these materials are made from non-chemically processed dry wood particles of various shapes and sizes and either synthethic or adhesive material.

The basic materials from which particle boards are made can be broken down into four basic types:

1. Chips – these are wood particles typically used in pulp manufacture.
2. Flakes – these are mechanically sliced wood particles and are usually one-tenth of an inch thickness.
3. Ribbons – these are wood particles of a specific thickness but varied in length.
4. Shavings – these are wafer-like particles and typically are known to be thin of great widy, short length and consisting some what of ruptured fibres.

There are other types of lingo-cellulosic materials that are used in wood polymer composite materials that sometimes are lumped in with particle boards, but differ mainly in the amount of processing involved in preparing the woody material fore composite manufacture. Fibre board is a good example in that the degree of fibreization is much higher than the above materials through chemical and physical treatment.

REVIEW QUESTIONS

1. What are the properties of a good timber?
2. What are the qualities of a good timber?
3. What is meant by seasoning of timber?
4. What are the objectives of seasoning of timber?
5. How is timber seasoned?
6. What is plywood and how is it manufactured?
7. Write short notes on
   a. Structure of wood
   b. Heartwood and sap wood
   c. Cambium and bark of wood
   d. Veneers and particle boards
   e. Artificial seasoning of timber
   f. Common timbers used for building construction

chapter 8

Steel

Steel is an alloy of iron and carbon. It is highly elastic, ductile, malleable and weldable. Steel has high tensile and compression strength and also stands wear and tear much better.

8.1 USES OF STEEL IN BUILDING WORKS

Steel can be used for various purposes in building works.

1. As structural material in trusses, beams, etc.
2. As non-structural material for grills, doors, windows, etc.
3. In steel pipes, tanks, etc.
4. In sanitary and sewer fittings, rainwater goods, etc.
5. Corrugated sheets.
6. As reinforcement for concrete.

8.2 STEEL AS A REINFORCEMENT IN CONCRETE

Although plain concrete is very strong in compression, it is very weak in tensile strength. So, steel is being used in concrete as a reinforcement as it is equally strong in compression and tension.

The steel for reinforcing is generally in the form of round bars varying in diameter from 5 to 40 mm, sometimes bars of other forms as mentioned above are also used.

Reinforced Cement Concrete (RCC) is more rigid, highly durable and fire resistant. It possess high tensile strength and it is economical in ultimate cost.

8.3 MARKET FORMS OF STEEL

The following are the various forms in which steel is available in the market:

1. Angle sections
2. Channel sections
3. Corrugated sheets
4. Expanded metal
5. Flat bars
6. I-sections
7. Plates

8. Ribbed-torsteel bars
9. Round bars
10. Square bars
11. T-section

8.3.1 Angle sections

Angle sections may be of equal legs or unequal legs.

Equal angle sections are available in sizes varying from 20 mm × 20 mm × 3 mm to 200 mm × 200 mm × 25 mm. The corresponding weights per metre length are respectively 0.95 kg and 73.60 kg. Unequal angle sections are available in sizes varying from 30 mm × 20 mm × 3 mm to 200 mm × 150 mm × 18 mm. The corresponding weights per metre length are respectively 1.10 kg and 46.90 kg. Figure 8.1 shows an equal angle section of size 100 mm × 100 mm × 10 mm with weight per metre length as 14.90 kg and an unequal angle section of size 90 mm × 60 mm × 10 mm with weight per metre length as 11.00 kg.

Angle sections are extensively used in structural steel work especially in the construction of steel roof trusses and filler joist floors.

8.3.2 Channel sections

Channel sections consist of a web with two equal flanges.

A channel section is designated by the height of the web and width of the flange. These sections are available in sizes varying from 100 mm × 45 mm to 400 mm × 100 mm. The corresponding weights per metre length are respectively 5.80 kg and 49.40 kg. Figure 8.2 shows a channel section of size 300 mm × 100 mm with weight per metre length as 33.10 kg. The ISI has classified channel sections as junior channel, light channel and medium channel and accordingly they are designated as ISJC, ISLC and ISMC, respectively. Channel sections are widely used as structural members in steel-framed structures.

8.3.3 Corrugated sheets

These are formed by passing steel sheets through grooves. These grooves bend and press steel sheets and corrugations are formed on the sheets. These corrugated sheets are usually galvanized and they are referred to as galvanized iron sheets or GI sheets. These sheets are widely used for roof covering (Figure 8.3).

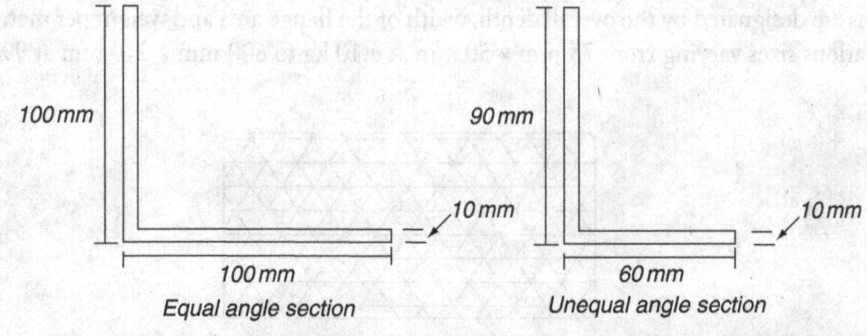

Figure 8.1 Equal angle and unequal angle sections

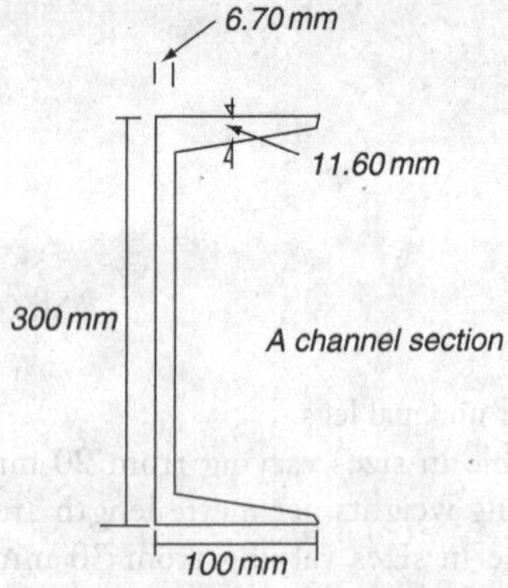

Figure 8.2 A channel section

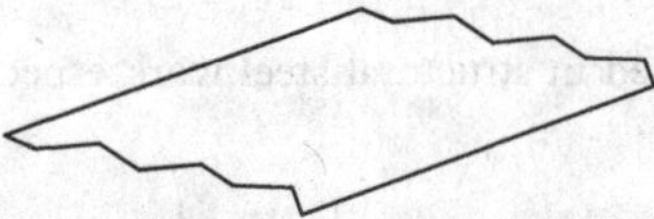

Figure 8.3 Corrugated sheets

8.3.4 Expanded metal

This form of steel is available in different shapes and sizes. It is prepared from sheets of mild steel, which are machine cut and drawn out or expanded. Expanded metal is widely used for reinforcing concrete in foundations, roads, floors, bridges, etc. It is also used as lathing material and for partitions (Figure 8.4).

8.3.5 Flat bars

These are available in suitable widths varying from 10 mm to 400 mm with thickness varying from 3 mm to 40 mm. They are widely used in the construction of steel grillwork for windows and gates.

8.3.6 I-sections

These are popularly known as rolled steel joists or beams. They consist of two flanges connected by a web.

I-sections are designated by the overall depth, width of the flange area and weight per metre length. They are available in various sizes varying from 75 mm × 50 mm at 6.10 kg to 600 mm × 210 mm at 99.50 kg. Figure 8.5

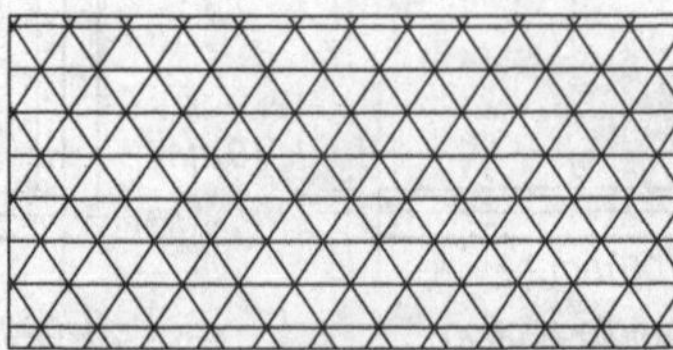

Figure 8.4 Expanded metal

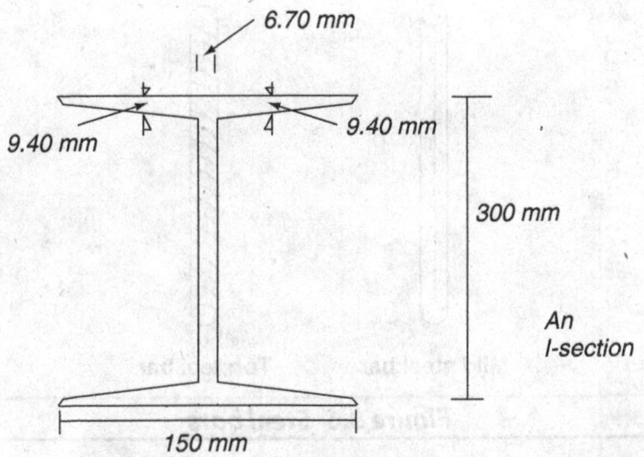

Figure 8.5 An I-section

shows a joist of size 300 mm × 150 mm at 37.70 kg. Wide flange beams are also available in sizes varying from 150 mm × 100 mm at 17.00 kg to 600 mm × 250 mm at 145.10 kg. Beams suitable for columns are available in H-sections which vary in sizes from 150 mm × 150 mm at 27.10 kg to 450 mm × 250 mm at 92.50 kg. The ISI has classified the I-sections into junior beams, light beams, medium beams, wide flange beams and heavy beams, and they are accordingly designated as ISJB, ISLB, ISMB, ISWB and ISHB respectively.

8.3.7 Plates

The plate sections of steel are available in different sizes with thickness varying from 5 to 50 mm. The corresponding weights per square metre are 39.20 kg and 392.50 kg, respectively. They are used mainly for the following purposes in structural steel work.

a. To connect steel beams for extension of the length.
b. To serve as tension members of steel roof truss.
c. To form built-up sections of steel.

8.3.8 Ribbed-torsteel bars

These bars are produced from ribbed-torsteel which is a deformed high-strength steel. These bars have ribs or projections on their surface and they are produced by controlled cold twisting or hot-rolled bars. Each bar is to be twisted individually and it is tested to conform to the standard requirements. Ribbed-torsteel bars are available in sizes varying from 6 to 50 mm diameter, with the corresponding weight per metre length as 0.222 kg and 15.41 kg. These bars are widely used as reinforcement in concrete structures, such as buildings, bridges, docks and harbour structures, roads, irrigation works, pile foundations and pre-cast concrete structures. The following are the advantages of ribbed-torsteel bars (Figure 8.6).

a. It is possible to bend these bars through 180 degrees without formation of any cracks or fractures on their outer surface.
b. It is possible to weld certain types of ribbed-torsteel bars by electric flash butt-welding or arc welding.
c. There is an overall reduction in reinforcement cost to the extent of about 30–40 per cent when these bars are used.

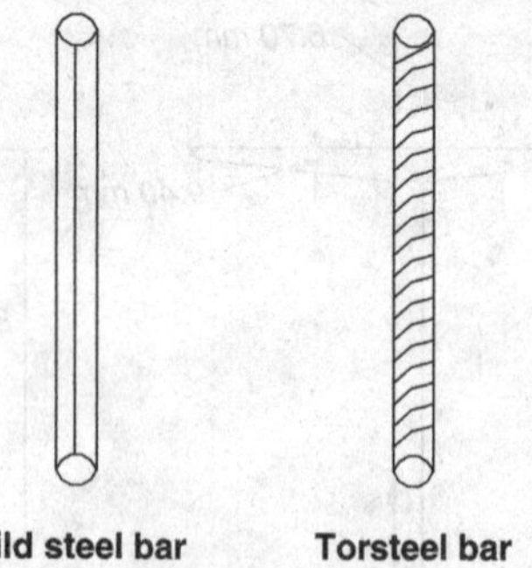

Figure 8.6 Steel bars

d. These bars are easily identified as they have got a peculiar shape.
e. These bars possess better structural properties than ordinary plain round bars. It is, therefore, possible to design them with higher stresses.
f. These bars possess excellent bonding properties and, hence, end hooks are not required.
g. They can be used for all major types of reinforced concrete structures.
h. They serve as efficient and economical concrete reinforcement.
i. When these bars are used, the processes of bending, fixing and handling are simplified to a great extent. It results in less labour charges.

8.3.9 Round bars

These are available in circular cross sections with the diameter varying from 5 to 250 mm. They are widely used as reinforcement in concrete structures, construction of steel grillwork, etc. The commonly used cross sections have diameters varying from 5 to 25 mm with the corresponding weights per metre length as 0.15 kg and 3.80 kg, respectively.

8.3.10 Square bars

These are available in square cross sections with sides varying from 5 to 250 mm. They are widely used in all construction of steel grillwork, for windows, gates, etc. The commonly used cross sections have varying sides from 5 to 25 mm with corresponding weights per metre length as 0.20 kg and 4.90 kg, respectively.

8.3.11 T-section

The shape of this section is like that of letter 'T' and it consists of a flange and web. Overall dimensions and thickness designate it. These sections are available in sizes varying from 20 mm × 20 mm × 3 mm to 150 mm × 150 mm × 10 mm. The corresponding weights per metre length are 0.90 kg and 22.80 kg, respectively. These sections are widely used as members of steel roof trusses and to form built-up sections.

In addition to the above standard shapes, rolled steel sections are also available in miscellaneous sections such as acute and obtuse angle sections, rail sections, tough sections and z-section. These miscellaneous sections are used to a limited extent in the structural steel work.

REVIEW QUESTIONS

1. What are the uses of steel in building works?
2. What are the different market forms of steel?
3. Write short notes on
   a. Steel-L-Section
   b. Steel I-Section
   c. Corrugated sheets
   d. Steel Channel sections
   e. Ribbed tor steel bars
4. What are the advantages of ribbed steel bars?
5. What are the main purposes of plates in structural steel work?

chapter 9

Aluminium

Aluminium occurs in abundance on the surface of the earth. It is available in various forms such as oxides, sulphates, silicates, phosphates, etc. But it is commercially produced mainly from bauxite (Al_2O_3, $2H_2O$) which is hydrated oxide of aluminium.

Bauxite deposits occur almost all over India, and particularly those at Balaghat, Kutni, Chhota Nagpur, Bhopal and Rewa are well known. The laterite deposits on the south-eastern and south-western coasts are also rich in bauxite content. Indian bauxite deposits compare favourably with those in any other parts of the world in the percentage of aluminium content.

9.1 MANUFACTURE

Aluminium is extracted from bauxite in two stages:

1. Purification of the ore to produce alumina (Al_2O_3)
2. Reduction of alumina by electrolysis

9.1.1 Purification of the ore to produce alumina (Al_2O_3)

The first process is called Bayer process, in which the alumina in the bauxite is dissolved in a solution of caustic soda, filtered and reprecipitated to obtain aluminium hydroxide and calcinated to form pure Al_2O_3.

9.1.2 Reduction of alumina by electrolysis

In the next process, called Hall-Hiroult's process, a steel tank is used, which is lined with refractory material with a second lining of carbon, which forms the cathode. Electric current is supplied through carbon electrodes with their lower ends dipped in the solution derived from Bayer process. These electrodes serve as anode. When a current of 8,000–20,000 amperes is switched on, the alumina, which is dissolved in the solution, is separated from the oxygen and deposits at the bottom of the steel tank in a molten state. The oxygen is liberated from carbon monoxide with carbon and burns off.

Pure aluminium is too soft for practical purposes. Hence, it is alloyed with copper, manganese, magnesium, zinc, silicon and nickel, which increases its tensile strength and hardness, while retaining the characteristics of lightness and durability. Aluminium is highly resistant to corrosion and is a very good conductor of heat and electricity. The commercial forms of aluminium are aluminium sheets, aluminium plates, aluminium foils, aluminium powder used as a pigment in the manufacture of paints, aluminium rods, aluminium bars and other aluminium structural parts.

9.2 PROPERTIES OF ALUMINIUM

1. It is a good conductor of heat and electricity.
2. It is a silvery white metal with a bluish tinge and it exhibits bright lustre on a freshly broken surface.

3. It is rarely attacked by nitric acid, organic acid or water. It is highly resistant to corrosion.
4. It is light in weight, malleable and ductile.
5. It is a soft metal.
6. The melting point of aluminium is about 658°C.
7. It possesses great toughness and tensile strength (1,960 kg/cm^2).
8. Its specific gravity is about 2.70.
9. It dissolves in hydrochloric acid.

9.3 USES

This metal is mainly used for making parts of aeroplanes, utensils, electric wires, window frames, door frames, corrugated sheets, structural members, foils, posts, panels, etc. Other uses of aluminium are as follows.

1. It is used as a reducing agent in the manufacture of steel.
2. It is used for making alloys and surgical instruments.
3. It is used in the casting of steel.
4. It is used in the manufacture of paints.
5. It is used in the manufacture of electrical conductors and capacitors.

After 1950s, this material has been tried for structural members. Its use as a structural material is, therefore, very recent. But it has made good progress during this period and research is yet going on to make maximum use of this metal as structural material.

9.4 ALUMINIUM AS A BUILDING MATERIAL

Aluminium is used as an important building material, especially in advanced countries. The following are the important properties of aluminium that make it useful as a building material:

1. *Air tightness:* A well-designed aluminium door, window, etc. is perfectly airtight and sealed for dust and rainwater when closed. This is a very important advantage in a modern air-conditioned building.
2. *Appearance:* The finished aluminium will be having a handsome appearance, and depending on the decorative style of the building, shades of various colours can be selected.
3. *Ease in fabrication and assembly:* As aluminium is comparatively soft and ductile, the fabrication of doors, windows, etc. can be easily carried out. An aluminium structure can easily be dismantled, transported and re-erected in different locations.
4. *Handling and transport cost:* Aluminium is very light and, hence, the handling and transport cost is very low.
5. *High corrosion resistance:* Aluminium has excellent corrosion resistance and it can resist weathering actions as well. It can also withstand extremely humid and hot dry conditions.
6. *High scrap value:* The scrap value of aluminium is very high, and as it hardly deteriorates, it enjoys a high resale value.

7. *Maintenance cost:* Due to the high corrosion resistance of aluminium, its maintenance cost is very low.

8. *Noise control:* Aluminium is an excellent reflector of electromagnetic and sound waves. An aluminium building is, therefore, less affected by external noises as compared to buildings made from other materials.

9. *Reinforcement:* Its use as reinforcement, instead of steel, will reduce foundation cost, increase durability and reduce the total cost.

10. *Roofing:* Corrugated sheets for roofing, window frames, etc. and many other general purpose building components can be made out of aluminium.

Duralumin is one of the most important alloys most extensively used in the aircraft industry, containing the following percentage of alloying metals, copper – 4.0 per cent, magnesium – 1.0 per cent, manganese – 0.5 per cent, silicon and iron – each 0.7 per cent and tin – 0.3 per cent. It has a specific gravity of 2.85 (compare this with 7.88 of iron) and possesses a tensile strength of 4,400 kg/cm^2, which is high as that of mild steel. It is susceptible to heat treatment and has the property of age hardening, i.e., once heated and quenched and left to itself, it acquires hardness by ageing in 2-3 days, when it offers considerably more resistance to corrosion.

REVIEW QUESTIONS

1. Write short notes on the extraction of aluminium using Hall-Hiroult's process.
2. Write short notes on duralumin.
3. Why is aluminium alloyed?
4. What are the properties of aluminium?
5. Why is aluminium used as a building material?

chapter 10

Paints and Varnishes

Paint is a substance used as the final finish to all surfaces and as a coating to protect or decorate the surface. Paint is a pigmented opaque material that completely covers and hides the surface to which it is applied. Paint is available in oil-based and water-based formulae.

It is used as a protective coating and is normally sprayed/brushed on. Paint prevents corrosion. It is a combination of pigments with suitable thinners or oils to provide decorative and protective coatings.

Painting protects a surface from weathering effects and also prevents corrosion of metals. Paint consists of two ingredients:

- A base of solid matter which helps to obscure the surface.
- A liquid vehicle, which carries the solid matter and allows it to be evenly spread on the surface. The vehicle also forms a binder for the solid matter and causes it to adhere to the surface.

10.1 CHARACTERISTICS OF AN IDEAL PAINT

1. When applied to the surface, paint should form a thin film of uniform nature.
2. The surface of the paint should not show cracks after drying.
3. The colour of the paint should withstand the adverse environmental conditions for a long time.
4. It should have an attractive and pleasing appearance.
5. It should be fairly cheap and economical.
6. It should be easily applicable with a brush or spraying devices.
7. It should possess maximum adhesion capacity to the material over which it is intended to be used.
8. It should have ideal resistance to corrosion and protect the material over which it is used.
9. The application of paint should not affect the health of workers.
10. It should be such that it dries in a reasonable time and not very rapidly.
11. It should possess good spreading or covering power, as it determines the cost.

10.2 INGREDIENTS OF PAINT

Paints are prepared by mixing the different ingredients in well-balanced proportions. Paint consists of the following ingredients.

10.2.1 Base

Base is a solid substance that forms the body of the paint and usually consists of finely divided suitable material. It determines the character of the paint and imparts durability to the surface. The function of the base material is to make the ultimate film harder and elastic, prevent formation of shrinkage cracks and protect the surface from moisture as well as ultra violet rays. Table 10.1 gives the common bases used for paints.

Table 10.1 Common Bases Used for Paints

No.	Name	Description
1.	White lead	It possesses good bulk and is the most commonly used base. It is dense, permanent and waterproof. This is a carbonate of lead and it forms the base of lead paints.
	Red lead	This is an oxide of lead and it forms the base of lead paints. It is suitable for painting iron surfaces and for providing a priming coat to the wood surfaces.
	Zinc white	This is an oxide of zinc and it forms the base of all zinc paints. It is smooth, transparent and non-poisonous. It is costly, less durable and difficult to work with.
	Oxide of iron	This is an oxide of iron and it forms the base of all iron paints. It is cheap and durable. It mixes easily with the vehicle. It is generally used for prime coat of iron surfaces.
	Titanium white or Titanium dioxide	This material possesses intense opacity. It is non-poisonous and provides a thin transparent film. It is used as an undercoat for enamel. It is unaffected by heat and acids and has very high covering power.
	Antimony white	This is nearly similar to the titanium white.
	Aluminium powder	This forms the bulk of aluminium paints. It is generally used for a priming coat to new woodwork. It prevents cracking and warping of wood.
	Lithophone	This is a mixture of zinc sulphide and barium sulphate. It is similar in appearance to the oxide of zinc. When exposed to daylight, it changes colour. Hence it is used for interior work of inferior nature.

10.2.2 Vehicles

Vehicles are the liquid substances which hold the ingredients of paint in liquid suspension (Table 10.2). They serve mainly two purposes:

a. To make it possible to spread the paint evenly and uniformly on the surface in the form of a thin layer.

b. To provide a binder for the ingredients of a paint so that they may stick or adhere to the surface.

10.2.3 Driers

These act as catalysts. These are substances that are used with a view of accelerating the rate of drying of the paint film. A drier absorbs oxygen from the air and transfers it to the vehicle, which in turn gets hardened. Driers may be either in the form of soluble driers or paste driers. Soluble driers are compounds of cobalt, lead, manganese, etc. dissolved in linseed oil or some other volatile liquid. Paste driers are compounds of the same metal.

Table 10.2 Materials Commonly Used as Vehicles and Their Properties

Number	Name	Description
1.	Linseed oil	This is the most common material used as a vehicle of paint. It is extracted from flax seeds. It is used in various grades.
	a. Raw linseed oil	This oil is thin and pale. It requires more time for drying and is used for interior work of inferior nature.
	b. Boiled linseed oil	This oil is thicker and darkly coloured than raw oil. It dries quickly. It is used for exterior surfaces.
	c. Pale boiled linseed oil	This is similar to boiled linseed oil except that it does not possess a dark colour. It is more suitable for painting plastered surfaces.
	d. Double boiled linseed oil	This oil dries very quickly and is suitable for external work. It however requires a thinning agent like turpentine.
	e. Stand oil	This oil dries slowly and provides a durable, clear and shining finish.
	f. Tung oil	This oil is far superior to linseed oil and is used for preparing paints of superior quality.
2.	Poppy oil	This oil is prepared from poppy seeds. It dries slowly but its colour lasts long. It is used for making paints of delicate colours.
3.	Nut oil	This oil is extracted from ordinary walnuts. It is nearly colourless and dries rapidly. It is used for ordinary work as it is cheap.

Litharge, red lead and sulphate of manganese can also be used as driers. Litharge is the most commonly used drier. Red lead is less effective than litharge. The following precautions should be taken while using driers:

a. A drier should not be added until the paint is about to be used.
b. More than one drier should not be used in a mixture.
c. The driers need not be used with pigments that dry well.
d. The driers should not be used unnecessarily nor in excess especially in the finishing coat as they may destroy the elasticity of the paint and injure its colour.

10.2.4 Colouring pigments

These are the colouring materials added to the paints in order to impart a desired shade and colour. They are added in the finely powdered state. Table 10.3 shows the colouring pigments that are used for a particular colour.

Table 10.3 Colouring Pigments Used for a Particular Colour

Colour of paint	Pigments
Black	Graphite, lamp black, ivory black, vegetable black
Blue	Indigo, Prussian blue
Brown	Burnt amber, raw amber
Green	Chrome green, copper sulphate
Red	Carmine, red lead, vermilion red
Yellow	Chrome yellow, yellow ochre, zinc chrome

10.2.5 Solvents

They are volatile substances that are added to the paint in order to make their application easy and smooth. Solvents help the paint in penetrating through the porous surface. The most commonly used solvents are spirit and turpentine. Solvents are also known as thinners from their function of thinning the original paint which is highly viscous. They are added to reduce the property of paint blistering.

They also help the paint to penetrate through the porous surface and they increase the spreading power of the paint. After application, the solvent evaporates and the resulting surface is more even and smooth.

10.3 TYPES OF PAINTS

1. *Enamel paints:* These paints are available in numerous shades. They mainly consist of white lead or zinc white, resinous matter and petroleum spirit. Their formation into hard, impervious, decay resistant enamel-like surface soon after application protects it from being affected by acids, alkalies, fumes and gas, hot and cold water, etc. They can be used for internal as well as external purposes.
2. *Cement paints:* These include a variety of paints in which cement is the main constituent responsible for the hardness and durability of the painted surface. They are available in dry, powder form. Cement paints are waterproof. It is desirable to provide cement paints on a rough surface rather than smooth surface because its adhesion power is more on rough surface than on smooth surface. It proves to be economical as compared to oil paints. They are suitable for painting fresh plasters having high alkalinity because cement paints are not likely to be attacked by the alkalinity of the masonry surface. It is not necessary to remove the existing paint for the application of new paint.
3. *Oil paints:* They are generally applied in three different layers with varying composition. These are termed as primes, undercoats and finishing coats. The dampness of the wall affects the life of the oil paint; hence, it must not be applied during damp weather. This is the ordinary paint which is cheap, fairly workable and possesses the qualities of opacity, good appearance, durability and resistance to weathering effects.
4. *Cellulose paints:* They are prepared from the nitro cotton, celluloid sheets, photographic films, etc. The cellulose paints harden by evaporation of thinning agents. The surface painted with cellulose can be washed and cleaned easily. They are a little more costly than other paints.
5. *Aluminium paints:* The finely ground aluminium is suspended in either quick drying spirit varnish or slow drying oil varnish as per requirement. As the spirit or oil evaporates, a thin film of aluminium is formed on the surface. These paints form a better protective surface over steel and iron. They are impervious to moisture and possess high electrical resistance. They have a good appearance and are visible in darkness.
6. *Emulsion paints:* These paints contain polyvinyl acetate, synthetic resins, etc. It is easy to apply and is retained for a long period and can be cleaned easily by water. For a rough plastered surface, a thin coat of cement paint may first be applied to smoothen the surface. It is necessary to have a sound surface to receive the paint.
7. *Anticorrosive paints:* They consist of oil and a strong drier. The pigments such as chromium oxide, lead or zinc chrome is taken and after mixing it with a small quantity of very fine sand is added to the paints. They are cheap and last for a long duration. They are black in colour, usually.
8. *Synthetic rubber paints:* These paints are prepared from resins. They have an excellent chemical resistant property. They can be applied to surfaces which may not be completely dry. They offer good resistance

to water and are not affected by heavy rains. They dry very quickly. They are not affected by weather and sunlight and are quite easy to apply.

10.4 USES OF PAINT

1. It protects the surface from weathering effects of the atmosphere and the actions by other liquids, fumes and gases.
2. It prevents decay and corrosion in metal.
3. It is used to give good appearance to the surface.
4. It provides a smooth surface.

10.5 VARNISH

Varnish may be defined as a homogeneous liquid containing essentially a resinous substance dissolved in suitable oil or volatile liquid. The objectives of varnishing are as follows:

1. It protects the unpainted wooden surfaces from the actions of atmospheric agencies.
2. It brightens the appearance of wooden surfaces.
3. It protects painted surfaces from atmospheric reactions.

10.6 INGREDIENTS OF VARNISH

The ingredients of varnish are as follows:

1. *Resins:* The natural resins include resin, copal and shellac. The synthetic resins that are prevalent in varnishes are phenolic resins, alkyl and vinyl resins.
2. *Driers:* The presence of driers accelerates the process of drying. The common types of driers used are white copper and lead acetate.
3. *Solvents:* Depending on the nature of resins the solvents are decided. The solvents used in the varnish industry include mainly boiled linseed oil and turpentine; alcoholic solvents like methyl and ethyl alcohol are also commonly used. Woodnaphtha is a cheap variety of resin.

10.7 CHARACTERISTICS OF IDEAL VARNISHES

1. It should not shrink or show cracks after drying.
2. It should make the surface glossy.
3. It should dry rapidly and present a finished surface which is uniform in nature and pleasing in appearance.
4. The colour of the varnish should not fade in course of time.
5. The varnish should be tough, hard and durable.

10.8 TYPES OF VARNISHES

Depending on the solvent, varnishes are classified into:

1. Oil varnish: Linseed oil is used as the solvent in this type of varnish. They are obtained by adding amber and copal in linseed oil. They form a hard and durable surface even though they dry slowly. They are specially adopted for exposed works which require frequent cleaning.
2. Turpentine: Turpentine is used as the solvent in this type of varnish. This varnish is quick drying but not resistant to weathering and, hence, it is used for interior work only. They possess light colours.
3. Spirit varnish: Spirit varnish is also known as French varnish. Methylated spirits of wine are used as solvents in this type of varnish. This varnish is quick drying but not resistant to weathering. They are generally used for furniture. They are obtained in the desired colour by addition of the required pigment.
4. Water varnish: Shellac is dissolved in hot water and enough quantity of ammonia or potash or soda is added so that the shellac is dissolved. They are used for delicate internal work.

REVIEW QUESTIONS

1. What do you mean by paint? Give the characteristics of an ideal paint.
2. What are the ingredients of paint? Explain in detail.
3. What is the role of a drier in paint? Give some examples for driers.
4. What are the general colouring pigments used in the manufacture of paints?
5. What are the types of paints available in market? Explain in detail.
6. What are the uses of paints and varnishes?
7. What are the ingredients of varnish? Also give its classification.

chapter 11

Miscellaneous Building Materials

There are various other materials used for the construction of buildings. An attempt is made here to introduce the various miscellaneous materials that are used in building works.

11.1 GLASS

Glass is being used as an engineering material. It is a mixture of metallic silicates, one of which is usually that of an alkali material. It is amorphous, transparent or translucent. It may also be considered as a solidified super-cooled solution of various metallic silicates having infinite viscosity. For the purpose of classification, glass may be grouped into the following three categories:

1. Soda-lime glass
2. Potash-lime glass
3. Potash-lead glass

Glass is not a single compound. It is very difficult to give any particular chemical formula for it. Numerous varieties of glass have been developed recently and it is possible to make glass lighter, safer and stronger as compared to any other building material. Glass is being used as a building material for purposes such as walls and ceilings, windows, furniture, bathroom fittings, looking glass, etc.

The chemical formulas for the three groups of glass, as classified above, are as follows:

Soda-lime glass: Na_2O, CaO, $6SiO_2$

Potash-lime glass: K_2O, CaO, $6SiO_2$

Potash-lead glass: K_2O, PbO, $6SiO_2$

11.1.1 Properties of glass

a. It absorbs, refracts or transmits light.
b. It can take up a high polish.
c. It has no definite crystalline structure.
d. It has no sharp melting point.
e. It is affected by alkalies.
f. It is an excellent electrical insulator.
g. It is available in beautiful colours.
h. It behaves more as a solid than most solids, in the sense that it is elastic. When the elastic limit exceeds, it fractures instead of deforming.

i. It is capable of being worked in many ways, such as blown, drawn and pressed.

j. It is difficult to cast it in large pieces.

k. It is extremely brittle.

l. It is usually unaffected by air or water.

m. Ordinary chemical reagents do not attack it.

n. It is possible to weld pieces of glass by fusion.

o. It is transparent and translucent.

p. When it is heated, it becomes softer and softer with the rise in temperature and is ultimately transformed into a mobile liquid.

11.1.2 Uses of glass as a building material

a. Glass can be used for window panels.

b. Glass blocks can be used for partitions up to 6 m for insulation.

c. Sheet glass can be used for glazing.

d. Structural glass can be used for insulation, panel walls, wall facings, enclosures, etc.

e. Potash lead glasses are used for making electric bulbs.

f. Tinted glass can be used for decorative glassworks.

g. Fibre glass reinforced plastics can be used to construct furniture, lampshades and bathroom fittings.

11.2 RUBBER

Rubber is an important engineering material and is widely used as a building material. The general uses of rubber are as follows:

1. Rubber is used as a gasket to make doors and windows airtight.
2. Rubber latex, a solution of rubber hydrochloride, is used for bonding rubber to metal, wood and similar surfaces.
3. Rubber is an important material used for repair of building works.
4. Rubber is used as a sealant in water-retaining structures.
5. Rubber can be used as a joint filler – cellular rubber and granulated cork compounds are used as fillers.
6. Rubber flooring and carpet mats are used in buildings.
7. Rubber coatings and linings are used for corrosion protection in offshore engineering.
8. Rubber can be used as sound absorbers.

11.3 PVC

PVC or polyvinyl chloride is the most versatile plastic and the use of PVC pipes in buildings is becoming popular day by day. It is possible to substantially change or modify the properties of PVC resin by the technique of compounding, i.e., addition of other additives to PVC. It is, thus, possible to prepare a PVC rigid pipe. PVC is obtained by polymerization of vinyl chloride. Water, kerosene oil, acids and chemicals have no effect on PVC. Hence, it is effectively useful for carrying sewage water and rainwater. PVC pipes are faster to install than other materials. PVC pipes have an expansion coefficient of 0.001°C. PVC sheets of thickness 0.1–12.5 mm can be used as a building construction material. They can also be used as a water proofing material.

11.4 PLASTER OF PARIS

To meet specific requirements in the finishing coats, special finishing materials like plaster of paris is used in building construction. The following are the main uses of plaster of paris in building construction.

1. It is used for repair of holes and cracks in plastered surfaces.
2. It is very effectively used for ornamental works.
3. It is used in combination with ordinary lime for building works(interior works).

11.5 DAMP PROOFING MATERIALS

Prevention of dampness in a building is achieved by using a suitable damp proofing material.

The properties of damp proofing materials are:

1. It should be impervious to moisture.
2. The damp proofing property should remain constant with lapse of time.
3. Damp proofing material should be stable in loaded and unloaded conditions.

11.5.1 Different types of damp proofing materials

a. Cement mortar: Cement mortar in the ratio 1:3, in layers of 2-3 cm with a suitable damp proofing compound, such as cico, impermo, etc.

b. Cement concrete: Cement concrete 1:2:4, 6 to 15 cm thick, with or without 3–5 per cent damp proofing compound.

c. Bitumen: A 3 mm thick layer of bitumen is placed on a bed of concrete or mortar. Bitumen is a by-product of petroleum distillation, which is used for roadwork.

d. Bituminous felt: This material is available in rolls of normal wall thickness.

e. Metal sheets: Copper and aluminium sheets are used as damp proofing material.

11.6 ADHESIVES

An adhesive is a substance which is used to join two or more parts so as to form a single unit. Adhesives are largely used in the manufacture of veneers, plywoods and other laminated wooden products and for fixing wall and ceiling linings, tiles and other floor coverings. The application of adhesives has many advantages over the conventional methods of bolting, riveting and welding. Glue is a general term which is used to indicate an adhesive substance. Following are the various types of adhesives.

1. *Albumin glue:* It is a glue of better quality. It is not attacked by water. It is used for making furniture.
2. *Animal protein glue:* It is obtained by boiling waste pieces of skins, bones, etc. of animals in hot water. Animal glue develops strong and tough joints and it is easy to apply. But it is affected by damp and moist conditions.

3. *Glue from natural resins:* It is prepared from natural resins. It is used for labelling, building paper, etc.
4. *Glue from synthetic resins:* This glue is based on synthetic resins. It may be either thermo-setting glue or thermo-plastic glue. The thermo-setting glue becomes permanent once it is set. The thermo plastic glue can be made plastic again by reheating.
5. *Nitrocellulose glue:* It is prepared from pyroxylin, which is a nitrated cellulose. It is derived by treating cellulose with nitric acid. It produces films, which strongly adhere to glass.
6. *Rubber glue:* It is prepared by dissolving rubber in benzene. It is used for joining rubber, plastics, glass, etc.
7. *Special glue:* It is specially prepared to join metals. Cycle weld is a modified form of rubber and it is used to join aluminium sheets. Araldite is another variety of special glue. It is used to join light metals.
8. *Starch glue:* It is prepared from vegetable starch. It has good strength in dry conditions. But it is not moisture resistant. It is cheap and is used from interior quality of plywood.
9. *Vegetable glue:* It is prepared from natural gums and starches. It is used for preparing paper board articles, labelling, etc.

11.6.1 Advantages of adhesives

a. A wide variety of combinations in joining is possible.
b. It can be used for bonding the surface of glass, metal, plastics and wood.
c. It creates a massive effort.
d. It is possible to prevent corrosion between different metals joined by adhesives.
e. It produces adequate strength.
f. The process of applying adhesives is easy, economical and speedy.

11.6.2 Disadvantages of adhesives

a. It is not possible to adopt any adhesive for all substances. Depending upon the properties of substances to be joined, a suitable adhesive has to be selected.
b. The adhesive substance does not become strong immediately after its application.
c. The adhesive substance generally does not remain stable at high temperatures.

11.7 COST-EFFECTIVE MATERIALS

11.7.1 Fly ash

Fly ash is an industrial waste from thermal power stations using coal as boiler fuel. Disposal of fly ash poses an operational constraint and is an environmental hazard. Annual generation of fly ash is 30 million tonnes. Fly ash can be used:

a. As a replacement of cement by 20 per cent by weight in RCC works.

b. In lean concrete mixes and cement mortars.

c. To make good burnt bricks by mixing 15–20 per cent of it with clay.

d. To make stabilized bricks by mixing it with asphalt and sand in proper proportions. Some quantity of sulphur can be added for bonding.

11.7.2 Soil-cement blocks

These are a mixture of pulverized soil and measured amount of Portland cement and water compacted to high density. These reduce cement consumption in construction.

11.7.3 Stabilized mud blocks

Any material used for wall construction should possess adequate compressive strength and erosion resistance. These properties can be imparted to the mud blocks by stabilizing it by cementitious admixtures, such as cement and lime and by compaction.

Lime is relatively cheap compared to cement and locally available in many places. So pressed soil lime blocks are better for rural applications.

11.7.4 Stone blocks

Random rubble masonry walls are widely used where building stone is abundantly available and cheaper than local bricks. The new method developed uses pre-cast stone masonry blocks of 30 cm × 20 cm × 15 cm nominal size. The blocks have one face with stone texture and weigh 18 kg. Their use reduces wall thickness from 30 to 20 cm, saving material cost and increasing the usable floor area.

a. Their use avoids the requirement of skilled labour.

b. They require only 1.5 cm plastering thickness instead of 2.5 cm required for random rubble masonry and that too only on one side.

11.7.5 Lime-based stone masonry blocks

Stone masonry building blocks made with lime-based binders possess many advantageous properties including compressive strength sufficient for use in load-bearing walls of 3–5 storey heights.

Stone masonry building blocks have the following advantages:

a. Convenient to use

b. Economic

c. Speed up construction work

d. Require less amount of mortar

e. Compressive strength increases continuously for long periods, even after 28 and 60 days

f. Can be used for the construction of load-bearing walls

g. Show good resistance to rain penetration and are subject to hardly any volume change or cracks with ageing.

11.7.6 Uses of lime

a. Easy to handle
b. Does not require any further processing at site
c. Good workability, less shrinkage, good strength resistance to moisture and freedom from major cracks
d. Replaces the use of cement by 50 per cent
e. Reduces the overall cost of construction

11.7.7 Concrete hollow blocks

Concrete hollow blocks have the following cost-effective characteristics:

a. Material economy
b. Lightweight material
c. Fast rate of construction
d. Good heat tightness
e. Good heat and sound insulation

Their advantages over solid masonry are:

a. Increased size of individual units reduce the construction time
b. The above implies few joints and, thus, increased compressive strength and saving in cement mortar
c. It reduces the self-weight of masonry

11.7.8 Fibre reinforced concrete (FRC)

It has superior fatigue, crack and impact resistance. It has greater durability and is used in roads, pavements, air fields, etc.

11.7.9 Sand lime bricks

These are made from silicious sand and hydrated lime with just sufficient water to allow them to be moulded under pressure. The bricks are cured in an autoclave under saturated steam.

The advantages are as follows:

a. Economical as less energy is used in their manufacture, i.e., 1/3 of fuel is used compared to that for a brick.
b. No need of clay and the land used for brick manufacture can be used for agricultural purpose.
c. In some places removal of sand leaves a good cultivable land.

11.7.10 Ferrocement

Highly versatile form of RCC in which cement mortar is reinforced with layers of continuous and relatively small wire meshes. It is considered as a cheap construction material. It has a unique combination of stiffness, high strength and serviceability. It is a homogeneous material and cracks only at higher loads.

The advantages are:

a. Easy to make, maintain and repair
b. This technique improves the engineering properties of the material such as failure, tensile and flexure strength, toughness, fatigue resistance, etc.
c. Mouldable and is of one-piece construction

11.7.11 Fibre reinforced polymers (FRPs)

FRPs are typically organized in a laminate structure, such that each lamina (or flat layer) contains an arrangement of unidirectional fibres or woven fibre fabrics embedded within a thin layer of light polymer matrix material. The fibres, typically composed of carbon or glass, provide the strength and stiffness. The matrix, commonly made of polyester, epoxy or nylon, binds and protects the fibres from damage and transfers the stresses between fibres.

The strength properties of FRPs collectively make up one of the primary reasons for which civil engineers select them in the design of structures. A material's strength is governed by its ability to sustain a load without excessive deformation or failure. The response of FRPs to axial compression is reliant on the relative proportion in volume of fibres, the properties of the fibre and resin and the interface bond strength. FRP composite compression failure occurs when the fibres exhibit extreme (often sudden and dramatic) lateral or sideways deflection called fibre buckling.

Among FRP's high strength properties, the most relevant features include excellent durability and corrosion resistance. Furthermore, their high strength-to-weight ratio is of significant benefit; a member composed of FRP can support larger live loads since its deadweight does not contribute significantly to the loads that it must bear. Other features include ease of installation, versatility, anti-seismic behaviour, electromagnetic neutrality, excellent fatigue behaviour and fire resistance.

There are three broad divisions into which applications of FRP in civil engineering can be classified: applications for new construction, repair and rehabilitation applications and architectural applications. FRPs have been used widely by civil engineers in the design of new construction. Structures such as bridges and columns built completely out of FRP composites have demonstrated exceptional durability and effective resistance to effects of environmental exposure. Pre-stressing tendons, reinforcing bars, grid reinforcement and dowels are all examples of the many diverse applications of FRP in new structures. One of the most common uses for FRP involves the repair and rehabilitation of damaged or deteriorating structures. The many applications for which FRP can be used have also been discovered. These include structures such as siding/cladding, roofing, flooring and partitions.

REVIEW QUESTIONS

1. How is glass classified? Give the chemical formula for each category?
2. What are the properties of a good glass?
3. What are the general uses of glass as a building material?
4. What are the general uses of rubber as a building material?
5. What are the main uses of plaster of paris in building construction?
6. What is damp proofing material? What are its properties?

7. What is an adhesive and explain the various types of adhesives?

8. Explain the advantages and disadvantages of adhesives.

9. Write short notes on

    a. Fly ash as a cost-effective material
    b. Stone blocks as a cost-effective material
    c. Ferrocement as a cost-effective material
    d. Fibre reinforced material
    e. Ferrocement
    f. Fibre reinforced polymers

Part-II

Building Construction

chapter 12

Component Parts of a Building

The basic requirements a building should satisfy in design and performance are:

i. It must be strong enough to withstand the loads coming on it including the self-weight, live load, wind load and earthquake load.

ii. It must not deflect under the loads.

iii. It must give comfort and convenience to the inhabitants.

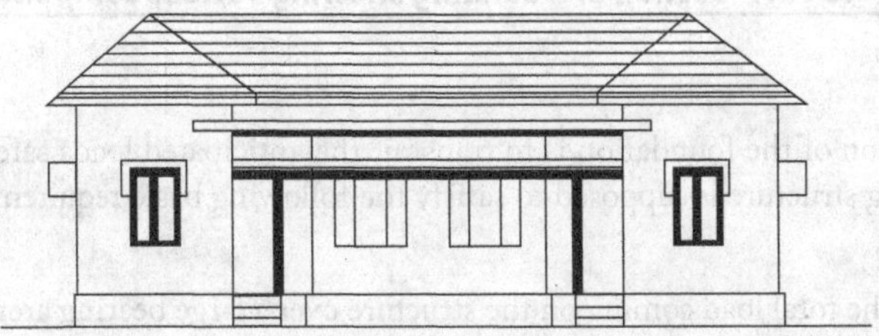

Front elevation drawing of a building

12.1 BUILDING COMPONENTS AND THEIR BASIC REQUIREMENTS

A building broadly consists of three parts:

1. Foundation
2. Plinth
3. Superstructure

12.1.1 Foundation

The foundation is the most critical part of any structure and most of the failures are probably due to faulty foundations. Hence, it is highly essential to secure good foundation to maintain the stability of the structure. A good foundation must remain in position without sliding, bending, overturning or failing in any other manner.

The foundation of any structure should be laid much below the surface of the ground in order to attain the following:

a. To secure a good natural bed.

b. To protect the foundation courses from atmospheric influences.

c. To increase the stability of the structure against overturning due to wind uplift.

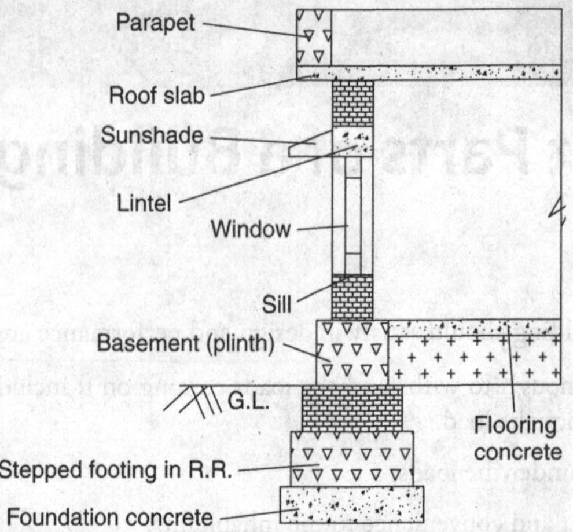

Figure 12.1 Section of a building showing various components

The primary function of the foundation is to transmit the anticipated loads safely to the soil below. The foundation in a building structure is supposed to satisfy the following basic requirements in their design and construction:

a. To distribute the total load coming on the structure over a large bearing area so as to prevent it from any movement.

b. To load the bearing surface or area at a uniform rate so as to prevent it from any movement.

c. To prevent the lateral escape or movement of the supporting material or alternatively to ensure the stability of the structure against sliding.

d. To secure a level or firm natural bed upon which to lay the course of masonry and also support the structure.

e. To increase the stability of the structure as a whole to prevent it from overturning or sliding against the disturbing forces such as wind, rain and frost.

12.1.2 Plinth

This is the portion of the structure between the surface of the surrounding ground and the surface of the floor immediately above the ground. The thickness of the plinth wall depends upon the weight of the superstructure and the width of the foundation concrete. The minimum height of the plinth is usually kept as not less than 4.5 cm.

The plinth wall should satisfy the following requirements in a building structure in its design and construction:

a. To transmit the load of the superstructure to the foundation.

b. To act as a retaining wall so as to keep the filling in position below the raised floor or the building.

c. To protect the building from dampness or moisture.

d. To enhance the architectural appearance of the building.

12.1.3 Walls and piers in superstructure

The primary function of the wall is to enclose or divide space. Piers are usually in the form of a thickened section of a wall, placed at intervals along the wall to take concentrated vertical loads or to provide lateral support to the wall. These walls may be built of different materials such as brick or stone masonry, plain concrete and reinforced masonry.

A load-bearing wall should satisfy the following requirements:

a. *Strength:* A wall should be strong enough to take up the loads safely. The loads coming in the wall include its own weight, weight by superimposed loads and bilateral pressures like wind.

b. *Stability:* It should be stable against overturning by lateral forces and buckling caused by excessive slenderness.

c. *Weather Resistance:* All the external walls whether load bearing or panel constructions should provide adequate resistance to rain, sun and wind.

d. *Fire Resistance:* The walls should offer sufficient resistance to fire as they behave as vertical barriers for spread of fire in the horizontal direction.

e. *Heat Insulation:* It should be possible for walls to attain insulation against heat.

f. *Sound Insulation:* The walls should be made of such materials and by such technique so as to insulate the building against sound.

g. *Privacy and Security:* The walls should provide sufficient privacy and afford security against theft.

12.1.4 Floors

The main function of a floor is to provide support for occupants, furniture and equipment of a building, and the function of providing different floors is to divide the building into different levels for creating more accommodation within the limited space (Table 12.1).

A floor should satisfy the following requirements:

a. *Strength and Stability:* All the floors, whether basement, ground or upper should be strong enough to support the floor covering and other superimposed loads.

b. *Durability and Damp Prevention:* The floors provide a clean, smooth, impervious, durable and wear-resisting surface.

c. *Heat Insulation:* Insulation against heat should be provided in case of ground and basement floors, especially when suspended and ventilated timber floors are used.

Table 12.1 Criteria for Calculating Floor Areas and Height of Structures

Type of building	Cubic contents per capita (m³)	Floor area per capita (m³)
Residential buildings	9	2.5–9
Dormitories	12–15	3-4
Educational buildings	4.5–7.5	1-2
Institutional buildings	30	8–10
Industrial buildings	7.5	2–2.5

d. *Sound Insulation and Fire Resistance:* The insulation against sound and fire should be provided in the case of upper floors as they act as horizontal barriers for the passage of sound and fire in a vertical direction.

12.1.5 Doors and windows

The main function of doors in a building is to serve as a connecting link between internal parts and to allow free movement to the outside of the building. Windows are generally provided for proper ventilation and lighting and their number should be determined according to the requirements.

Doors and windows should satisfy the following requirements:

a. *Weather Resistance:* They should be strong enough to resist the adverse effects of weather.

b. *Sound and Thermal Insulation:* They should be capable of being made air tight to achieve insulation against sound and heat.

c. *Damp Prevention and Termite Prevention:* They should not be affected by white ants and the moisture penetration as this will reduce the strength and durability.

d. *Fire Resistance and Durability:* They should offer fire resistance and should be durable.

e. *Privacy and Security:* They should offer sufficient privacy without inconvenience or trouble and security against theft.

12.1.6 Sills, lintels and weather shades

Window sills are provided between the bottom of the window frame and the wall below to protect the top of the wall from wear and tear. The openings are provided in the wall of a building to accommodate the doors and windows. The actual frame of a door or window is not strong enough to support the weight of the wall above the opening and a separate structural element is, therefore, introduced between the top of the window frame and the wall coming over it. This is known as the ***lintel***. Weather shades are generally combined with lintels of windows to protect them from the weathering agencies.

12.1.7 Roofs

A roof is the uppermost part of a building whose main function is to enclose the space and to protect the same from the effects of weather elements. A good roof is just as essential as a safe foundation. As a well-designed foundation secures the building against destruction starting at the bottom, similarly a good roof affords protection for the building itself and what the building contains and prevents destruction from the top.

A roof should satisfy the following requirements:

a. *Strength and Stability:* The roof structure should be strong and stable enough to take up the anticipated loads safely.

b. *Weather Resistance:* The roof covering should have adequate resistance to resist the effects of weather elements.

c. *Heat Insulation:* The roof should provide adequate insulation against heat.

d. *Sound Insulation:* The roof should have adequate insulation against sound from external sources.

e. *Fire Resistance:* The roof should offer an adequate degree of fire resistance in order to give protection against the spread of fire from any adjacent buildings and to prevent early collapse of the roof.

The form of construction should also be such that the spread of fire from its source to other parts of the building by way of roof cannot occur.

f. *Day Lighting:* The roof provides daylight in buildings with large floor area.

12.1.8 Steps and stairs

A step usually consists of a tread and riser supported by strings. A stair is a structure consisting of a number of steps leading from one floor to another. Location of stairs in all types of residential and public buildings should be such as to afford the easiest and quickest service possible to the building. The main function of the stairs is firstly to provide a means of communication between the various floors. Secondly, it also acts as an escape from the upper floors in the event of fire.

Steps and stairs should satisfy the following requirements:

a. *Strength and Stability:* The stairs should be designed like floors such that they are strong and stable enough to carry the anticipated loads safely due to the weight of the people using them and also the weight of the furniture or equipment being carried up or down through them.

b. *Fire Resistance:* The stairs should be made of fire-resisting material and should be connected to different floors, such that they provide safe means of escape in the event of fire.

c. *Sound Insulation:* The stairs should have adequate insulation against sound from external sources.

d. *Weather Resistance:* The stairs if exposed to open air should offer sufficient resistance to weather elements such as rain and heat.

e. *Comfort:* The proper design of steps and proper location of stairs in a building offer several advantages such as comfort and efficiency in vertical movement, natural light and ventilation and safety in emergency.

12.1.9 Finishes for walls

Finishes of several types such as pointing, plastering, painting and distempering and decorative colour washing are applied on the walls. The main functions of these finishes are as follows:

a. They protect the structure, particularly the exposed surfaces, from the effects of weather.

b. They provide a true, even and smooth finished surface and also improve the aesthetic appearance of the structure as a whole.

c. They cover up the unsound and porous materials used in the construction.

12.2 PLANNING REGULATIONS

12.2.1 Building line and control line

The 'building line' refers to the line of building frontage, i.e., the line up to which the plinth of the building adjoining the street or an extension of street or on a future street may lawfully extend. This line is often known as set back or front building line and is laid down in each case, parallel to the plot boundaries, by the authority, beyond which nothing can be constructed towards the plot boundaries. Certain buildings, such as factories and business centres that attract large number of vehicles, should further set back a distance apart from the building line. The line that accounts for this extra margin is known as 'control line'(Table 12.2).

Table 12.2 Distances of Building Lines and Control Lines

Type of road	In open and agricultural country		Ribbon development along approaches		Actual limits in urban area	
	Building line (m)	Control line (m)	Building line (m)	Control line (m)	Building line (m)	Control line (m)
National and state highways	30	36	18	30	30	45
Major district roads	24	45	9	15	15	24
Other district roads	15	24	6	9	9	25
Village roads	12	18	6	9	9	15

12.2.2 Built-up area

The built-up area of a plot means the plot area minus the area of open spaces in it (Table 12.3).

12.2.3 Open space requirements

For attaining natural ventilation, open spaces are necessary around the building. As per the Kerala Building Rules 1984, the minimum open space requirements are given as below.

a. For buildings upto 10 m height.
   i. Every building shall have at least front and backyards of a minimum of 3 m.
   ii. A clear open space of not less than 1.5 m shall be there on the sides other than front and rear.

b. For buildings above 10 m and upto 25 m height, there shall be an increase in the minimum open spaces at the rate of 0.5 m per every 3 m exceeding 10 m or fraction of it.

12.2.4 Size of the rooms

The minimum area required for the individual rooms is also specified. For example, the minimum area required for a kitchen is specified as 4.5 m^2. The bathroom is to have a minimum size of 1.5 m × 1.2 m or an area of 1.8 m^2, while a combined bathroom and water closet is to have a minimum floor area of 2.8 m^2, with a minimum width of 1.2 m.

The minimum height of an individual room is specified as (i) 2.75 m for habitable rooms, (ii) 2.2 m for bathrooms and WC and (iii) 2.55 m for kitchens. The specifications for maximum height of buildings are given in Table 12.4.

Table 12.3 Limitation of Built-up Area for Residential Buildings

Area of plot (m^2)	Maximum permissible built-up area
Less than 200	60% with two-storeyed structure
200–500	50% of the site
500–1,000	40% of the site
More than 1,000	33.33% or 1/3rd of the site

Table 12.4 Maximum Height of Buildings

Width of street (m)	Height of the building
Say W	Height = 1.5 W + front open space
Upto 8	1.5 times the width of street
8–12	Not more than 12 m
Above 12	Not more than the width of street and in no case more than 24 m

12.2.5 Lighting and ventilation of rooms

The aggregate area of doors and windows is to be not less than one-fourth of the room area. The total area of windows is to be one-tenth of the floor area for dry hot climate and one-sixth of the area for wet climate.

In addition to the above, ventilators are to be provided for at least 0.3 m^2 area or at the rate of 0.1 m^2 for every 10 m^3 of space.

The depth of the foundation is to be 0.7–1 m below the finished ground for single-storeyed buildings and for double-storeyed buildings it is to be 1–1.3 m. The plinth of the building is to be kept higher than the surrounding ground level by 30 cm or greater.

REVIEW QUESTIONS

1. What are the various basic requirements a building should satisfy in design and performance?
2. What are the main components of a building?
3. What are the objectives of having a good foundation for a building?
4. What are the basic requirements of a foundation?
5. What do you mean by plinth wall of a building and what are its requirements?
6. What are the requirements of a load-bearing wall?
7. What are the requirements of doors and windows of a building?
8. What do you mean by roof of a building and what are its requirements?
9. Write short notes on planning regulation of a building explaining built-up area, room size and open space requirements.

chapter 13

Foundation

13.1 PURPOSE OF PROVIDING A FOUNDATION

A foundation is that part of the structure that is in direct contact with the ground. It is that part of the structure that transmits the weight of the structure to the ground. The foundation is, therefore, a connecting link between the structure proper and the ground that supports it. Foundation transfers the load of the structure to the soil below on a large area. It prevents the differential settlement by evenly loading the substrata. Foundations are generally built of bricks, stones, concrete, steel, etc. The selection of the material and type of foundation depends upon the type of the structure above and the underlying soil.

Foundation design must take into account the effects of construction on the environment, such as pile driving vibration, pumping and discharge of groundwater, the disposal of waste materials and operation of heavy mechanical plants. Foundations must be durable to resist the attack of harmful substances.

A foundation is designed such that:

1. The soil below does not fail in shear.
2. The settlement is within a safe limit.

The pressure that the soil can safely withstand is known as the 'allowable bearing pressure'.

Foundations may be broadly classified into two categories.

1. *Shallow foundation:* A shallow foundation transmits the load to the strata at a shallow depth.
2. *Deep foundation:* A deep foundation transmits the load at a considerable depth below the ground surface.

13.2 BEARING CAPACITY OF SOIL

The bearing capacity of soil is the maximum load per unit area which the soil or material in foundation, may be rock or soil, will support without displacement. Very often, a structure fails by unequal settlement or differential settlement.

The allowable bearing capacity or the safe bearing capacity of a soil is obtained by dividing the ultimate bearing capacity by a certain factor of safety and is used in the design of foundation. It is suggested that a factor of safety of 2 for buildings in ordinary construction and a factor of safety of 2.5 or 3 for heavy constructions be adopted.

13.2.1 Bearing capacity of various types of soil

Types of rocks and soils	Presumptive safe bearing capacity	Remarks
1) Rocks	***Kg/cm^2***	–
a) Rocks (hard) without lamination and defects, e.g., granite, trap and diorite	33	
b) Laminated rocks, e.g., sandstone and limestone in sound condition	16.5	

(continued)

c) Residual deposits of shattered and broken bed rock and hard shale, cemented material	9.0	
d) Soft rock	4.5	
2) Non-cohesive soils		
a) Gravelly sand and gravel, compact and offering high resistance to penetration when excavated by tools	4.5	Dry means that the groundwater level is at a depth not less than a width of foundation below the base of the foundation
b) Coarse sand, compact and dry	4.5	
c) Medium sand, compact and dry	2.5	
d) Fine sand, silt (dry lumps easily pulverized by the fingers)	1.5	
e) Loose gravel or sand gravel mixture, loose coarse to medium sand, dry	2.5	
f) Fine sand, loose and dry	1.0	
3) Cohesive soils		
a) Soft shale, hard or stiff clay in deep bed, dry	4.5	This group is susceptible to long-term consolidation settlement
b) Medium clay, readily indented with thumb nail	2.5	
c) Moist clay and sand clay mixture which can be intended with strong thumb pressure	1.5	
d) Soft clay indented with moderate thumb pressure	1.0	
e) Very soft clay which can be penetrated several centimetres with the thumb	0.5	

13.2.2 Methods for improving the bearing capacity of soil

It happens sometimes that the required safe bearing capacity of the soil is not available at shallow depth or it is so low that the dimensions of the footings work out to be very large and uneconomical. Therefore, on such circumstances, depending on the site conditions it becomes necessary to improve the safe bearing capacity.

a. *By increasing the depth of foundation:* In most of the cases, the bearing capacity increases with the depth due to the confining weight of the overlying material. This method is not economical because the cost of construction increases with the depth and the load on the foundation increases with increase in depth. This method should not be used on silts where the subsoil material grows wetter as the depth increases.

b. *By draining the soils:* The presence of water decreases the bearing capacity of the soil. The studies show that around 50 per cent of bearing capacity is lost in sandy soils due to the presence of excess water. Suitable drains should, therefore, be provided in the foundation channel to drain off the excess water.

c. *By compacting the soil:* The compaction of soils results in increase in density and strength and, hence, the bearing capacity. Better compaction is achieved in two ways. (1) By hand packing the rubble boulders or spreading broken stone gravel or sand and thereafter ramming well in the bed of trenches. (2) By driving piles either of wood or concrete or driving and withdrawing the piles and filling the holes with sand and concrete.

d. *By confining the soil:* The movement of soil under the action of load can be prevented by confining the ground by the use of sheet piles. These confined soils can be further compacted for better strength. This method is especially useful for sand soils underlying shallow foundations.

e. *By increasing the width of the foundation:* By increasing the width of the foundation the bearing area increases and, hence, the intensity of pressure decreases. This method has limited use, since the width of the foundation cannot be increased indefinitely.

f. *By replacing the poor soils:* The poor soil is first removed and then the gap is filled by superior materials such as sand, rubble stone, gravel or other hard materials. First the foundation trenches are excavated to a depth of 1.5 m, then filled in stages of 30 cm by hard material and finally rammed.

g. *By grouting:* In poor soil bearing strata, sufficient number of boreholes are driven. Then the cement grout is injected under pressure, because it scales off any cracks or pores or fissures which otherwise reduce the bearing capacity of the soil. This method is employed for materials having pores, fissures or cracks underneath the foundation.

h. *By chemical treatment:* The chemicals like silicates of soda and calcium chloride with soil particles form a gel-like structure and develop into a compact mass. This is called chemical stabilization and is used to impart additional strength to soft soils at deeper depth. However, the chemicals are added in traces only, but even then it has proved to be costly and, hence, is adopted in exceptional cases.

i. *By using geotextiles:* This is a method of reinforcing weak soils to improve their bearing capacity. Coir geotextiles are found to be very useful in this context.

13.3 PLATE LOAD TEST

This is one of the most commonly used methods for determining the allowable bearing capacity of soils.

13.3.1 Working principle

In this test the loading platform, consisting of a bearing plate of steel or cast iron or composite material made of wooden sleepers and steel joist, is subjected to a gradual increment of load and the corresponding settlement values are noted. The load settlement curve is then plotted from which the ultimate bearing capacity is found as based on settlement considerations for cohesive and non-cohesive soils. Finally, by dividing this ultimate bearing capacity by the suitable factor of safety, the allowable bearing capacity of soil is found out.

13.3.2 Test set-up

The test pit should be at least five times as wide as the test plate. The test plate is made to rest in the centre of the pit in a depression which is of the same size as that of the test plate and the bottom level of which shall correspond to the level of the actual foundation. The depth of the hole shall be such that the ratio of depth to the width of loaded area is the same as for the actual foundation. The test plate is 2.5 cm in thickness and the following are their sizes for different soils.

a. Clayey soils, sandy and silty soils – size, 60 cm square
b. Gravelly and dense sandy soils – size, 30 cm square

Larger sizes up to 75 cm square can be used depending upon practical considerations.

The test plate should be machined on the sides and edges. The sides of the pit are lined with wooden sheeting if the soil is soft. The test plate shall be bedded to the soil by plaster of paris, aluminous cement slurry or fine sand. At the start of the test, the platform will be preloaded with a load of 0.7 kg/cm^2 and released.

13.3.3 Testing procedure and observations

The load is applied through a column by means of deadweights such as sand bags or pig iron on lead bars or by a reaction frame which may be a truss frame anchored to the soil by anchors or jacking against a loaded platform with a steel joist placed centrally underneath. The hydraulic jack should butt against the joist with a ball bearing placed in between. The load is increased in regular increments of 200 kg or one-fifth of the approximate ultimate bearing capacity or until the ultimate load is reached.

The settlements should be recorded at least to an accuracy of 0.02 mm with the help of at least two dial gauges to take care of any differential settlement that may occur. Settlements should be observed for each increment of load after an interval of 1, 4, 10, 20, 40, 60 minutes and thereafter at hourly intervals. In the case of clayey soils, the time-settlement curve should be plotted at each load stage and the load should be increased to the next stage either when the curve indicates that the settlement has exceeded 70–80 per cent of the expected ultimate settlement at that stage or at the end of a 24-hour period. For soils other than clayey soils, each load increment shall be kept for not less than 1 hour and up to the time when no further measurable settlement occurs. The next increment of load shall then be applied and the observation is repeated (Figure 13.1).

13.3.4 Limitations of plate load test

a. *Size effect:* The results of the plate load test reflect the strength and the settlement characteristics of the soil within the pressure bulbs. As the pressure bulb depends on the size of the loaded area, it is much deeper for the actual foundation as compared to that of the plate. The plate load test does not truly represent the actual conditions if the soil is not homogenous and isotropic to a large depth.

b. *Scale effect:* The ultimate bearing capacity of saturated clays is independent of the size of the plate, but for cohesionless soils it increases with the size of the plate. To reduce the scale effect it is desirable to repeat the plate load test with plates of two or three different sizes, extrapolate the bearing capacity for the actual foundation and take the average of the values obtained.

c. *Time effect:* A plate load test is essentially a test of short duration. For clayey soils, it does not give the ultimate settlement. The load settlement curve is not truly representative.

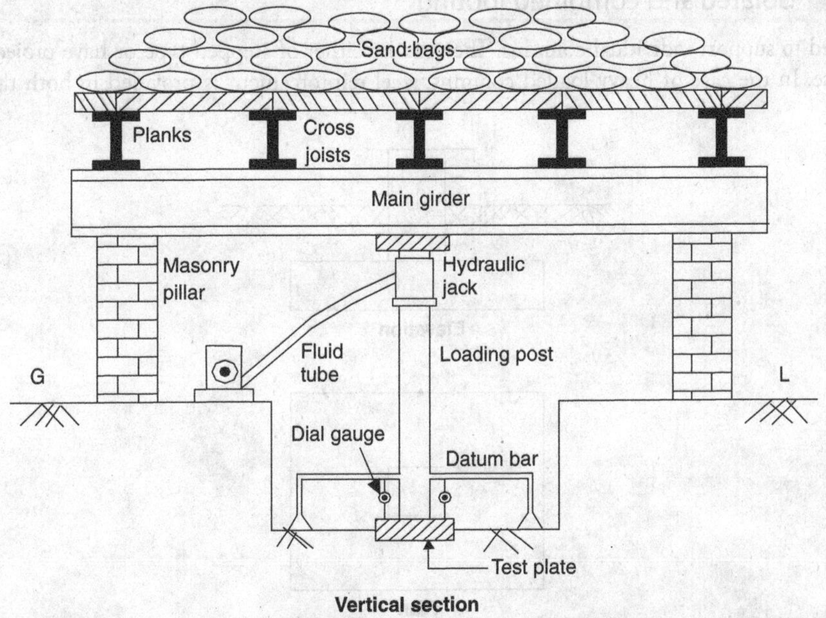

Figure 13.1 Plate load test

d. *Interpretation of failure load:* A failure load is well defined, except in the case of a general shear failure. An error of personal interpretation may be involved in other types of failure.

e. *Reaction load:* It is not practical to provide a reaction of more than 250 KN and, hence, the test on a plate of size larger than 0.6 m width is difficult.

f. *Water table:* The level of water table affects the bearing capacity of the sandy soils. If the water table is above the level of the footing, it has to be lowered by pumping before placing the plate. The test should be performed at the water table level if it is within 1 m below the footing.

13.4 VARIOUS TYPES OF FOUNDATIONS WITH SKETCHES

13.4.1 Spread footing

This is the most common type of foundation and can be laid using open excavation by allowing natural slopes on all sides. This type of foundation is practicable for a depth of about 5 m and is normally convenient above the water table. The base of the structure is enlarged or spread to provide individual support. This type of footing is given for structures of moderate height built on sufficiently firm ground and for light structures. They have only one projection beyond the width of the wall on either side (Figure 13.2).

13.4.2 Stepped footing

Here, we have more than one projection on either side of the width of the wall as shown in Figure 13.3.

The depth of each layer is at least twice the projections and its base width should be twice the width of the layer above that. Generally, the projections provided are kept as 15 cm on either side. The depth is generally limited to 0.9 m in general cases.

13.4.3 Isolated and combined footing

They are used to support individual columns. They can be either of stepped type or have projections in the concrete base. In the case of heavy loaded columns, steel reinforcement is provided in both the directions

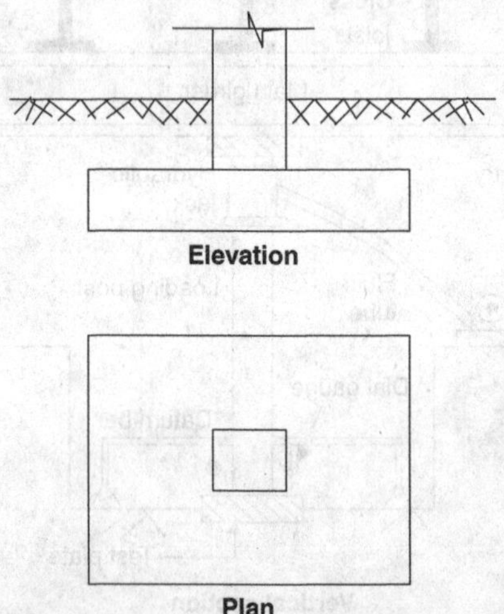

Figure 13.2 Spread footing

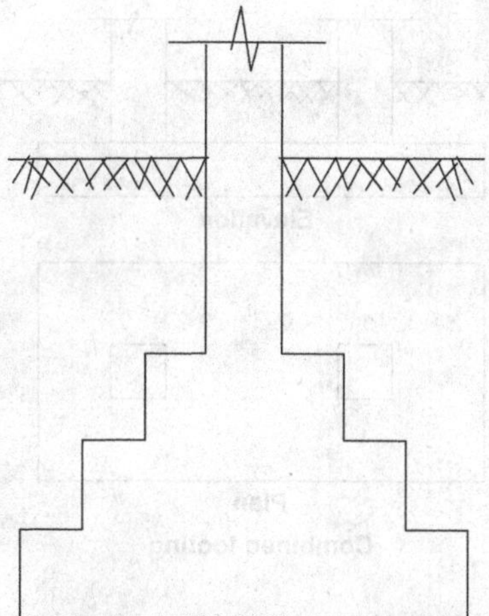

Figure 13.3 Stepped footing

in the concrete bed. Generally, 15 cm offset is provided on all sides of the concrete bed. In the case of brick masonry columns, an offset of 5 cm is provided on all the four sides in regular layers. The footing of concrete columns may be slab, stepped or sloped ones.

A combined footing supports two or more columns in a row. The combined footing can be rectangular if both the columns carry equal loads or can be trapezoidal if there are space limitations and carry unequal loads. Generally, they are constructed of reinforced concrete (Figure 13.4).

13.4.4 Mat or raft foundation

The raft or mat foundation is a combined footing that covers the entire area beneath the structure and supports the columns. If required the beam and slab construction in reinforced cement concrete (RCC) can also be carried out. When the allowable soil pressure is low or the structural loads are heavy, the use of spread footings would cover more than one-half of the area and it may prove more economical to use raft foundation. A raft may undergo large settlements without causing harmful differential settlement. For this reason almost double the settlement of that permitted for footings is acceptable for rafts. Usually, when hard soil is not available within 1.5–2.5 m, a raft foundation is adopted. The raft foundations are useful for public buildings, office buildings, school buildings, residential buildings, etc. (Figure 13.5).

13.4.5 Pile foundation

The pile foundation is a construction for the foundation supported on piles. A pile is an element of construction composed of timber, concrete or steel or a combination of them. Pile foundation may be defined as a column support type of foundation, which may be cast in situ or precast. The piles may be placed separately or they may be placed in the form of a cluster throughout the length of the structure. This type of construction is adopted when the loose soil extends to a great depth. The load of the structure is transmitted by the piles to the hard stratum below or it is resisted by the friction developed on the sides of piles.

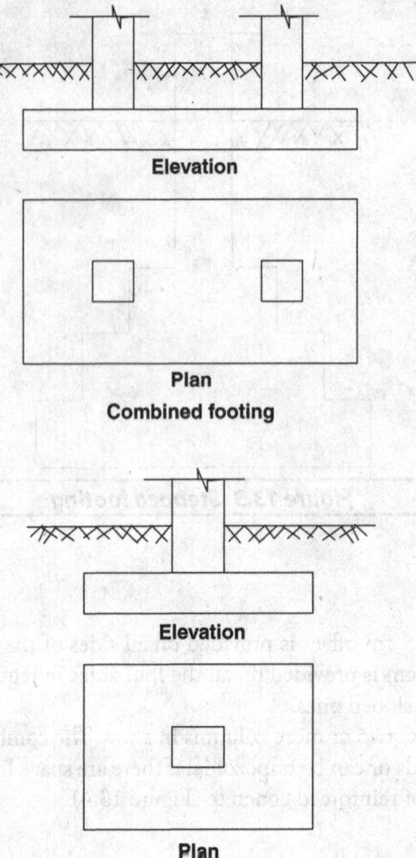

Figure 13.4 Isolated and combined footing

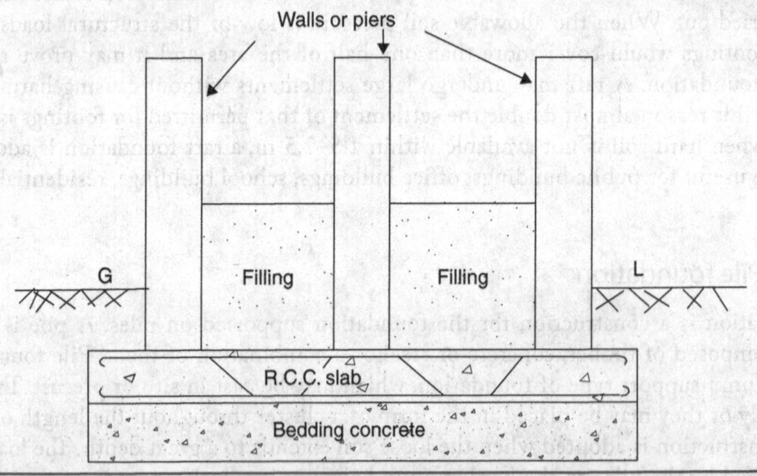

Figure 13.5 Raft foundation

13.4.5.1 *Classification of piles*

Piles can be broadly classified into two categories.

Classification based on the function

i. *Bearing piles:* These piles penetrate through soft soils and their bottom rests on a hard stratum. The soft ground through which the piles pass also gives lateral support and increases the load carrying capacity of the bearing piles.

ii. *Friction piles:* When loose soils extend to a great depth, piles are driven up to such a depth that frictional resistance developed at the sides of the piles equals the load coming on the piles.

iii. *Screw piles:* Screw piles consist of a hollow cast-iron or steel cylinder with one or more blades at the bottom. The blades are generally made of cast iron. The bottom end of the screw pile with a blunt point is useful when the ground to be penetrated consists of clay or sand.

iv. *Compaction piles:* They are used to compact loose granular soils in order to increase their bearing capacity. These piles do not carry any load.

v. *Uplift piles:* These piles anchor down the structure subjected to uplift due to overturning movement.

vi. *Batter piles:* These are used to resist large horizontal or inclined forces.

vii. *Sheet piles:* These are used as bulk heads or as an impervious cut off to reduce seepage and uplift.

Classification based on materials and composition

i. *Cement concrete piles:* Cement concrete piles possess excellent compressive strength. These piles can be reinforced or prestressed.

ii. *Timber piles:* Timber piles are prepared from trunks of trees. They may be circular or square. They are 30–50 cm in diameter with a length not exceeding 20 times its top width. At the bottom, a cast iron shoe is provided and at the top a steel plate is fixed. If a group of timber piles is driven, the top of each member of the group is brought at the same level and then a concrete cap is provided to have a common platform. It is difficult or even impossible to drive these piles into the hard stratum or boulders.

iii. *Steel piles:* The different forms of steel piles are H piles, box piles and tube piles. These piles can easily withstand the stresses due to driving.

- These piles can be easily lengthened by welding and also can be cut off easily.
- These piles can resist lateral forces in a better way.
- The bearing capacity of these piles is comparatively high.
- These piles can take up the impact of stresses and can resist lateral forces.

The disadvantage in steel piles is corrosion. To prevent corrosion they should be coated with paints or may be encased with cement concrete.

iv. *Sand pile:* These piles are made by boring holes on the ground and filling them with sand. The top of the sand piles is filled with concrete to prevent the sand from coming upwards due to lateral force. Sand piles are spaced 2-3 m and the length is about 12 times the diameter. Even though they are easy to construct and can be used irrespective of any position of water table, they are not suitable for earthquake regions.

v. *Composite piles:* A composite pile is formed when it is a combination of piles of two different materials. They are suitable where the upper part of a pile is to project above the water table. They are economical and easy to construct.

13.4.6 Well foundations

These foundations are used in underwater constructions. They are foundations with a hollow portion. These are the most commonly used deep foundations, especially for structures like bridges. They are generally built of timber, metal, reinforced concrete, masonry, etc.

REVIEW QUESTIONS

1. What do you mean by foundation and what are its main objectives?
2. How are foundations classified according to their depth?
3. What is the bearing capacity of soil?
4. Explain the different methods for improving the bearing capacity of soil.
5. Explain Plate Load Test with the help of a neat diagram.
6. What are the limitations of Plate Load Test?
7. Explain the various types of foundation with neat sketches.
8. How are piles classified based on their function?
9. How are piles classified based on the material used?
10. What is well foundation?

chapter 14

Mortar

The term mortar is used to indicate a paste prepared by adding a required quantity of water to a mixture of binding material like cement or lime and fine aggregate like sand. This is used to bond masonry or other structural units.

14.1 FUNCTIONS OF MORTAR IN BUILDING WORKS

The following are the major functions of mortar:

1. To bind building materials such as bricks and stones into a solid mass.
2. To carry out pointing and plasterwork on exposed surfaces of masonry.
3. To form an even and soft bedding layer for building units.
4. To form joints of pipes.
5. To improve the general appearance of a structure.
6. To prepare moulds for coping, corbels, cornice, etc.
7. To serve as a matrix or cavity to hold the coarse aggregates, etc.
8. To distribute uniformly the super-incumbent weight from the upper layer to the lower layer of bricks or stones.
9. To hide the open joints of brickwork and stonework.
10. To fill up the cracks detected in the structure during maintenance process, etc.

14.2 TYPES OF MORTARS AND THEIR PREPARATION

The following major types of mortars are recognized based on the type of the cementing material used in its preparation: lime mortars, cement mortars and gauged mortars are the most common types.

14.2.1 Lime mortars

Lime mortars are defined as mixes of lime with fine sand and/or pozzolanic materials like surkhi, pumice, ash and cinder in water.

14.2.1.1 Preparation

The following are the main stages in the preparation of lime mortars: selection of the raw materials, proportioning of the raw materials and mixing of the raw materials.

Selection of the raw materials

Lime, sand, surkhi, pumice, ash and cinder form the major raw materials commonly used in the preparation of lime mortars.

Lime: All types of lime can be used for making mortars, although all lime mortars cannot be used for different situations. The main types of lime, namely the fat lime and hydraulic lime, yield mortars of different qualities, the former being suitable only for limited purposes like jointing and light loaded masonry work whereas the latter being more useful in other situations as well. It is essential that the lime to be used be free from impurities like silica, iron oxide and, especially, gypsum.

Sand: The sand for making a strong and durable mortar must be of good quality, i.e., it must be clean and sharp grained. The recommended fineness modulus of sand for mortar is 2.00–3.00.

Sand is added to the lime for at least two purposes:

i. To avoid shrinkage and cracking of mortar on hardening, because lime used alone shrinks and cracks on hardening.

ii. To increase the bulk of the mortar; this is essential to make it more economical.

When used in appropriate proportions, sand facilitates the hardening of the lime paste by yielding a porous structure and allowing the atmospheric carbon dioxide access to the inner lime.

Pozzolanic materials: These include a variety of lightweight materials like surkhi, cinder, pumice and ash, which have been found to increase the strength of mortars when used in place of sand or even along with sand.

Water: In general, water suitable for drinking purposes can be used in the preparation of mortar. The water must be essentially free from alkalies, acids and organic residues.

Sometimes, cement is also added in small quantities to lime mortars to enhance the strength and setting properties of mortar. Depending on the type of the aggregate used, lime mortars are further distinguished into lime–sand mortars, lime–surkhi mortars, lime–sand–surkhi mortars and so on.

Proportioning of the raw materials

The main aim of proportioning is to fix such ratios of the raw materials that will result in a mortar of desired quality, i.e., required strength, durability and finish. Since the mortars used in different situations in construction are not subjected to the same forces and conditions, no single rule for proportioning a mortar can be framed. The widely recommended and commonly adopted proportions for different conditions are summarized in Table 14.1.

Mixing of the raw materials

It is of paramount importance that the ingredients of mortar be mixed as thoroughly as possible. Two common methods adopted are manual mixing and mortar mill mixing.

Manual mixing: This is done on a watertight platform made of masonry or in a tank of suitable dimensions. A spade is the main tool required. Measured volumes of slaked lime and the aggregate are placed on the platform or on the tank. These are first mixed in dry state by giving turns with the help of the spade. Water is added gradually and mixing continued simultaneously until a mortar of uniform consistency is obtained.

Mortar mill mixing: A mortar mill is a mechanical device for grinding the mortar ingredients in the presence of water and is run either by animals or by power. In the traditional method, animal-driven mortar mills are used.

Table 14.1 Various Lime Mortar Compositions for Different Construction Works

Situation	Mortar composition recommended	Remarks
(A) Foundations		
1. Foundation concrete in dry subgrade with water level below 2.4 m of the foundation level	Any one of the following mixes: 1 lime, 2 sand 1 lime, 1 sand, 1 surkhi 1 lime, 2 surkhi 1 cement, 3 lime, 12 sand	For moist subgrade where water table is within 2.4 m of the foundation level, only cement–sand mortar of 1:3 mix should be used for all foundation work
2. Foundation masonry with loading less than 4 tons/ ft²	1 lime, 2 sand 1 lime, 1 sand, 1 surkhi 1 lime, 2 surkhi 1 cement, 3 lime, 12 sand	
3. Foundation masonry in heavy and medium loading in dry subgrade	1 cement, 1 lime, 6 sand	
(B) Superstructures		
4. Load-bearing walls with brick masonry.	1 cement, 1 lime	
i) Light loading	1 lime, 1 sand, 1 surkhi 1 lime, 1 sand, 1 cinder	Loading less than 4 tons/ft²
ii) Medium loading	1 cement, 3 lime, 12 sand	Loading between 4 and 6 tons/ft²
iii) Heavy loading	1 cement, 2 lime, 9 sand to 1 cement, 1 lime, 6 sand	Loading more than 8 tons/ft²
5. Non-load bearing partition wall with concrete slabs or hollow blocks	1 lime, 3 sand	
(C) Plasters		
6. i) External plasters below damp proof course	1 cement, 1 lime, 6 sand	
ii) External plasters on all walls	1 cement, 2 lime, 9 sand	
iii) Internal plasters on all walls	All mixes are suitable	
General purpose mortar	1 cement, 1 lime, 6 sand	Suitable for most of masonry work and plasterwork

Hardening of lime mortars

Mortars made of fat lime or hydrated lime hardens by a simple process of crystallization preceded by a loss of water due to evaporation. The lime takes up the carbon dioxide from the atmosphere and forms a crystal mass of calcium carbonate that is quite hard and responsible for the strength of the mortar.

14.2.2 Cement mortar

The most common type of mortars of the present day generally consists of mixtures of cement with sand in the presence of suitable quantity of water. In all situations where a strong, durable and resistant mortar is desired, the cement mortar becomes indispensable.

Table 14.2 Proportions Recommended for Cement–Sand Mortar

Type and situation of the work	Recommended proportions
1. For ordinary masonry work with bricks or stones as structural units.	1 cement, 3 sand to 1 cement, 6 sand
2. For reinforced brickwork and for all work in moist situations.	1 cement, 2 sand to 1 cement, 3 sand

14.2.2.1 *Preparation*

The preparation of cement mortar involves the same stages as of lime mortar, namely selection of raw materials, their proportioning and thorough mixing. Among the materials, cement, sand and water are the essential ingredients. Cement used for preparing mortar must be in perfectly undamaged and undeteriorated condition. Sand in addition to possessing its usual properties must be free from impurities like oxides of iron, clay and mica and must be thoroughly cleaned before being used for good quality mortar. Generally, no other aggregate can replace sand completely in a cement mortar, although nowadays several other alternatives are being used. The functions of sand in the mortar include increasing the bulk of the mortar, providing resistance against shrinkage and cracking on the setting of cement and also making the mortar more strong.

14.2.2.2 *Proportioning*

Only very carefully proportioned volumes of the ingredients are to be used in the preparation of cement mortars. The proportioning is usually by volume. Different mix ratios are specified for different works, the most commonly used are shown in Table 14.2.

14.2.2.3 *Mixing of the ingredients*

For small jobs, the manual mixing of mortar is commonly practised and for large-scale construction where larger batches of the mortar are required the mortar mill mixing is indispensable.

14.2.3 Gauged mortars

These mortars contain both cement and lime as the essential ingredients besides the fine aggregate, which is generally sand. The addition of cement to lime mortar improves considerably the quality of mortar, especially in respect of its workability, time of setting, hardening and ultimate strength.

Since the gauged mortars consist of cement as one of the essential ingredients, it is important that such mortars be consumed within 2 hours of their preparation to avoid any deterioration in quality.

REVIEW QUESTIONS

1. What do you mean by mortars and explain its main functions in building works?
2. Explain briefly the different types of mortars and their preparation.
3. What are the raw materials used in the preparation of different types of mortars?
4. What are the general proportions of raw materials in the preparation of mortar used for brick and stone masonry?
5. What is gauged mortar?

chapter 15

Masonry Works

Masonry is used to indicate the art of building a structure in either stones or bricks. The masonry wall is built of individual blocks of materials such as stones, bricks, concrete, hollow blocks, cellular concrete and laterite, usually in horizontal courses cemented together with some form of mortar. The binding strength of mortar is usually disregarded as far as the strength of the wall is concerned.

Masonry can be classified into the following categories:

1. Stone masonry
2. Brick masonry
3. Hollow block concrete masonry
4. Reinforced masonry
5. Composite masonry

Comparison Between Brick and Stone Masonry

Description	Stone masonry	Brick masonry
1. Uses	Construction of piers, dams, docks, marine structures, residential and monumental buildings	Residential buildings
2. Strength	High crushing strength	Much less
3. Durability	Excellent	Excellent
4. Source	Natural	Artificial
5. Danger from dampness	No such danger	Causes disintegration
6. Cost	High cost. It is restricted to areas where stone is plenty	Much less cost, easily available
7. Bond	It requires a great deal of time and extra labour in maintaining proper bond	Regular shape and size result in quick construction bond
8. Construction	Requires high skill	Ordinary skill required
9. Moulding into desired shape	Not convenient	Convenient
10. Handling	Requires lifting device	Easy to handle

15.1 DEFINITION OF TERMS

The definitions of some important terms used in masonry are given below.

1. *Course:* A course is a horizontal layer of bricks or stones.
2. *Bed:* This is the surface of a stone perpendicular to the line of pressure. It indicates the lower surface of bricks or stones in each course.

3. *Back:* The inner surface of a wall that is not exposed is called a back. The material forming the back is known as backing.
4. *Face:* The exterior of a wall exposed to weather is known as face. The material used in the face of a wall is known as facing.
5. *Hearting:* It is the interior portion of a wall between the facing and backing.
6. *Stretcher:* This is a brick laid with its length parallel to the face or front or direction of a wall. The course containing stretchers is known as stretcher course.
7. *Header:* This is a brick laid with its breadth or width parallel to the face or front or direction of a wall.
8. *Arrises:* The edges formed by the intersection of plane surfaces of a brick are called the arrises and they should be sharp, square and free from damage.
9. *Perpends:* The vertical joints separating the bricks in either length or cross direction are known as the perpends; for a good bond the perpends in alternate courses should be vertically one above the other.
10. *Lap:* The horizontal distance between the vertical joints in successive courses is termed as a lap; for a good bond it should be one-fourth of the length of a brick.
11. *Closer:* A piece of brick which is used to close up the bond at the end of brick courses is known as the closer. It helps in preventing the joint of successive courses to come in a vertical line. Generally, the closer is not specially moulded.
12. *Queen closer:* This is obtained by cutting the brick longitudinally in two equal parts. It can also be made from two quarter bricks, known as the quarter closers, to minimize the wastage of bricks.
13. *King closer:* This is obtained by cutting a triangular portion of the brick such that half a header and half a stretcher are obtained on the adjoining cut faces. A king closer is used near door and window openings to get satisfactory arrangements of the mortar joint.
14. *Frog:* A frog is a mark of depth about 10–20 mm, which is placed on the face of a brick to form a key for holding the mortar.
15. *Bat:* This is a piece of brick, usually considered in relation to the length of a brick and accordingly known as half bat or three-quarter bat.

15.2 STONE MASONRY

It is the art of building a structure with stones. The selection of material depends on the following:

1. The availability
2. Ease of working
3. Appearance
4. Strength and stability
5. Polishing characteristics
6. Economy
7. Durability

Table Showing Different Stones Used for Different Purposes

Purpose	Stone used
Heavy engineering works like dock, breakwater, lighthouse, bridges, etc.	Granite and gneiss
Buildings situated in industrial towns	Granite and compact sandstone
Pavements and railway	Granite and basalt
Electrical switchboards	Marble slab and slate
Fire resistance works	Sandstone
Carving and ornamental works	Granite and marble

The general principles in stone masonry construction are as follows:

1. The stones to be used for stone masonry should be hard, tough and durable.
2. The pressure acting on stones should be vertical. The pressure acting along the direction of bedding planes causes splitting of stones.
3. The stones should be properly dressed as per the requirement.
4. The headers and bond stones should not be of a dumb-bell shape.
5. In order to obtain uniform distribution of load, under the ends of girders, roof trusses, etc. large stones should be used.
6. The beds of the stones and the plan of the courses should be at right angles to the slope in the case of sloping retaining walls.
7. Wood boxing should be fitted into walls having fine dressed stonework to protect it during further construction.
8. The mortar to be used should be of good quality and in the specified proportion.
9. The construction work of stone masonry should be raised uniformly.
10. Plumb bob should be used to check the verticality of erected walls. In addition, wooden templates should be used to check the battered faces.
11. The stone masonry section should always be designed to take compression and not the tensile stresses.
12. The masonry work should be properly cured after the completion of work for a period of 2-3 weeks.
13. As far as possible, broken stones or small chips should not be used.
14. Double scaffolding should be used for working at higher levels.
15. Properly wetted stones should be used along with mortar for construction work.

15.3 CLASSIFICATION OF STONE MASONRY

Based on the arrangement of stones in the construction and the degree of refinement in the surface finish, the stone masonry can be classified broadly into the following two categories:

1. **Rubble masonry**
   a. Coursed rubble
   b. Uncoursed rubble

c. Random rubble
d. Dry rubble
e. Polygonal rubble
f. Flint rubble

2. **Ashlar masonry**

a. Ashlar fine
b. Ashlar rough tooled
c. Ashlar rock or quarry faced
d. Ashlar chamfered
e. Ashlar block-in course

15.3.1 Rubble masonry

In this type of construction, stones of irregular sizes are used. The stones as obtained from the quarry are taken to be used in the same form or are broken and shaped into suitable sizes by means of a hammer as the work proceeds. The strength of the rubble masonry mainly depends on:

a. The quality of mortar.
b. The use of the long stones at frequent interval for proper bonding.
c. The proper filling of the mortar between the spaces of stones.

15.3.1.1 Coursed rubble masonry

In this type of rubble masonry, the heights of stones vary from 50 mm to 20 cm. The stones are stored before the work commences. The masonry work is then carried out in courses such that the stones in a particular course are of equal heights. This type of masonry is used for the construction of public buildings and residential buildings. The coursed rubble masonry is further divided into three categories.

i. *Coursed rubble masonry I sort:* In this type, stones of the same height are used and the courses are of the same heights. The face stones are dressed by means of a hammer and the brushings do not project by more than 40 mm. The thickness of the mortar joint does not exceed 10 mm (Figure 15.1).

ii. *Coursed rubble masonry II sort:* The stones to be used are of different heights. The courses need not be of equal heights. Only two stones are to be used to make up the height of one course. The thickness of the mortar joint is 12 mm (Figure 15.2).

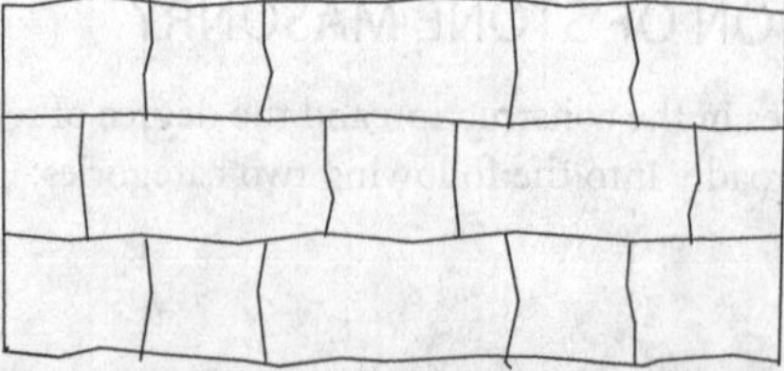

Figure 15.1 Coursed rubble masonry I sort

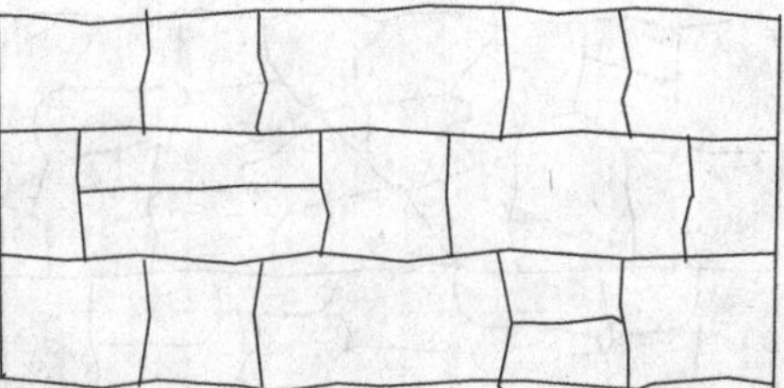

Figure 15.2 Coursed rubble masonry II sort

iii. *Coursed random rubble III sort:* The stones to be used are of different heights, the minimum being 50 mm. The courses need not be of equal heights. Only three stones are to be used to make up the height of one course. The thickness of the mortar joint is 16 mm (Figure 15.3).

15.3.1.2 Uncoursed rubble masonry

In this type of rubble masonry, the stones are not dressed. However, they are used as they are available from the quarry, except by knocking out some corners. The courses are not maintained regularly. The larger stones are laid first and the spaces between them are then filled by means of spalls. The wall is brought to a level every 30–50 cm. This type of rubble masonry being cheaper is used for the construction of compound walls, garages, labour quarters, etc. (Figure 15.4).

15.3.1.3 Random rubble masonry

In this type of rubble masonry, stones of irregular sizes and shapes are used. The stones are arranged to have a good appearance. It is to be noted that more skill is required to make the masonry structurally stable. If the face stones are chisel dressed and the thickness of mortar joints does not exceed 6 mm, it is known as random

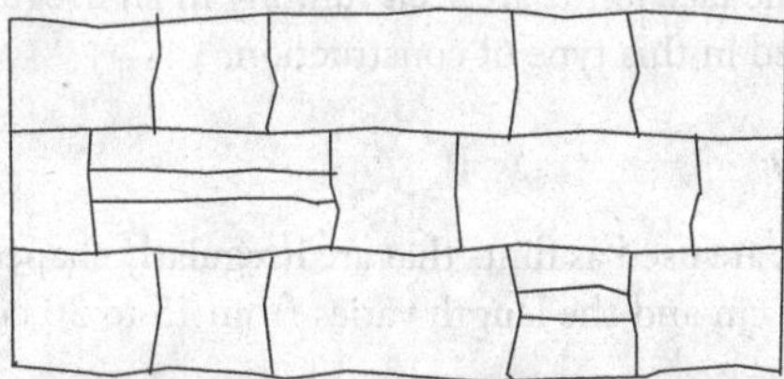

Figure 15.3 Coursed rubble masonry III sort

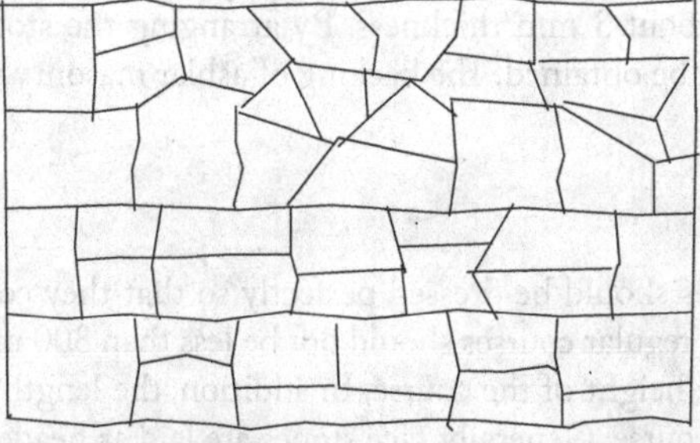

Figure 15.4 Uncoursed rubble masonry

Figure 15.5 Random rubble masonry

rubble masonry I sort. If the face stones are hammer dressed and the thickness of mortar joints does not exceed 12 mm, it is known as random rubble masonry II sort. This type of masonry is used for construction of residential buildings (Figure 15.5).

15.3.1.4 Dry rubble masonry

This is just similar in construction to the coursed rubble masonry III sort except that no mortar is used in the joints. This type of construction is the cheapest, but it requires more skill in construction. It is extensively used for compound walls pitching on bridge approaches, retaining walls, etc. In order to prevent the displacement of stones and to make the work more stable, the two courses at the top and about 50 cm length at the ends are sometimes built in mortar.

15.3.1.5 Polygonal rubble masonry

In this type of masonry, the stones are hammer dressed and the stones selected for face work are dressed in an irregular polygonal shape. Thus, the face joints are seen running in an irregular shape in all directions. It is to be noted that more skill is required in this type of construction.

15.3.1.6 Flint rubble masonry

In this type of masonry, the stones are used as flints that are irregularly shaped nodules of silica. The width and thickness vary from 80 mm to 15 cm and the length varies from 15 to 30 cm (Figure 15.6).

15.3.2 Ashlar masonry

This is costlier, high grade and superior quality masonry. This is built from accurately dressed stones with uniform and very fine joints of about 3 mm thickness. By arranging the stone blocks in various patterns, different types of appearances can be obtained. The backing of ashlar masonry walls may be built of ashlar or rubble masonry.

15.3.2.1 Ashlar fine masonry

At all beds, joints and faces stones should be dressed perfectly so that they conform to the desired pattern. The size of the stones to be laid in regular courses should not be less than 300 mm in height. The width of the stones should not be less than the height of the course. In addition, the length of the stones should be more than two times the height of the course. Generally, face stones are laid as headers and stretchers alternatively. The header comes under the middle portion of the stretchers. In order to break the continuous vertical joints,

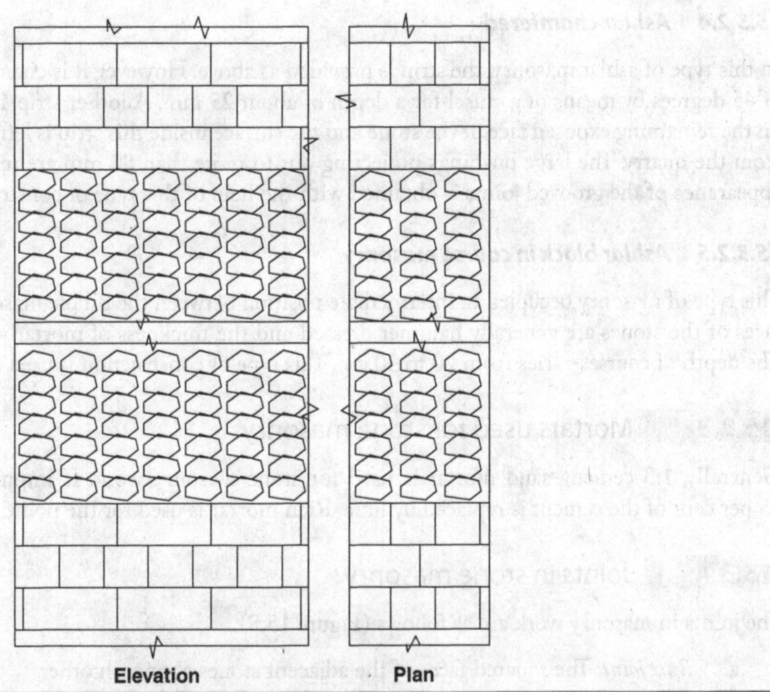

Figure 15.6 Flint rubble masonry

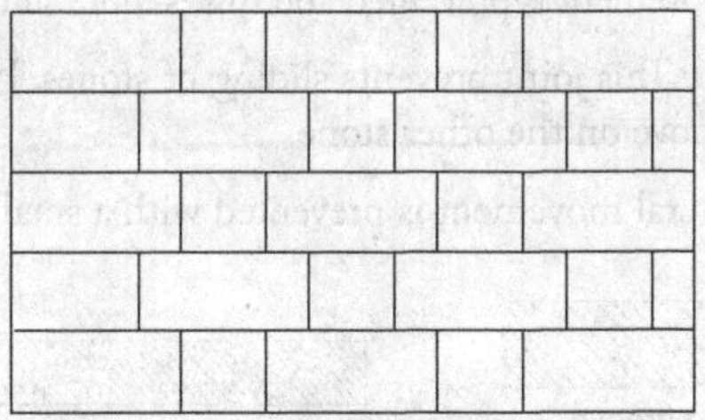

Figure 15.7 Ashlar fine masonry

the stones in the adjacent layers should have a lap of more than half of the height of the course. All the joints, either horizontal or vertical, should be made of fine mortar with a maximum thickness of 3 mm. The broken edges of the stones should not be visible at the joints (Figure 15.7).

15.3.2.2 Ashlar rough tooled

The exposed faces of this type of masonry should be given a fine dressed chisel drafting of about 25 mm width. In between the drafts, portions should be roughly tooled. The joint thickness should be 6 mm.

15.3.2.3 Ashlar rock or quarry faced

In this case, the exposed faces of the facing stones between the chisel draftings all around are left undressed. However, the projections of sizes more than 8 cm are broken. All other specifications regarding the sizes of the stones and the bond are kept similar to that of ashlar rough tooled masonry.

15.3.2.4 *Ashlar chamfered*

In this type of ashlar masonry, the strip is provided as above. However, it is chamfered or bevelled at an angle of 45 degrees by means of a chisel for a depth of about 25 mm. Another strip 12 mm wide is then provided on the remaining exposed face of the stone and the surface inside this strip is left in the same form as received from the quarry. The large bushings projecting out to more than 80 mm are removed by a hammer. A neat appearance of the grooved joints is obtained with the help of this type of construction.

15.3.2.5 *Ashlar block in course masonry*

This type of masonry occupies an intermediate position between the rubble masonry and ashlar masonry. The faces of the stones are generally hammer dressed and the thickness of mortar joints does not exceed 6 mm. The depth of courses varies from 20 to 30 cm. This type of construction is used for heavy engineering works.

15.3.3 Mortars used for stone masonry

Generally, 1:3 cement–sand mortar is used for stone masonry work. If improved workability is required, 15 per cent of the cement is replaced by lime. Rich mortar is used for the pointing works.

15.3.4 Joints in stone masonry

The joints in masonry work are as follows (Figure 15.8):

a. *Butt joint:* The squared faces of the adjacent stones abut each other.
b. *Lapped or rebated joint:* This is used for arches, stones laid on slopes, etc.
c. *Table joint:* Here, lateral movement is prevented and this is more suitable where lateral pressure is high.
d. *Tongued and grooved joint:* This joint prevents sliding of stones. Here, a projection of one stone fits into the depression or groove on the other stone.
e. *Dowel joint:* Here also lateral movement is prevented with a small key as in Figure 15.8.

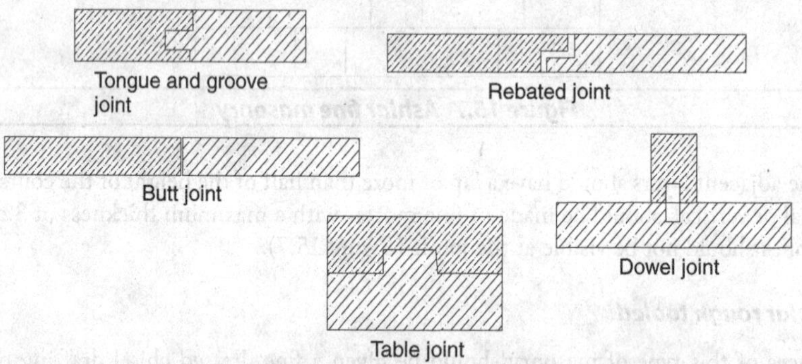

Figure 15.8 Joints in stone masonry

15.4 BRICK MASONRY

The bricks are obtained by moulding clay in rectangular blocks of uniform size and then by drying these blocks. They are of uniform size, light in weight and require no lifting appliances. The bricks are of various sizes. The actual size of a modular brick is 19 cm × 9 cm × 9 cm. With mortar thickness size of such a brick becomes 20 cm × 10 cm × 10 cm and it is known as the nominal size of the modular bricks. Common bricks must have a minimum compressive strength of 50 kg/m² and the average water absorption should not be more than 20 per cent by weight.

15.5 TYPES OF BRICK MASONRY

The types of brick masonry are as follows:

1. **Brickwork in mud:** In this type of brickwork, mud is used to fill up the joints. The mud is prepared by intimately mixing sand and clay. The thickness of the mortar joints is 12 mm. This type of brickwork is adopted in cases of low-cost construction and the maximum height up to which a wall can be constructed in this type of brickwork is 4 m.
2. **Brickwork in cement mortar or lime mortar – I class:** In this type of brickwork, cement or lime mortar is used. The bricks are table-moulded, are of standard shape and are burnt in kilns. The surfaces and edges of the bricks are sharp, square and straight. They comply with all the requirements of a good standard brick. The thickness of mortar joints does not exceed 10 mm.
3. **Brickwork in cement mortar or lime mortar – II class:** The bricks to be used in this type of brickwork are moulded on ground and are burnt in kilns. The surface of these bricks is somewhat rough and their shape is slightly irregular. The bricks may have hair cracks and their edges may not be sharp and uniform. These bricks are commonly used at places where the brickwork is to be provided with a coat of plaster. The thickness of mortar joints is 12 mm.
4. **Brickwork in cement mortar or lime mortar – III class:** This type of brickwork is the same as II class except that the bricks to be used are burnt in clamps. These bricks are not hard and they have rough surfaces with irregular and distorted edges. These bricks give a dull sound when struck together. They are used for unimportant and temporary structures and at places where rainfall is not heavy.

15.6 BONDS IN BRICK MASONRY

15.6.1 Stretcher bond

In this type of bond, all the bricks are laid with their length in the direction of the wall. The stretcher bond is useful for one-brick partition walls as there are no headers. This bond does not develop proper internal bond and it should not be used for walls having thickness greater than that of one-brick wall (Figure 15.9).

Figure 15.9 Stretcher bond

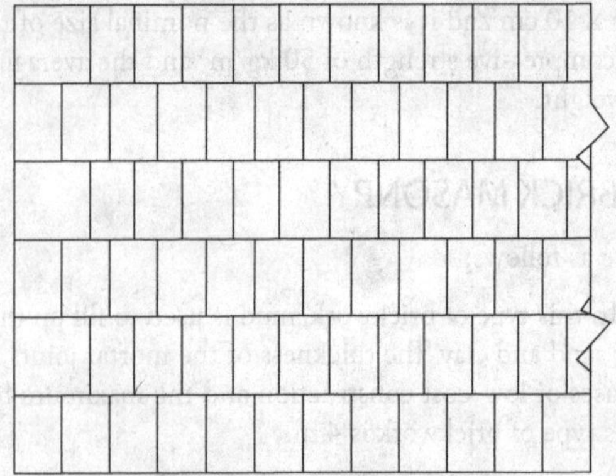

Figure 15.10 Header bond

15.6.2 Header bond

In this type of bond, all the bricks are laid with their ends towards the face of the wall. Thus, the bond does not have the strength to transmit pressure in the direction of the length of the wall. This bond is used for curved surfaces (Figure 15.10).

15.6.3 English bond

In this type of bond, alternate courses of headers and stretchers are laid. It is necessary to place queen closers after the first header in the heading course for breaking the joints vertically (Figures 15.11, 15.12 and 15.13).

a. A queen closer must be provided after a quoin header or first header. A header course should never start with a queen closer.

b. Each alternate header should be centrally placed over a stretcher.

c. Continuous vertical joints should not be allowed except at the stopped end.

d. In case the wall thickness is equivalent to an even number of half bricks, the wall shall present similar appearance in both faces.

e. In case the wall thickness is equivalent to an odd number of half bricks, the same course shall have stretcher on one face and header on the other face.

f. Only headers should be used for the hearting of the thicker walls.

g. The joints on the header course should be made thinner than those in the stretcher course. This is because of the fact that the number of vertical joints in the stretcher course is half the number of joints in the header course.

15.6.4 Flemish bond

In this type of bond, the headers are distributed evenly and, hence, it creates a better appearance than the English bond.

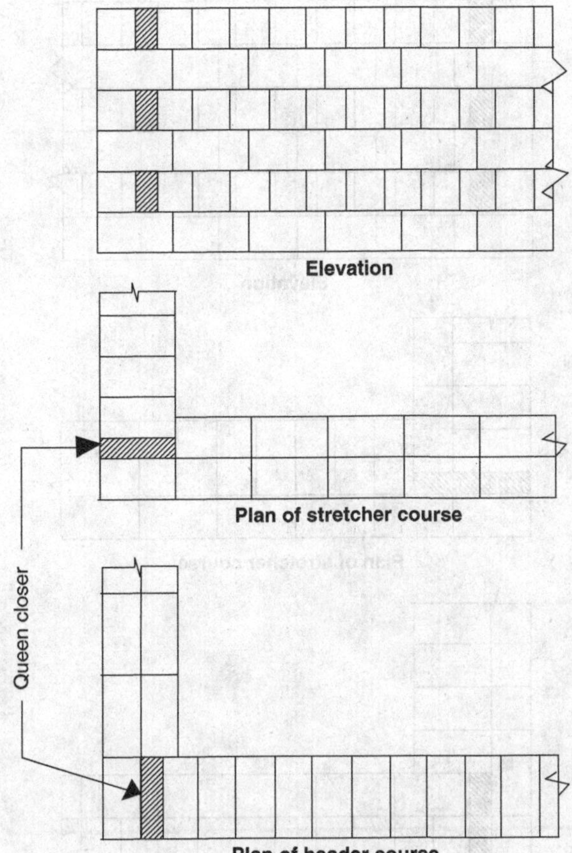

Figure 15.11 English bond – 1 brick thick

In the flemish bond, for every course the headers and stretchers are placed alternatively. The queen closer is put next to the quoin header in alternate courses to develop the face lap. Every header is centrally supported over a stretcher below it.

The flemish bond is divided into two groups.

15.6.4.1 Double Flemish bond

In this bond, alternate headers and stretchers are laid to each course. This type of bond is better in appearance than the English bond. The facings and the bracings are of the same appearance. Brickbats are used in the case of walls having thickness equivalent to an odd number of half bricks. The queen closer is placed next to the quoin header in alternate courses in order to break the continuity of the vertical joints (Figure 15.14).

15.6.4.2 Single Flemish bond

The face elevation is of flemish bond and the filling as well as backing are of the English bond. This type of bond is an attempt to combine the strength of the English bond with the appearance of the flemish bond.

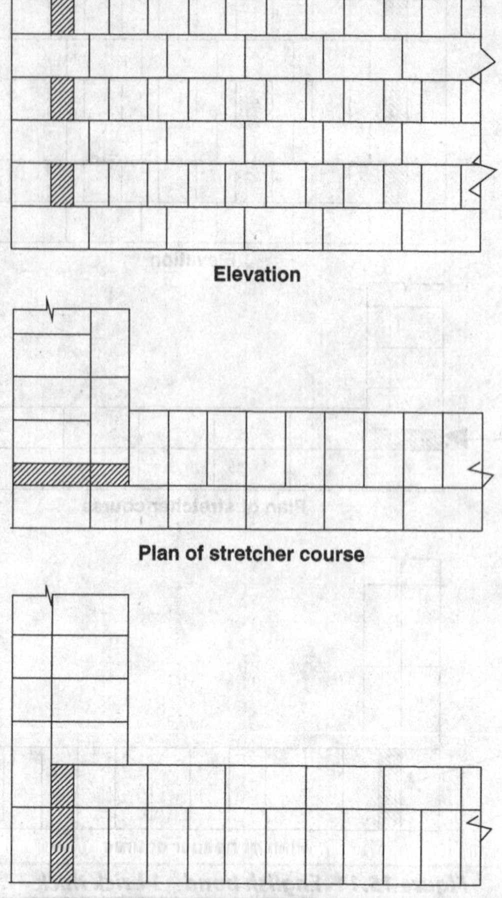

Figure 15.12 English bond – 1½ brick thick

In order to construct this bond, a wall of minimum thickness of 1½ bricks is required. The bricks in the same course do not break joints with each other. The joint is straight. In this bond, short continuous vertical joints are formed. The following table gives the comparison between English and Flemish bonds.

A Comparison Between English and Flemish Bonds

English bond	Flemish bond
More compact and strong for walls having thickness more than 1½ bricks	Less compact and less strength
Less pleasing in appearance from facing	Better appearance in the facing
Strict supervision and skill are not required	Good workmanship and careful supervision required
More in cost	Cheaper in cost

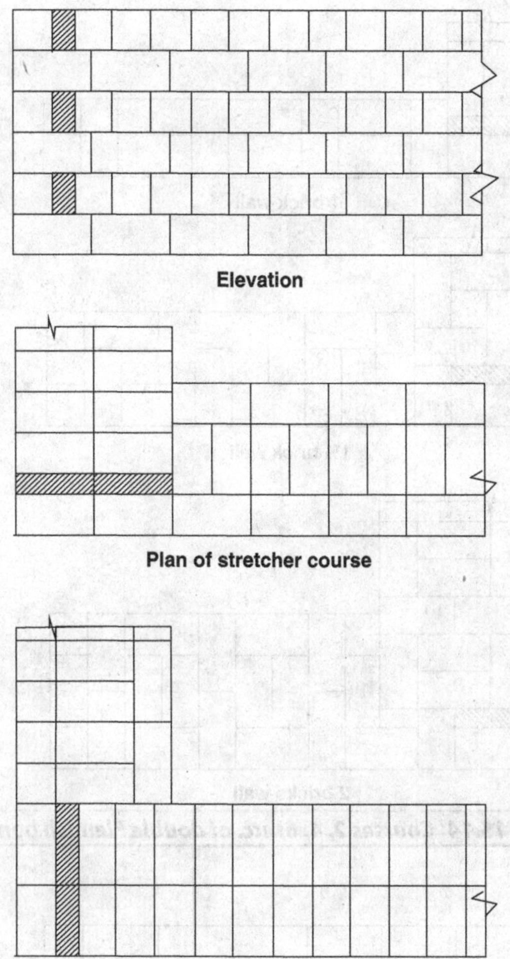

Figure 15.13 English bond – 2 bricks thick

15.6.5 Garden-wall bond

This type of bond is employed for the construction of garden walls, compound walls, boundary walls, etc.

15.6.5.1 English garden-wall bond

This type of bond comprises of one course of header to three or five courses of stretchers. In order to break the continuity of vertical joints, a queen closer is laid next to the header of the heading course and the middle course of stretchers is started with a header (Figure 15.15).

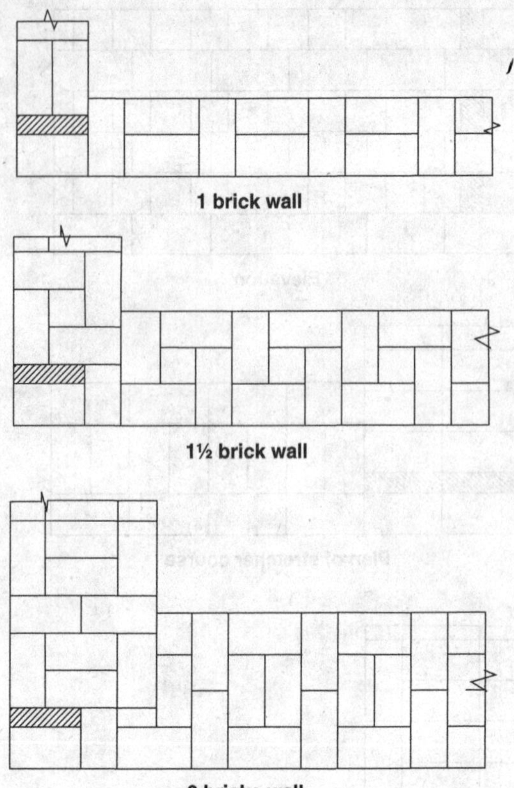

Figure 15.14 Courses 2, 4, 6, etc. of double Flemish bond

Figure 15.15 Garden-wall English bond

15.6.5.2 Flemish garden-wall bond

In this type, each course contains one header to three or five stretchers. A three-fourth brickbat is placed next to the quoin header in every alternate course to develop the necessary lap. A header is placed centrally over each middle stretcher (Figure 15.16).

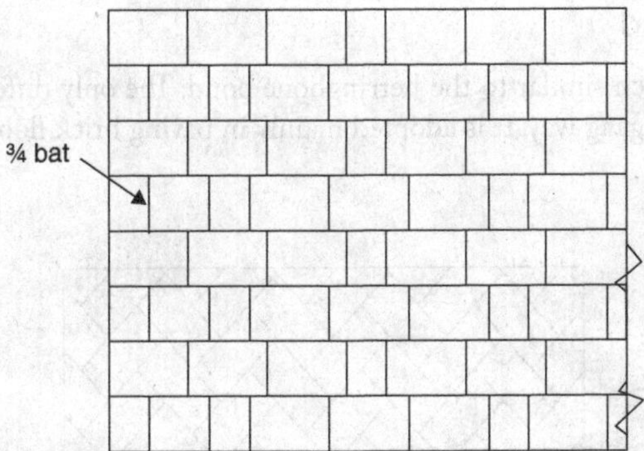

Figure 15.16 Garden-wall Flemish bond

15.6.6 Dutch bond

This type of bond is a modified form of English bond. The corners of the wall provided with the Dutch bond are quite strong. The alternate courses in this type of bond are headers and stretchers. In the stretchers course, a three-fourth bat is used as quoin. A header is placed next to the three-fourth bat in every alternate stretcher (Figure 15.17).

Figure 15.17 Dutch bond

15.6.7 Herringbone bond

In this bond, the bricks are placed at an angle of 45 degrees from the central line in both the directions. This type of bond is used in the case of walls having thickness of more than four bricks or for paving (Figure 15.18).

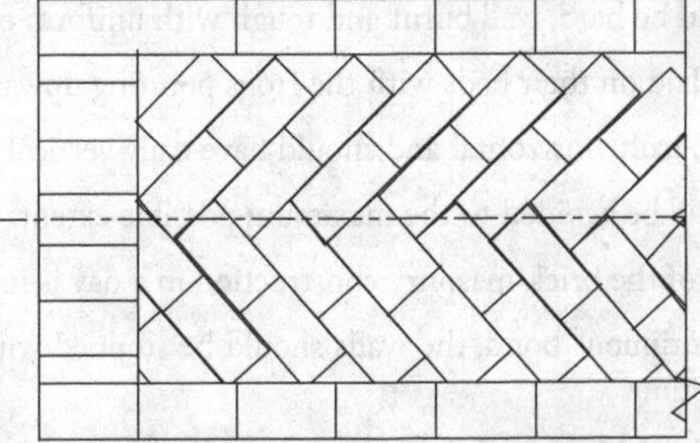

Figure 15.18 Herringbone bond

15.6.8 Zigzag bond

This type of bond is very much similar to the herringbone bond. The only difference in this type of bond is that the bricks are laid in a zigzag way. It is adopted mainly in paving brick floors (Figure 15.19).

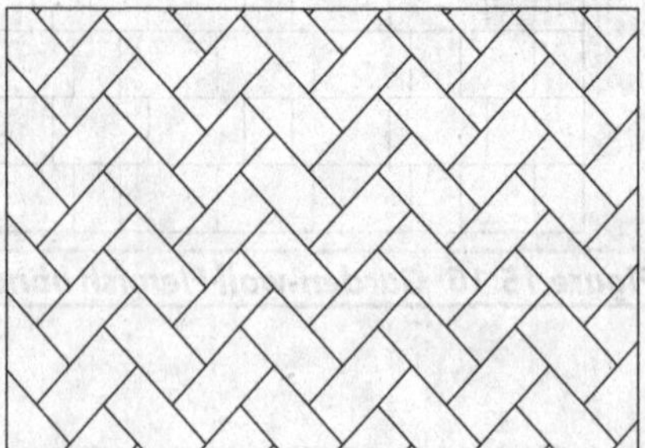

Figure 15.19 Zigzag bond

15.6.9 Brick on edge bond

In this type of bond, the bricks are laid on edge instead of bed. This bond is economical as it consumes less number of bricks and less quantity of mortar. However, it is not strong and, hence, is used for the construction of garden walls, compound walls, partition walls, etc. In this bond, the bricks are laid as headers and stretchers in alternate courses in such a way that the stretchers are laid at the edge.

15.6.10 Facing bond

In this type of bond, bricks of different thickness are used in the facing and backing of the walls. In this case, a header course is placed after several stretcher courses. The distance between the successive heading courses is equal to the least common multiple of the thickness of backing and facing bricks.

15.7 GENERAL PRINCIPLES IN BRICK MASONRY

1. The bricks used should be hard, well burnt and tough with uniform colour, shape and size.
2. The bricks should be laid on their beds with the frogs pointing upwards.
3. The courses should be truly horizontal and should have truly vertical joints.
4. Use of brickbats should be avoided to the maximum possible extent.
5. Generally, the height of the brick masonry construction in a day is limited to 1.5 m.
6. In order to ensure continuous bond, the walls should be stopped with a toothed end at the end of each stage of construction.
7. Finished brickwork should be cured for at least 2-3 weeks where lime mortar is used and for 1-2 weeks where cement mortar is used.

Estimated Quantities of Materials Required Per Cubic Metre of Brickwork

No.	Nominal mix Cement	Nominal mix Sand	Water cement ratio	Cement By weight in kg	Cement By number of bags	Sand in litres	Bricks 19 cm × 9 cm × 9 cm (nos)
1	1	3	0.40	151.5	3.03	315.6	450
2	1	4	0.53	118.5	2.37	329.1	450
3	1	5	0.62	101.3	2.03	339.6	450
4	1	6	0.70	84.0	1.68	350.1	450
5	1	8	0.90	66.0	1.32	366.6	450

Assumption is that one cubic metre of brick masonry is made of 70 per cent bricks and 30 per cent mortar

REVIEW QUESTIONS

1. How are masonry works classified?
2. Define course, phase, stretcher, header and lap of masonry works.
3. What is the difference between king closer and queen closer in a masonry work?
4. What is the difference between frog and bat in masonry work?
5. What are the criteria for the selection of material for stone masonry?
6. What are the general principles of stone masonry?
7. How is stone masonry classified?
8. On what factors does the strength of rubble masonry depend?
9. With a neat sketch explain uncoursed rubble masonry and random rubble masonry.
10. Write short notes on
    a. Flint rubble masonry
    b. Ashlar masonry
    c. Flemish bond
    d. Zigzag bond
11. What are the different types of joints used in stone masonry?
12. What are the different types of brick masonry?
13. What are the different bonds used in brick masonry? Explain about English bond 1-brick thick and 2-brick thick wall masonry using a neat sketch.
14. What are the general principles in brick masonry?

chapter 16

Concrete

Cement concrete is an artificial building material that is obtained by mixing together cement, water and some other inert materials. The mixture in a plastic condition when allowed to set becomes as hard as stone. By suitably adjusting the proportions of various ingredients, concrete with sufficient compressive strength for various uses can be developed. The strength of concrete depends mainly on its ingredients, their relative quantities and the manner in which they are mixed and placed.

Because of its high strength, it is used extensively for construction of roads, heavy structural member-like columns, gravity dams, etc., and also for foundations.

In addition to its strength, concrete also possesses other qualities such as high durability, better appearance, ease of construction, greater fire resistance and economy.

As plain concrete is weak in tensile strength, reinforcing with steel is done to increase the tension carrying capacity.

16.1 INGREDIENTS OF CONCRETE AND THEIR FUNCTIONS

Cement concrete is a composite mixture that consists mainly of relatively inert mineral matter in the form of particles or fragments held together by a binding medium, which gives concrete its solidity and strength. Sand and stone chips or boulders are inert materials and a combination of Portland cement and water is the binder. The inert material, called 'aggregate', is normally graded in size from fine sand to boulders or fragments of stone.

There are four ingredients, which make up the composite material of cement concrete:

1. Cement
2. Sand (fine aggregate)
3. Stone chips or boulders (coarse aggregate)
4. Water

1. **Cement:** The function of cement in the concrete is to bind the coarse and fine aggregate particles together by setting and hardening around such particles. There are different types of cement and each type is used under certain conditions due to its special properties. However, for ordinary construction, generally Ordinary Portland Cement is used. When water is mixed with cement, a chemical reaction takes place because of which the cement paste first loses its plasticity and becomes stiff, at the same time it acquires hardness and strength.

2. **Fine aggregate:** This is the inert or chemically inactive material, most of which passes through a 4.75 mm IS sieve and contains not more than 5 per cent coarser material. The fine aggregates serve the purpose of filling all the open spaces in between the coarse particles, and thus by decreasing the porosity of the final mass, its strength is considerably increased. Sand is universally used as a fine aggregate although many other materials have been developed for special-purpose concretes.

3. **Coarse aggregate:** The inert material, most of which is retained on a 4.75 mm sieve and contains not more than 0–10 per cent of finer materials, is known as coarse aggregate. The function of the coarse aggregate is to act as the main load-bearing component of the concrete. When a good number of coarse aggregate fragments are held together by a binding material, their behaviour towards the imposed loads is just like a rock mass. Gravels and crushed stones are commonly used for this purpose.
4. **Water:** This is the least expensive but most important ingredient of concrete. It governs the important properties related to cement concrete such as durability, strength and watertightness. The purposes of mixing water are (a) to damp the aggregates and prevent them from absorbing the water vitally necessary for the chemical combination between cement and water which is called 'hydration' (b) to flux the cementing material over the surface of the particles of aggregates and (c) to make the concrete workable so that it can be placed easily and uniformly between the reinforcing bars and in the corners.

16.2 PROPORTION OF MIX USED FOR DIFFERENT WORKS

The process of selection of relative proportions of cement, sand, coarse aggregate and water to obtain a concrete of desired quality is known as proportioning the concrete. There are various methods for determining the volumetric proportions of various components, like the arbitrary method, fineness modulus method, minimum voids method and maximum density method.

The recommended mixes of concrete for various types of construction are given in the following table. The maximum sizes of aggregates are also mentioned in the table. The proportions are by volume.

Proportion of concrete mix	Maximum size of aggregate	Nature of work
1:1:2	12–20 mm	Heavily loaded RCC columns and RCC arches of long span
1:2:2	12–20 mm	Small precast members of concrete, such as poles for fencing telegraphs, long piles, watertight constructions and heavily stressed members of the structures.
1:1½ :3	20 mm	Water-retaining structures, piles, precast products, etc.
1:2:3	20 mm	Water tanks, concrete deposited under water, bridge construction and sewers
1:2½:3½	25 mm	Footpaths and roadworks
1:2:4	40 mm	For all general RCC works in building, such as stair, beam, column, weather shed, slab and lintel, machine foundation subjected to vibration and RCC piles.
1:3:6	50 mm	Mass concrete works in culverts, retaining walls, etc.
1:4:8 or 1:5:10 or 1:6:12	60 mm	Mass concrete work for heavy walls, foundation, footings, etc.

16.3 FINE AGGREGATE AND COARSE AGGREGATE

16.3.1 Fine aggregate

Fine aggregate is the inert or chemically inactive material, most of which passes through a 4.75 mm IS sieve and contains not more than 5 per cent coarser material. They may be classified as follows:

a. Natural sand: Fine aggregate resulting from the natural disintegration of rocks and which has been deposited by streams or glacial agencies.

b. Crushed stone sand: Fine aggregate produced by crushing of hard stone.

c. Crushed gravel sand: Fine aggregate produced by crushing of natural gravel.

The fine aggregates serve the purpose of filling all the open spaces in between the coarse particles. Thus, it reduces the porosity of the final mass and considerably increases its strength. Usually, natural river sand is used as a fine aggregate. However, at places, where natural sand is not available economically, finely crushed stone may be used as a fine aggregate.

16.3.2 Coarse aggregate

The inert material, most of which is retained on a 4.75 mm sieve and contains not more than 0–10 per cent of finer materials, is known as coarse aggregate. They may be put under the following categories:

a. Uncrushed gravel or stone which results from the natural disintegration of rocks.

b. Crushed gravel or stone which results from crushing of gravel or hard stone.

c. Partially crushed gravel or stone which is a product of the mixture of the above two types.

The function of the coarse aggregate is to act as the main load-bearing component of the concrete. The nature of work decides the maximum size of the coarse aggregate. For thin slabs and walls, the maximum size of the coarse aggregate should be limited to one-third the thickness of the concrete section. The aggregates to be used for cement concrete work should be hard, durable and clean. The aggregates should be completely free from lumps of clay, organic and vegetable matter, fine dust, etc. The presence of all such debris prevents adhesion of aggregates and, hence, reduces the strength of concrete.

16.4 SIGNIFICANCE OF SAND IN CONCRETE

Sand or the fine aggregates form an important constituent of concrete. It helps to increase the bulk or volume of concrete, which results in the reduction of cost. It helps in the adjustment of the strength of concrete by variation of its proportion with cement. It also increases the resistance to crushing. The aggregates reduce shrinkage and affect economy. Earlier, aggregates were considered as inert materials but now it has been recognized that some of the aggregates are chemically active and also that certain aggregates exhibit chemical bond at the interface of the aggregate and paste. The sand together with the coarse aggregate forms 70–80 per cent of the volume of concrete. The fine aggregate also helps in filling the voids formed by the coarse aggregates.

16.5 WATER–CEMENT RATIO

Cement and water are the only two chemically active elements in concrete. By their combination they form a glue-like binder paste, which surrounds and coats the particles of the inert mineral aggregates, sets and upon hardening binds the entire product into a composite mass. Next only to cement, water is the most important element in concrete governing all the properties of cement concrete like durability, strength and watertightness.

The functions of mixing water are (a) to damp the aggregates and prevent them from absorbing the water vitally necessary for the chemical combination between cement and water which is called 'hydration' (b) to flux the cementing material over the surface of particles of aggregates and (c) to make the concrete workable so that it can be placed easily and uniformly between the reinforcing bars and in the corners.

One of the most recent improvements in concrete manufacture is the control of water in the mixture. The ratio of the amount of water to the amount of cement by weight is known as 'water–cement ratio', and the strength of concrete depends on this ratio.

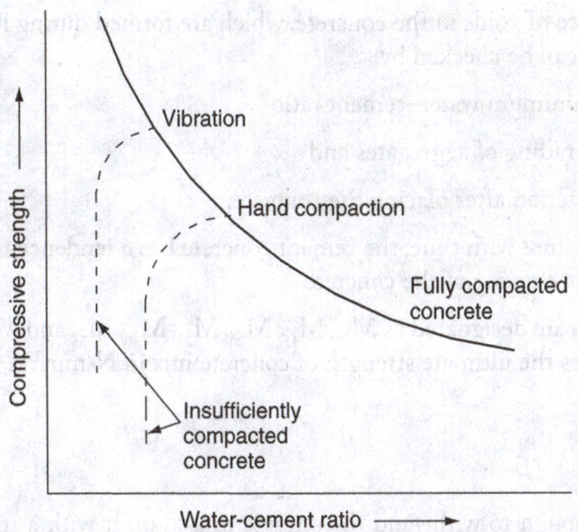

Figure 16.1 Relation between strength and water–cement ratio of concrete

Abram's water/cement ratio law states that the strength of concrete is dependent upon water–cement ratio, provided the mix is workable. The relation between water–cement ratio and the strength of concrete is shown in Figure 16.1.

It can be seen that lower water–cement ratio could be used when the concrete is vibrated to achieve higher strength, whereas comparatively higher water–cement ratio is required when concrete is hand compacted. In both cases, when the water–cement ratio is below the practical limit, the strength of concrete falls rapidly due to introduction of air voids. The graph showing the relationship between the strength and water–cement ratio is approximately hyperbolic in shape.

16.6 PROPERTIES OF CONCRETE

Cement concrete possesses the following important properties:

1. It possesses a high compressive strength.
2. It is free from corrosion and there is no appreciable effect of atmospheric agents on it.
3. It gives a hard surface capable of resisting abrasion.
4. It is more economical than steel, as sand and coarse aggregate, which constitutes the bulk of concrete, are generally available at a cheaper rate. Formwork can be reused for other construction works.
5. Improved appearance and various types of finishes can be given to the concrete surface.
6. It continues to harden and attains more strength as time passes. It is this property of cement concrete that gives it a distinct place among the building materials.
7. It can develop good bondage with steel. Steel reinforcement is usually placed in cement concrete at suitable intervals to take up the tensile stresses, as plain concrete is weak in tension. This is termed as Reinforced Cement Concrete or RCC.

8. Due to the presence of voids in the concrete, which are formed during its placing, it has a tendency to be porous. This can be checked by
   i. The use of minimum water–cement ratio
   ii. The proper grading of aggregates and
   iii. Better compaction after placing the concrete.
9. Due to loss of moisture with time, the cement concrete has a tendency to shrink. The shrinkage can be reduced by proper curing of the concrete.
10. The concrete mixes are designated as M_{10}, M_{15}, M_{20}, M_{25}, M_{30}, M_{35} and M_{40}. 'M' refers to the mix and the number denotes the ultimate strength of concrete mix in N/mm^2 at the end of 28 days.

16.6.1 Strength

Concrete is to be strong enough to withstand the stresses caused on it with a required factor of safety. The strength of the concrete is measured in N/mm^2 as said earlier and it is the ultimate compressive strength of 15 cm cubes after 28 days of curing (sometimes 7-day curing strength is also found out).

The tensile strength of concrete is about 8–12 per cent and shear strength is 8–10 per cent of its compressive strength.

16.6.2 Durability

Concrete should be able to resist the forces of disintegration owing to natural and chemical causes. The durability of concrete can be increased by using good quality materials, adopting optimum water–cement ratio, using dense graded aggregates, careful mixing and placing through compaction and adequate curing.

16.6.3 Workability

Workability is the easiness with which the concrete mix can be mixed, handled, transported, placed, moulded and compacted. A workable concrete should not show any segregation or bleeding, i.e., the materials should not separate out or the excess water should not come up to the surface.

The workability of concrete can be measured by two tests, namely the slump test and the compacting factor test.

16.6.3.1 Slump test

The slump test is carried out to have a rough estimate of the workability of concrete. It does not measure all factors contributing to workability, nor is it always representative of the placeability of the concrete. However, it is conveniently used as a control test and gives an indication of the uniformity of the concrete from batch to batch.

The apparatus for conducting the slump test essentially consists of a metallic mould in the form of a frustum of a cone having internal dimensions as follows:

Bottom diameter	-	20 cm
Top diameter	-	10 cm
Height	-	30 cm

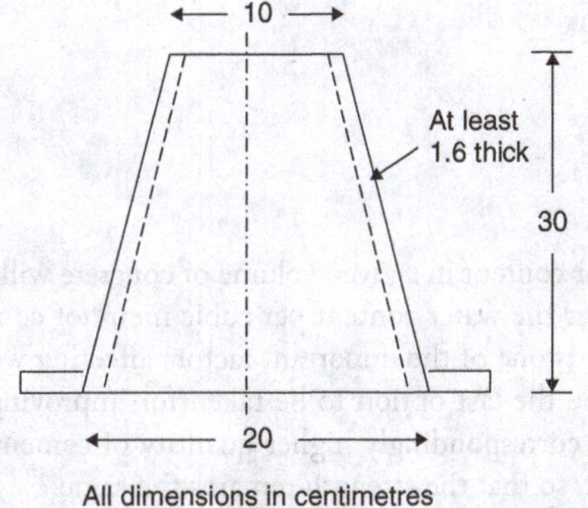

Figure 16.2 Typical mould for slump test

Figure 16.2 shows the details of the slump apparatus.

The internal surface of the mould is thoroughly cleaned and freed from any superfluous moisture. The mould is placed on a smooth, horizontal, rigid and non-absorbent surface. The mould is then filled in four layers, each approximately one-fourth of the height of the mould. For tamping the concrete, a steel tamping rod of 16 mm diameter and 0.6 m length with a bullet end is used. Each layer is tamped 25 times by the tamping rod. After the top layer has been rodded, the concrete is struck off level with a trowel and tamping rod. The mould is removed from the concrete immediately by raising it slowly and carefully in a vertical direction. This allows the concrete to subside. The difference between the height of the mould and that of the subsided concrete is measured in mm and this is referred to as the slump of concrete.

It is seen that the slump test gives good consistent results for a plastic mix. This test is not sensitive for a stiff mix. Despite many limitations, the slump test is very useful on site to check the day-to-day or hour-to-hour variation in the quality of the mix.

16.6.3.2 Slumps for different works

The recommended slump for concrete for different types of works is shown in the table.

Number	Type of concrete	Slump
1	Concrete for road construction	20–40 mm
2	Concrete for top of curbs, parapets, piers, slabs and walls that are horizontal	40–50 mm
3	Concrete for canal linings	70–80 mm
4	Concrete for arch and side walls of tunnels	90–100 mm
5	Normal RCC work	80–150 mm
6	Mass concrete	25–50 mm
7	Concrete to be vibrated	10–25 mm

16.6.3.3 Factors affecting workability of concrete

The major factors affecting the workability of concrete are given below:

i. Water content

ii. Mix proportions

iii. Size of aggregates

iv. Shape of aggregates

v. Grading of aggregates

vi. Use of admixtures

i. **Water content:** Water content in a given volume of concrete will have a significant influence on the workability. The higher the water content per cubic metre of concrete, the higher will be the fluidity of concrete, which is one of the important factors affecting workability. However, increasing the water content must be the last option to be taken for improving the workability. More water can be added, provided a correspondingly higher quantity of cement is also added to keep the water–cement ratio constant, so that the strength remains the same.

ii. **Mix proportions:** Aggregate–cement ratio is an important factor affecting workability. The higher the aggregate–cement ratio, the leaner is the concrete. In the case of lean mix, less quantity of paste is available for providing lubrication and, hence, the mobility of aggregate is restrained. On the other hand, in the case of rich concrete with lower aggregate–cement ratio, more paste is available to make the mix cohesive and fatty to give better workability.

iii. **Size of aggregates:** The bigger the size of aggregates, less quantity of water and paste will be required. Hence, for a given quantity of water and paste, bigger size of aggregates will give greater workability. The above will be true within certain limits.

iv. **Shape of aggregates:** The shape of aggregates greatly influences the workability. Angular, elongated or flaky aggregates make the aggregate very harsh when compared to round- or cubical-shaped aggregates. Contribution to greater workability of rounded aggregates is due to the fact that for a given volume or weight it will have less surface area and less voids than angular or flaky aggregates.

v. **Grading of aggregates:** This is one of the factors which will have maximum influence on workability. A well-graded aggregate is the one which has least amount of voids in a given volume. Other factors being constant, when the total voids are less, excess paste should be available to give better lubricating effect.

vi. **Use of admixtures:** Of all the factors, which affect workability, the most important factor is the use of admixtures. Admixture is defined as a material, other than cement, water and aggregates, that is used as an ingredient of concrete and is added to the batch immediately before or during mixing. Plasticizers and superplasticizers are admixtures that greatly improve the workability many folds. The use of air entraining agents reduces the internal friction between the particles and gives easy mobility to the particles. Similarly, the fine glassy pozzolana materials, in spite of increasing the surface area, offer better lubricating effects for increasing the workability.

16.7 MIXING OF CONCRETE

Thorough mixing of the materials is necessary for the production of uniform concrete. The mixing should ensure that the mass becomes homogeneous, uniform in colour and consistency. There are two methods adopted for mixing of concrete.

1. Hand mixing
2. Machine mixing

16.7.1 Hand mixing

Hand mixing is practised for small-scale unimportant concrete works. As the mixing cannot be thorough and efficient, it is desirable to add 10 per cent more cement to cater for the inferior concrete produced by this method.

Hand mixing should be done on an impervious concrete or brick floor of sufficiently large size to take one bag of cement. Spread out the measured quantity of coarse and fine aggregate in alternate layers. Pour the cement on top of it and mix them dry by shovel, turning the mixture repeatedly until uniformity of colour is achieved. Water is then sprinkled over the mixture and simultaneously turned over. This operation is continued until a good, uniform and homogeneous concrete is obtained. It is of particular importance to see that the water is not poured but only sprinkled. Water in small quantity should be added towards the end of the mixing to get just the required consistency. At this stage, even a small quantity of water makes a difference.

16.7.2 Machine mixing

Mixing of concrete is invariably carried out by machine for reinforced concrete work and for medium- or large-scale concrete work. Machine mixing is not only efficient and fast but also economical when the quantity of concrete to be produced is large.

The mixers for mixing concrete can be classified as batch mixers and continuous mixers. Batch mixers produce concrete batch by batch with intervals whereas continuous mixers produce concrete without stoppage. Continuous mixers are used in large works such as dams. In normal concrete work, the batch mixers are used. Batch mixers may be of pan type or drum type. The drum type may be further classified as tilting, non-tilting and reversing or forced action type.

To get better efficiency, the sequence of charging the loading skip is as described below. Firstly, about half the quantity of coarse aggregate is placed on the skip over which about half the quantity of fine aggregate is poured. On that the full quantity of cement is poured, over which the remaining portion of coarse and fine aggregate is deposited in sequence. This prevents spilling of aggregate while discharging into the drum and also the blowing away of cement in windy weather.

Before the loading skip is discharged into the drum, about 25 per cent of the total water required for mixing is introduced into the mixer drum to wet the drum and to prevent any cement from sticking to the blades or the bottom of the drum. Immediately on discharging the dry materials, the remaining 75 per cent of water is added to the drum.

Concrete mixers are generally designed to run at a speed of 15–20 revolutions per minute. On the site, the normal tendency is to reduce the mixing time to increase the outturn. This results in poor quality of concrete. On the other hand, if the concrete is mixed for a comparatively longer time, it is uneconomical from the point of view of rate of production of concrete and fuel consumption. Therefore, it is of importance to mix the concrete for such a duration that will give optimum benefit. It is seen that the quality of concrete in terms of compressive strengths will increase with increase in the time of mixing, but for mixing time beyond 2 minutes the increase in compressive strength is not very significant.

Concrete mixers are often used continuously without stopping for several hours for mixing and placing. It is of utmost importance that a mixer should not stop in between the concreting operation. For this, the mixer must be well maintained. The mixer is placed on the site at a firm and levelled platform. The drum and blades must be kept clean at the end of the concreting operation. The drum must be kept in tilting position or kept covered when not in use to prevent the collection of rainwater. The skip is operated carefully and it must be provided with proper cushion such as sand bags.

16.8 TRANSPORTING AND PLACING OF CONCRETE

Soon after mixing, the concrete has to be transported and placed in the moulds. A maximum time limit of 1½ hours is allowed between the moment of actual mixing and placing of concrete. It should never be disturbed once the setting has commenced.

While transporting and placing of concrete, care should be taken to avoid segregation and loss of water added.

The surface on which the concrete is to be placed should be cleaned, prepared and well watered. The formwork or moulds should be given a coating of grease or some other material to prevent adhesion of concrete. Concrete should always be laid in even and thin layers and each layer should be well compacted before the next is placed, usually layers of 15–45 cm thickness is adopted. Concrete should never be thrown from a height. Concrete pumps are also now available to place the concrete. The reinforcement should be kept in a fixed position and should never be disturbed.

16.9 COMPACTION OF CONCRETE

Compaction of concrete is the method adopted for expelling the entrapped air from the concrete. If the air is not removed fully, the concrete loses strength considerably. Figure 16.3 shows the relation between loss of strength and air voids left due to lack of compaction.

It can be seen from Figure 16.3 that 5 per cent voids reduce the strength of concrete by about 30 per cent and 10 per cent voids reduce the strength by over 50 per cent. Therefore, it is imperative that 100 per cent compaction is one of the most important points to be kept in mind in good concrete-making practices.

16.9.1 Methods of compaction

The following methods are generally adopted for the compaction of concrete:

a. ***Hand compaction***
   i. Rodding
   ii. Ramming
   iii. Tamping

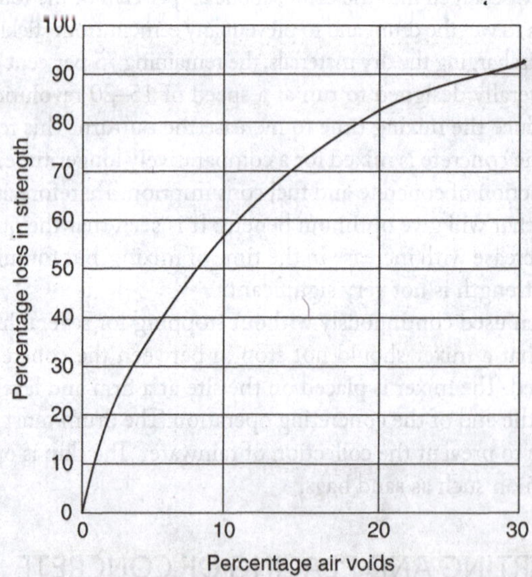

Figure 16.3 Relation between percentage loss in strength and percentage air voids

b. ***Compaction by vibration***

   i. Internal Vibrator (Needle vibrator)
   ii. External Vibrator (Formwork vibrator)
   iii. Table vibrator
   iv. Platform vibrator
   v. Surface vibrator (Screed vibrator)
   vi. Vibratory roller

c. ***Compaction by pressure and jolting***

d. ***Compaction by spinning***

16.9.1.1 Hand compaction

Hand compaction is adopted in the case of small unimportant works. This method is also used in situations where a large quantity of reinforcement is used, which cannot be normally compacted by mechanical means. When hand compaction is adopted, the consistency of concrete is maintained at a higher level. Hand compaction consists of rodding, ramming or tamping.

Rodding is nothing but poking the concrete with a rod about 2 m long and 16 mm diameter to pack the concrete between the reinforcement and sharp corners and edges. Rodding is done continuously over the complete area to effectively pack the concrete and drive away entrapped air.

Ramming should be done with care. Light ramming can be permitted in unreinforced foundation concrete or in ground floor constructions. Ramming should not be permitted in the case of reinforced concrete or in upper floor constructions, where the concrete is placed on formwork placed on struts.

Tamping is one of the usual methods adopted in compacting roof or floor slabs or road pavement where the thickness of concrete is comparatively less. Tamping consists of beating the top surface by a wooden cross beam. Since the tamping bar is sufficiently long, it not only compacts but also levels the top surface across the entire width.

16.9.1.2 Compaction by vibration

To compact concrete with low water–cement ratio, mechanically operated vibratory equipments must be used. Compaction of concrete by vibration has almost completely revolutionized the concept of concrete technology, making possible the use of low slump stiff mixes. Different methods for vibrating the concrete can be adopted.

Internal vibrator

Of all the vibrators, the internal vibrator is the most commonly used. It is also called 'needle vibrator', 'immersion vibrator' or 'poker vibrator'. This essentially consists of a power unit, a flexible shaft and a needle. The needle's diameter varies from 20 to 75 mm and its length varies from 25 to 90 cm. A bigger needle is used in the construction of mass concrete dam.

External vibrator (Formwork vibrator)

Formwork vibrators are used for concreting columns, thin walls or in the casting of precast units. The machine is clamped to the external wall surface of the formwork. The vibration is given to the formwork so that the concrete in the vicinity of the shutter gets vibrated. Since the vibration is given to the concrete indirectly through the formwork, it consumes more power and its efficiency is less than the internal vibrator.

Table vibrator

This is a special type of formwork vibrator, where the vibrator is clamped to the table or the table is mounted on springs, which are vibrated transferring the vibrations to the table. It is commonly used for vibrating concrete cubes. It is adopted mostly in laboratories and in making small, precise precast RCC members.

Platform vibrator

Platform vibrator is nothing but a table vibrator, but it is larger. It is used in the manufacture of large prefabricated concrete elements such as electric poles, railway sleepers and prefabricated roofing elements.

Surface vibrator

Surface vibrators are also known as 'screed board vibrators'. A small vibrator placed on the screed board gives an effective method of compacting and levelling thin concrete members. Mostly, floor slabs and roof slabs are so thin that an internal vibrator or any other type of vibrator cannot be used. In such cases, the surface vibrator can be effectively used.

Vibratory roller

One of the recent developments of compacting very dry and lean concrete is the use of vibratory roller. Such concrete is known as roller compacted concrete. This method is mainly used for the construction of dams and pavements. Heavy roller, which vibrates while rolling, is used for the compaction of dry lean concrete.

16.9.1.3 Compaction by pressure and jolting

This is one of the effective methods of compacting very dry concrete. This method is often used for compacting hollow blocks, cavity blocks and solid concrete blocks. The stiff concrete is vibrated, pressed and given jolts. By employing great pressure, a concrete of very low water–cement ratio can be compacted to yield very high strength.

16.9.1.4 Compaction by spinning

Spinning is one of the recent methods of compaction of concrete. This method of compaction is adopted for the fabrication of concrete pipes. The plastic concrete when spun at a very high speed gets well compacted by centrifugal force.

16.10 CURING OF CONCRETE

Concrete derives its strength by the hydration of cement particles. The quality of the product of hydration and consequently the amount of gel formed depend on the extent of hydration. Theoretically, water–cement ratio of 0.38 is required to hydrate all the particles of the cement and to occupy the space in the gel pores. In the field, even though higher water–cement ratio is used, since the concrete is open to atmosphere, the water used in the concrete evaporates and the water available in the concrete will not be sufficient for effective hydration to take place, particularly in the top layer.

Curing can be considered as creation of a favourable environment during the early period for an uninterrupted hydration. The desirable conditions are a suitable temperature and ample moisture content. Concrete while hydrating releases heat of hydration. This heat is harmful from the point of view of volume stability. The heat generated can also be reduced by means of water curing.

The curing methods may be broadly divided into the following categories:

1. Water curing
2. Membrane curing
3. Application of heat

16.10.1 Water curing

This is considered as the best method of curing as it satisfies all the requirements of curing, namely promotion of hydration, elimination of shrinkage and absorption of the heat of hydration. Water curing can be done in the following ways:

a. Immersion
b. Ponding
c. Spraying or Fogging
d. Wet covering

The precast concrete items are normally immersed in curing tanks for a certain duration. Pavement slabs, roof slabs, etc. are covered under water by making small ponds. Vertical retaining walls or plastered surfaces or concrete columns, etc. are cured by spraying water. In some cases, wet coverings such as wet gunny bags, jute matting and straw are wrapped to the vertical surface for keeping the concrete wet. For horizontal surfaces, saw dust, earth or sand are used as wet coverings to keep the concrete in wet condition for a longer time.

16.10.2 Membrane curing

The quantity of water normally mixed for making concrete is more than sufficient to hydrate the cement, provided this water is not allowed to go out from the body of concrete. For this reason, concrete could be covered with a membrane that will effectively seal off the evaporation of water from concrete. In addition, if concrete works are carried out in places where there is acute shortage of water, the lavish application of water for water curing is not possible due to reasons of economy.

Sometimes, the concrete is placed in some inaccessible, difficult or far off places. The curing of such concrete cannot be properly supervised. In such cases, it is much safer to adopt membrane curing than to leave the responsibility of curing to workers.

Large number of sealing compounds have been developed in recent years. The idea is to obtain a continuous seal over the concrete surface by means of a firm impervious film to prevent moisture in the concrete from escaping by evaporation. Some of the materials that have been used for this purpose are bituminous compounds, polyethylene or polyester film, waterproof paper, rubber compounds, etc.

16.10.3 Application of heat

The development of the strength of concrete is a function of not only time, but also of temperature. When concrete is subjected to higher temperature, it accelerates the hydration process resulting in faster development of strength. The exposure of concrete to higher temperature is done in the following manners:

a. Steam curing at ordinary temperature
b. Steam curing at high temperature
c. Curing by infra red radiation
d. Electrical curing

16.10.3.1 Steam curing at ordinary temperature

This method is often adopted for prefabricated concrete elements. Application of steam to in situ construction will be a difficult task. For steam curing, the concrete elements are stored in a chamber. The chamber should be large enough to hold a day's production. The door is closed and steam is applied. The steam may be applied either continuously or intermittently. An accelerated hydration takes place at this higher temperature and concrete attains the 28-day strength of normal concrete in about 3 days. In large prefabricated factories, they have tunnel curing arrangements. However, concrete subjected to higher temperature at the early period of hydration is found to lose some of the strength gained at a later stage.

It has been emphasized that a very young concrete should not be subjected suddenly to high temperature. A certain delay period after casting the concrete is desirable. In India, steam curing is often adopted for precast elements, especially precast concrete sleepers.

16.10.3.2 Steam curing at high temperature

The high-pressure steam curing is something different from ordinary steam curing, in that the curing is carried out in a closed chamber. The superheated steam at high temperature and high pressure is applied on the concrete. This process is also called 'autoclaving'. The following advantages are derived from the high-pressure steam curing process:

i. High pressure steam cured concrete develops in 1 day or less, the strengths developed at 28 days of normally cured concrete. In addition, it does not lose the strength at a later stage.

ii. High-pressure steam cured concrete exhibits higher resistance to sulphate attack, freezing and thawing action and chemical action. It also shows less efflorescence.

iii. High-pressure steam cured concrete exhibits lower drying shrinkage and moisture movement.

16.10.3.3 Curing by infra red radiation

Curing of concrete by infra red radiations has been practised in very cold climatic regions of Russia. It is claimed that much more rapid gain of strength can be obtained than with steam curing and does not cause a decrease in the ultimate strength as in the case of steam curing at ordinary pressure.

16.10.3.4 Electrical curing

Another method of curing concrete, which is applicable mostly to very cold climatic regions, is by the use of electricity. This method is not likely to find much application in ordinary temperatures due to economic reasons.

16.11 FORMWORK

When concrete is placed, it is in plastic state. It requires to be supported by temporary supports and casings of the desired shape till it becomes sufficiently strong to support its own weight. This temporary casing is known as the formwork.

16.11.1 Requirements of a good formwork

a. ***Easy removal:*** The design of formwork should be such that it can be removed easily with least amount of hammering. This will also prevent the possible injury to the concrete, which has not become sufficiently hard. Further, if the removal of formwork is easy it can be made fit for reuse with little expenditure. The operation of removing the formwork is commonly known as stripping, and when stripping takes place, the components of the formwork are removed and then reused for another part of the structure. Such forms whose components can be reused several times are known as panel forms.

b. ***Economy:*** It is noted that the formwork does not contribute anything to the stability of the finished structure, and hence it will be desirable to bring down its cost to a minimum, consistent with safety. The various steps such as reduction in the number of irregular shapes of forms, standardizing the room dimensions, use of component parts of commercial size and putting the formwork in use again as early as possible may be taken to affect the economy in the formwork. The formwork should be constructed of that material which is easily available at low cost and which can safely be reused several times.

c. ***Less Leakage:*** The formwork should be so arranged that there is minimum leakage through the joints. This is achieved by providing tight joints between sections of the formwork.

d. ***Quality:*** These forms should be designed and built accurately so that the desired size, shape and finish of the concrete are attained.

e. ***Rigidity:*** The formwork should be rigid enough to retain the shape without any appreciable deformation. For visible surfaces in the completed work, the deflection is limited to 1/300 of span and that for hidden surface is limited to 1/150 of span. It should be noted that a rigid formwork would be robust and stiff enough to allow repeated use.

f. ***Smooth Surface:*** The inside surface of formwork should be smooth to turn out a good concrete surface. This is achieved by applying crude oil or soft soap solution to the inside surface of formwork. This also makes the removal of formwork easy.

g. ***Strength:*** The formwork should be sufficiently strong enough to bear the dead load of wet concrete as well as the weights of equipments, labour, etc. required for placing and compacting the concrete. This requires careful design of the formwork. The overestimation of loads results in expensive formwork and the underestimation of loads results in the failure of formwork. The loads on vertical forms are to be assessed from various considerations such as density of concrete, dimension of section, concrete temperature, slump of concrete, reinforcement details, stiffness of forms and rate of pouring of concrete.

16.11.2 Steel formwork

Steel is used for formwork when it is desired to reuse the formwork several times. The initial cost of steel formwork is very high. However, it proves to be economical for large work requiring many repetitions of the formwork. The erection and removal of steel formwork are simple and it presents a smooth surface on removal.

16.11.2.1 Advantages

i. It can be reused several times, nearly ten times more than timber formwork.

ii. It does not absorb water from concrete and, hence, the chances of formation of honey-combed surface are brought down to the minimum level.

iii. It does not shrink or distort and, hence, it is possible to achieve higher degree of accuracy and workmanship by its use as compared to timber formwork.

iv. It is easy to install and dismantle and, hence, there is saving in the labour cost.

v. It gives excellent concrete surfaces requiring no further finishing treatment. The surface obtained by the use of timber formwork invariably requires plastering for getting the desired finish of the concrete surface.

vi. It possesses more strength and is more durable than timber formwork.

vii. The design calculations for the steel formwork system can be made precisely because of the known characteristics of steel.

16.11.3 Timber formwork

When formwork is required for small works requiring less repetition, timber is preferred to steel. Timber formwork is cheap in initial cost and it can be easily adopted or altered for new use. The timber to be used as formwork should be well seasoned, free from loose knots, light in weight and easily workable with nails without splitting.

a. The timber formwork should be neither too dry nor too wet. If it is too dry, the timber will swell and get distorted when wet concrete is laid on it and honey-combed surface will appear on removal of the formwork. On the other hand, if it is too wet, the timber will shrink on hot weather resulting in gaps in the formwork through which concrete will flow out. Hence, ridges will be formed on the concrete surface. It is found that a moisture content of about 20 per cent is appropriate for the timber formwork.

b. The dimensions of components of the timber formwork will depend upon the loads to be carried and the availability of timber sections. However, generally the latter is the governing factor as the former can be adjusted by suitable spacing of the supports.

c. Minimum number of nails should be used in timber formwork and the nail heads should be kept projecting so as to facilitate easy removal.

d. The timber formwork proves to be economical for buildings with minimum number of variations in the dimensions of the rooms. Thus, the cutting of timber pieces is brought down to the minimum.

e. It is common practise to support formwork for slab in buildings with the timber ballies, which are cut to approximate sizes with wedges below them for final adjustments. It leads to the formation of weak points, which are seldom prevented from displacement. The timber ballies are generally not straight and they do not transmit the load axially.

Plywood as formwork is becoming popular at present over the timber formwork because:

a. It can be reused several times as compared to ordinary timber formwork. Under normal conditions the plywood formwork can be used 20–25 times and the timber formwork can be used 10–12 times.

b. It gives surfaces which are plain and smooth and, hence, it may not require any further finishing treatment.

c. It is possible to cover up more area by using large size panel and, hence, there is considerable reduction in the labour cost of fixing and dismantling formwork.

16.11.4 Failures of formwork

Safety must be given importance in the design, construction, erection and stripping of formwork systems. The general rules to be observed to avoid the failure of formwork for concrete structures are as follows:

a. If high shoring is not suitably strengthened by diagonal braces, there are chances for formwork failure to occur.

b. It should be remembered that the forms are continuously supported structures and as such they must be provided with uniform bearing at each support.

c. The entire work should be carried out under the strict and direct supervision of skilled persons or engineers only.

d. The design of formwork should provide for possible shocks and vibrations.

e. The details that are difficult to perform should be avoided as in many cases such details will not be satisfactorily performed and may become the starting point for causing a formwork failure.

f. The stripping of form and reshoring should not be carried out in an unbalanced way. It will otherwise lead to unnecessary stresses in freshly laid concrete.

g. The wedging of posts to counterbalance load compression must be carried out with extreme care so that the assembled form support remains undisturbed.

16.11.5 Formwork for columns

The column formwork consists of a box prepared for four separate sides (Figure 16.4). The four sides of the box are held in position by wooden blocks, bolts and yokes. The important features in the RCC column are:

a. The formwork should be designed to resist the high pressure resulting from the quick filling of the concrete.

b. The spacing of yokes is about 1 m. However, it should be carefully determined by working out the greatest length of the formwork, which can safely resist the load coming on the formwork.

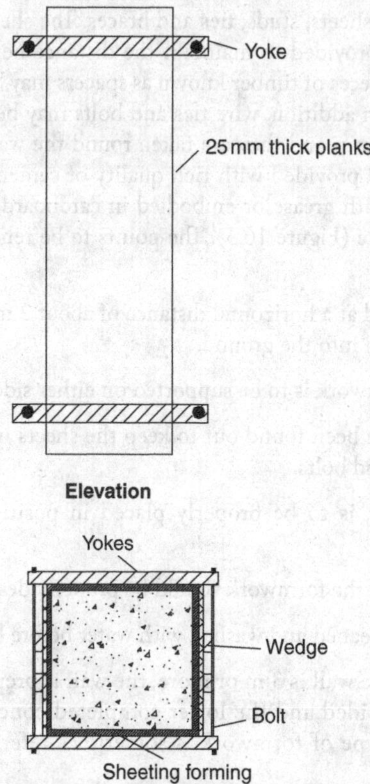

Figure 16.4 Wooden formwork for rectangular or square column

c. Depending upon the shape of the column, the box can be suitably prepared.

d. A hole is generally provided at the bottom of the formwork of column to remove the debris, which might have fallen before the concrete is placed. This hole is termed as the cleanout hole or washout hole and it is filled up before the placing of the concrete starts.

e. A wash with water is given to the inside of the formwork just before starting the laying of concrete.

f. The boxes should be designed in such a way that with little alterations they can be reused for columns with smaller cross sections on upper floors.

g. In order to make the dislocation of boxes easy, the nails are kept projecting instead of being firmly driven. Thus, they can be removed easily by the claw hammer.

h. The wooden yokes being efficient and cheap are wisely used, but they can be replaced by metal clamps of suitable design.

i. The formwork for circular columns is made of narrow vertical boards. These are known as the staves and they are correctly shaped to the required curvature. The staves in turn are fixed to the yokes, which are also suitably curved.

16.11.6 Formwork for walls

The formwork for walls consists of sheets, studs, ties and braces. The sheets are supported by vertical studs and horizontal wales. The ties are provided to maintain the distance between the sheets and to resist the bursting action of concrete. Small pieces of timber known as spacers may be used and they are to be removed as the concrete reaches that level. In addition, wire ties and bolts may be provided. The wires are placed at a horizontal distance of about 600 mm and they are taken round the wales. When formwork is struck off, the ends of wire ties are cut off and provided with rich quality of cement mortar to avoid rusting. If bolts are used, they are to be provided with grease or embodied in cardboard tubes to make their removal easy after 2 or 3 days of pouring concrete (Figure 16.5). The points to be remembered in the case of formwork for walls are as follows:

a. The braces may be provided at a horizontal distance of about 2 m and they are supported at ground level by stakes firmly driven into the ground.

b. If the wall is high, the formwork is to be supported on either side by guy wires instead of braces.

c. Several patent devices have been found out to keep the sheets in the correct position. They can be used in place of wire ties and bolts.

d. The reinforcement, if any, is to be properly placed in position before the laying of concrete starts.

e. The various components of the formwork should be properly designed.

f. The formwork should be cleaned and washed with water before laying concrete in it.

g. If the vertical height of the wall is 3 m or more, the wall is prepared in vertical lifts of about 1 m height. The braces are avoided and the lower completed concrete wall works as a platform for the upper portion. This type of formwork is sometimes referred to as the climbing formwork (Figure 16.5).

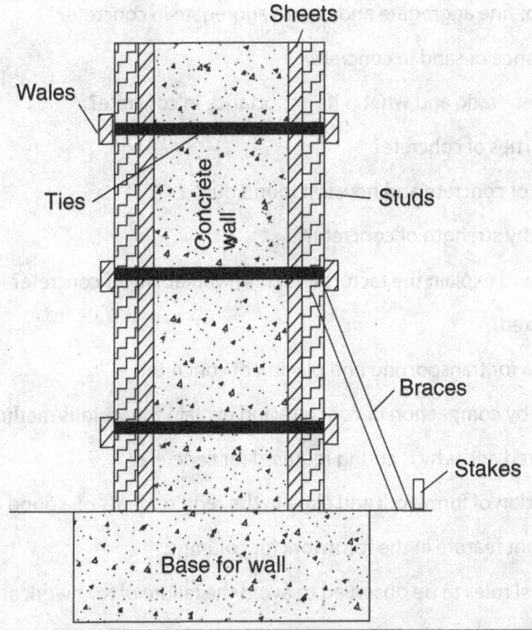

Figure 16.5 Formwork for wall

16.11.7 Stripping time of formwork

In normal circumstances and where Ordinary Portland Cement is used, forms may generally be removed after the expiry of the following periods.

a. Walls, column and vertical faces on all structural members	24–48 hours as may be decided by the engineer-in-charge
b. Slabs (props left under)	3 days
c. Beam soffits (props left under)	7 days
d. Removal of props under slabs:	
1. Spanning up to 4.5 m	7 days
2. Spanning over 4.5 m	14 days
e. Removal of props under beams and arches:	
1. Spanning up to 6 m	14 days
2. Spanning over 6 m	21 days

For other cements, the stripping time recommended for Ordinary Portland Cement may be suitably modified.

REVIEW QUESTIONS

1. What are the ingredients of concrete and their functions?
2. What are the general proportions of concrete mix used for different works?

3. What are the roles of fine aggregate and coarse aggregate in concrete?
4. What is the significance of sand in concrete?
5. What is water–cement ratio and what is its importance in concrete?
6. What are the properties of concrete?
7. What is workability of concrete and how is it found out?
8. What do you mean by strength of concrete?
9. What is slump test and explain the factors affecting workability of concrete?
10. How is concrete mixed?
11. What are the criteria for transporting and placing of concrete?
12. What do you mean by compaction of concrete and explain the various methods of compaction?
13. How is concrete cured and why is curing required for concrete?
14. What is the application of formwork and explain the requirements of a good formwork?
15. Explain the important feature in the formwork for columns.
16. What are the general rules to be observed to avoid the failure of formwork of concrete structures?
17. Write short notes on
    a. Formwork for walls
    b. Stripping time for formwork
    c. Timber formwork
    d. Steel formwork
    e. Curing of concrete by infra red radiation.
    f. Water curing of concrete
    g. Hand compaction of concrete
    h. Compaction of concrete by vibration.

chapter 17

Doors and Windows

A door is a movable barrier secured in an opening, known as the doorway, through a building wall or partition for the purpose of providing access to the inside of a building or rooms of a building. A door is held in position by doorframes, the members of which are located at the sides and top of the opening or doorway. Sills may or may not be provided at the bottom of doorways. A window is defined as an opening in a wall of a building to serve one or more of the functions like natural light, natural ventilation and vision.

The main function of a door in a building is to serve as a connecting link between the internal parts and to allow free movement to the outside of the building. Windows are generally provided for the proper ventilation and lighting of a building and their size and number should be properly determined as per the requirements. To perform their basic functions, the following functional requirements should be satisfied in their design and construction.

i. *Weather resistance:* They should be strong enough to resist the adverse effects of weather such as wind and rainfall.

ii. *Sound and thermal insulation:* They should be capable of being made airtight to achieve insulation against sound and heat. They should act as vertical barriers like walls for the passage of sound, heat and fire.

iii. *Damp prevention and termite proofing:* They should not be affected by termite and moisture penetration as they reduce their strength and durability.

iv. *Fire resistance and durability:* They should offer fire resistance and should be durable.

v. *Privacy and security:* They should offer sufficient privacy without inconvenience or trouble and security against burglars.

17.1 LOCATION OF DOORS AND WINDOWS

The designer or planner should observe the following rules while deciding the location of doors and windows:

1. The number of doors should be kept minimum for each room because larger numbers cause obstruction and decrease the utility of the accommodation. The location and size of the doors should be based on their functional requirement.
2. From the viewpoint of utility and privacy of the occupants, doors should preferably be located near the corner of a room.
3. For good ventilation and free air circulation inside the room, the doors should be located in opposite walls facing each other.
4. The location, number and size of the windows are decided considering various factors, like desired daylight, vision, privacy, ventilation and heat loss.

5. The sill of a window should be located at a height of 0.75–1 m above the floor level. However, windows when exposed to public places like shopping centres and cinema theatres are located at a higher level say about 2 m. This is essential for achieving privacy in buildings on the ground floor.
6. Doors and windows should be located by keeping in view the interior decoration of the room and views of the building owner.

17.2 DOORS

From the operational point of view, the doors are classified as below.

17.2.1 Swinging doors

In these doors, the shutter is hung to the door frame on hinges or butts fixed to one side of the shutter, so that they swing on a vertical axis. The doors may be single swinging, double swinging or double acting type.

In single swinging type doors, if a person is standing on the outside of a door and the hinges are at his left, the door is a left-hand door, but if they are at his right it is a right-hand door.

In double swinging doors, the shutters are hinged at opposite sides of an opening. These doors are extensively used at the entrances of buildings. In double acting doors, the shutters are provided with special hinges, which keep the door closed when it is not held open. Doors of this type can be easily pushed in either direction.

17.2.2 Folding doors

These doors are usually single or as folding partitions so that two rooms may be used together as a single room or separately. They are made of wood or metal and are used for very large openings. Doors are also hinged together.

17.2.3 Sliding doors

Sliding doors that slide sideways were extensively used in the past for residences. The door shutters can also slide either upward or downward. These doors do not cause any obstruction in movements. The vertical sliding doors are pulled up by cables or chains and are used for large openings in industrial and freight elevator doors. The right angle doors are suspended from an overhead track and are used to a very limited extent for garages.

17.2.4 Rolling doors

This is a modification over sliding doors. These doors are generally made of steel or slates of sheet metal and can be easily closed or opened by slightly pulling or pushing the shutter. They do not require much space and are commonly used for garages, show rooms, shops, etc.

17.2.5 Revolving doors

These doors are extensively used where frequent opening and closing of doors are to be avoided due to heavy foot traffic, like markets, public buildings, hotels, stores, theatres and hospitals. The arrangements are made to rotate the door to about one side of the shutter and get it closed automatically whenever pushed and left.

17.2.6 Collapsible doors

These doors consist of a mild steel frame, which is made up of light steel channel sections. They are provided with rollers at the bottom and top to roll on rails when they open or collapse. These doors work without hinges and can be opened or closed by a slight pull or push. These doors are extensively used for residential buildings, public buildings, schools, etc.

17.3 WINDOWS

17.3.1 Fixed window

It is fixed in the wall and makes no provision for natural air circulation.

17.3.2 Double hung window

In this type, both sashes slide vertically with the weight balanced by the sash weight, spiral springs or tape spring balances.

17.3.3 Horizontal sliding window

In this type, either one or both sashes are arranged to slide horizontally. Sashes are sometimes suspended from rollers operating on overhead tracks. Heavy sashes are often provided with nylon rollers at the bottom for ease in operation.

17.3.4 Casement window

Any hinged window, which may swing out or in like doors, is termed as casement window. These windows usually swing on extension hinges provided on the sides. There are out-swinging and in-swinging casement windows with two sashes. Extension hinges are used to make the sash swing clear of the inside surface of the wall.

17.3.5 Folding window

It is a form of out-swinging casement window with the two sashes hinged together on its meeting stile, rather than each to its outside stile. The projection arms are so arranged that the sashes operate symmetrically.

17.3.6 Pivoted window

Horizontally pivoted sash windows are often arranged in a row to form a continuous window in a sawtooth roof or monitor and they operate in harmony from the floor by a mechanical operation.

17.3.7 Top- and bottom-hinged window

Sash windows may be top-hinged out-swinging or top-hinged in-swinging or bottom-hinged in-swinging type.

17.3.8 Projected window

A window with a ventilation sash that projects outwards or inwards is called a projected window. In this type, the ends of the arms are pivoted to the side of the sash and to the frame.

17.3.9 Hopper window

Any inward projecting window when located at or near the bottom of a window is termed as hopper window.

17.4 TYPES OF DOORS

The doors commonly used in buildings are classified into various types depending upon several factors or aspects like materials used in the manufacture of doors, arrangement of door components, method of construction and their working operation.

17.4.1 Battened and ledged door

This is the simplest form of door, which is frequently used for narrow openings. The use of this type of door is preferred where cost is the main factor rather than the strength and appearance. This door consists of vertical boards known as battens, which are secured by horizontal pieces known as ledges. Usually, there are three ledges, namely top ledge, middle ledge or lock edge and bottom edge. The outer edges of the ledges are generally chamfered. The bottom and middle ledges are sometimes wider than the top ledges. The battens secured by means of tongued and grooved joint are either V-jointed or beaded (Figure 17.1).

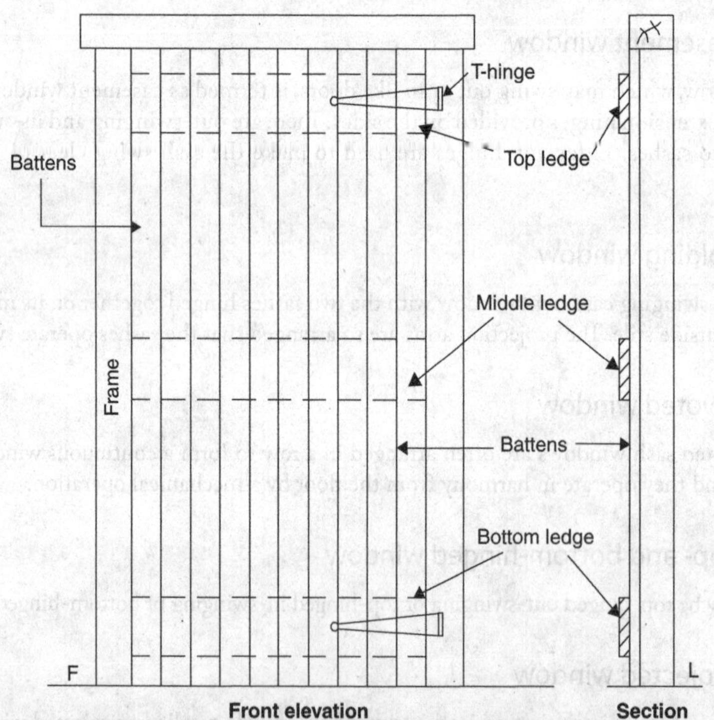

Figure 17.1 Battened and ledged door

The sizes of the door components are as follows:

a. Vertical battens: width is 10–20 cm; thickness is 2–4 cm
b. Top ledge: width is 10 cm; thickness is 3-4 cm
c. Middle and bottom ledge: width is 15–20 cm; thickness is 3-4 cm

17.4.2 Battened, ledged and braced door

This door is a modification over battened and ledged doors in which additional diagonal members called braces are provided to increase its rigidity and, hence, the strength. These braces act as struts as they are made to incline upwards from the hanging edge. By doing so, the tendency of dropping at the nose, in the case of wider doors, is prevented. Thus, these types of doors can be used for wider openings.

Braces: width is 10–15 cm; thickness is 3-4 cm. All other members are of the same size as in the above case.

This door is commonly adopted for bathrooms where the appearance is not so important as economy (Figure 17.2).

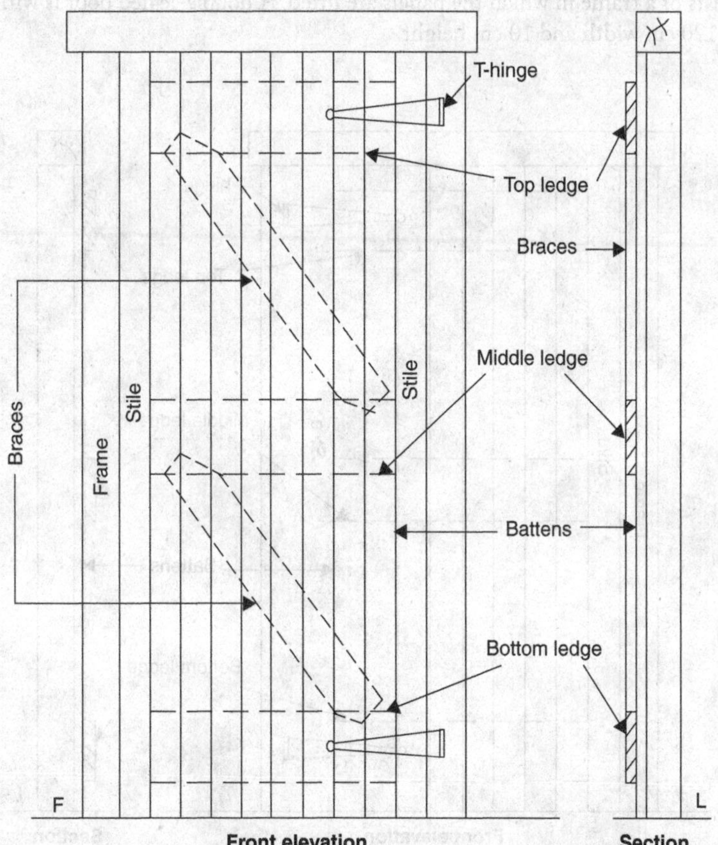

Figure 17.2 Battened, ledged and braced door

17.4.3 Battened, ledged and framed door

This door is provided with a framework for shutters to have better strength and appearance than the ordinary battened and ledged doors. This door consists of two vertical styles, three ledges or rails, bottom, top and middle and battened fixed in the framework. Battens and ledges are provided as usual. Stiles are generally 10 cm in width and 4 cm in thickness (Figure 17.3).

17.4.4 Battened, ledged, framed and braced door

This door is a modification over the previous type in which additional members known as braces have been introduced to increase its strength, durability and appearance. This type of door is largely used for external work. This door has a framework consisting of two stiles, three ledges on rails, battens and two inclined braces. Generally, the thickness of the top rail and the stiles is same and equal to that of the braces and batten together. The braces are 1.5 × 12 cm (Figure 17.4).

17.4.5 Framed and panelled door

This type of door is very commonly used and is constructed in various designs. The object of using such a door is to obtain a framework in which the tendency of shrinkages is reduced and the appearance enhanced. This type of door consists of a frame in which the panels are fitted. A double-leafed door is with modular dimensions of frame as 120 cm width and 10 cm height.

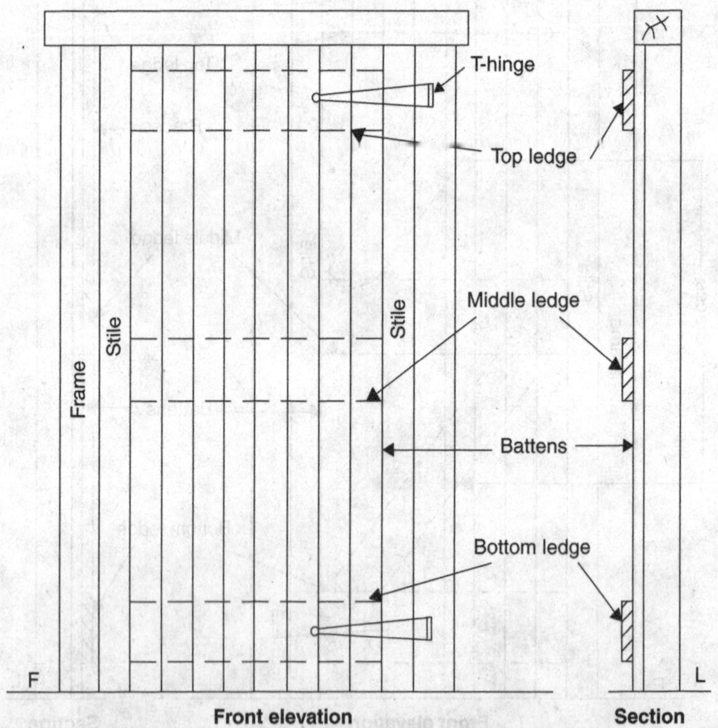

Figure 17.3 Battened, ledged and framed door

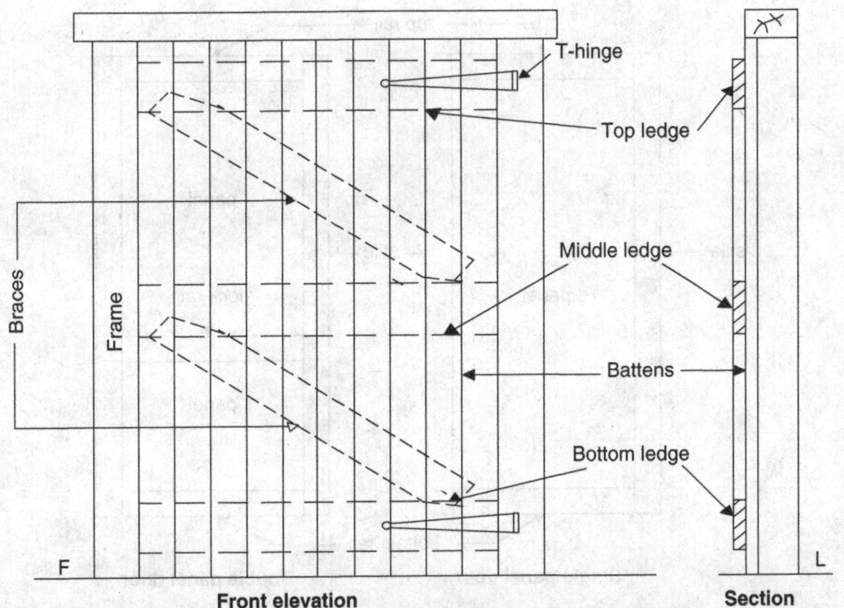

Figure 17.4 Battened, ledged, framed and braced door

a. The stiles are continuous from top to bottom for the full height.

b. The rails, top rail, bottom rail and lock rail, are jointed to the stiles.

c. The frame consists of narrow pieces, mortised and tenoned to each other and grooved on all the inside faces to receive the panels.

d. Bottom and lock rails are made of bigger size and are stronger than top and frieze rails.

e. It is generally recommended that the minimum width of stiles should not be less than 10 cm and for lock and bottom rails not less than 15 cm. The thickness of the shutter frame is usually kept 4-5 cm, but the actual value depends upon several factors like door size, situation of door, type of work, thickness of panels and size of the moulding (Figure 17.5).

17.4.6 Glazed or sash door

Sometimes, the doors either fully glazed or partly glazed and partly panelled are used to supplement the natural lighting provided by windows or to make the interior of one room visible from another (Figure 17.6). The glazed or sash door is extensively used these days in residential as well as public buildings. When sufficient light is required to be admitted through doors then fully glazed doors are provided (Figure 17.7). In the case of partly glazed and partly panelled doors, the usual proportion of glazed portion to the panelled portion is kept 2:1.

17.4.7 Flush door

Flush doors are becoming more popular these days for residential and public buildings because of several good characteristics like pleasing appearance, simplicity of construction, economy, strength and high durability.

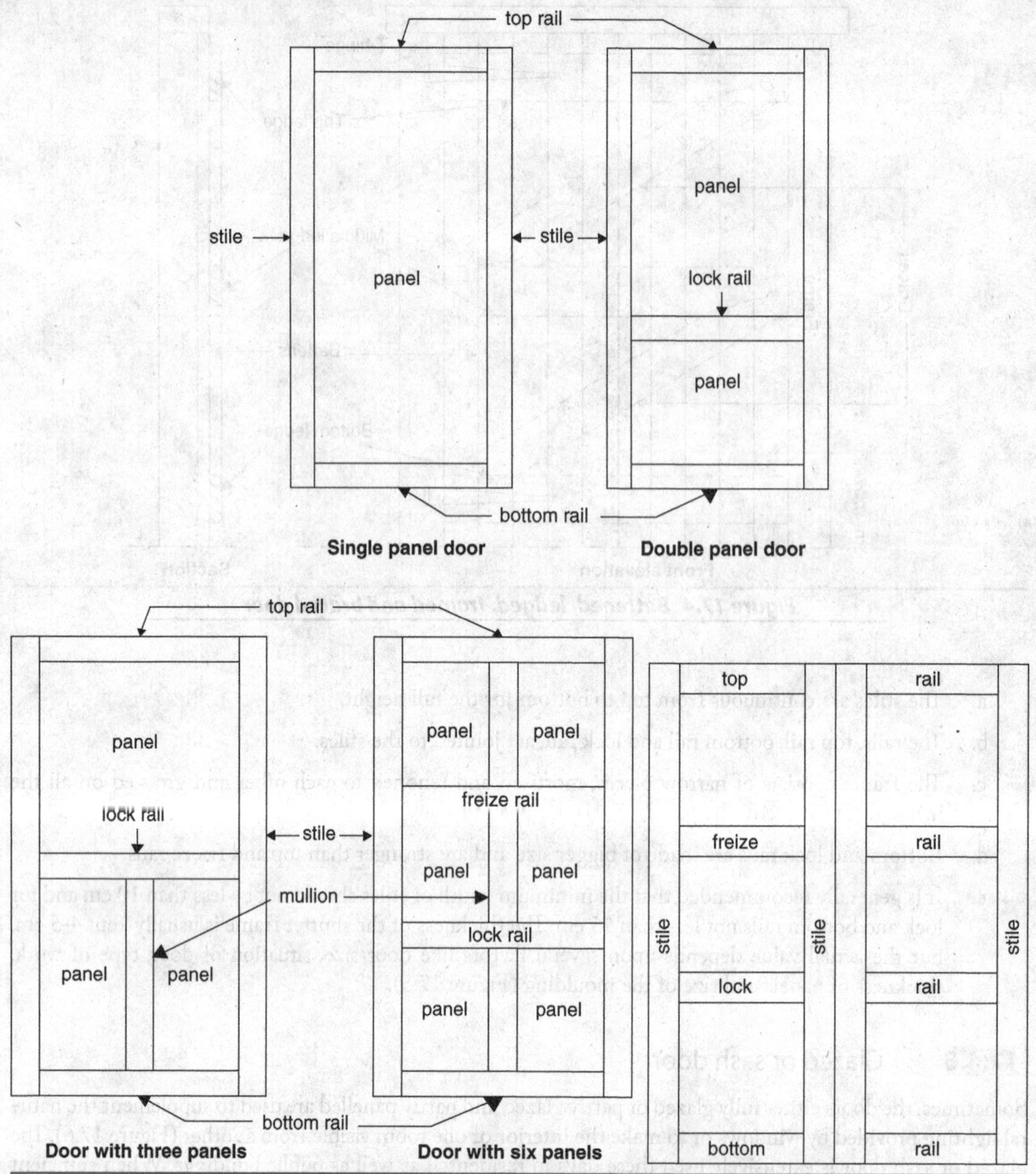

Figure 17.5 Framed and panelled door

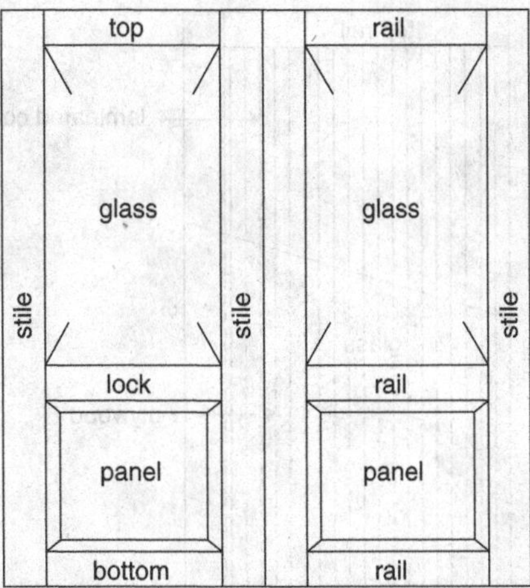

Figure 17.6 Partly panelled and partly glazed door

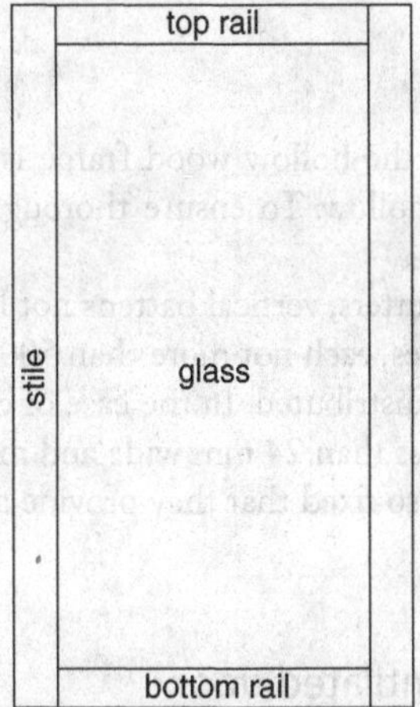

Figure 17.7 Fully glazed door

A flush door consists of a skeleton or a hollow framework of rails and stiles and it is covered on both the sides with laminated boards or plywood. This door has, therefore, fewer perfectly flush and joint surfaces on both the sides. The nominal thickness of flush door shutters varies from 25 to 40 mm depending upon the type of door (Figure 17.8).

a. *Solid Core Flush Door:* This door consists of a core of strips of a wood glued together under great pressure and faced on each side by plywood sheets. The laminated strips are not less than 20 mm in width and are glued edge to edge. The solid core or limited flush doors are heavy and require more material for their construction.

b. *Hollow and Cellular Core Flush Door:* In this type, the frame consists of stiles and top, bottom and intermediate rails, each not less than 7.5 cm wide, and this frame is covered on both the sides by

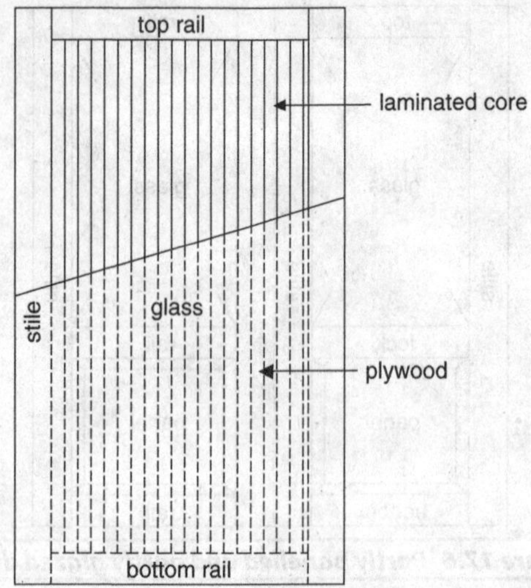

Figure 17.8 Flush door

sheets of plywood. Sometimes, the hollow wood frame is filled with granulated cork or any light material instead of being left hollow. To ensure thorough circulation of air within the framing, ventilation holes are provided.

In the case of hollow core shutters, vertical battens not less than 25 mm wide are so fixed on the rails that they provide void spaces, each not more than 500 cm². Moreover, their void spaces or hollow portions should be equally distributed. In the case of cellular core shutters, vertical battens and horizontal battens or ribs not less than 24 mm wide and made up of strips of wood and plywood or blocks of compressed wood are so fixed that they provide a grid of void spaces, each not more than 25 cm² in area.

17.4.8 Louvered door or ventilated door

The use of this door is recommended when privacy combined with natural ventilation and quietness for rest is desired, because it allows free passage of air even when closed. These doors are not much favoured since they are difficult to clean. The louvers are made either movable by connecting them to the pivot by means of hinges or fixed into the stile. In the case of movable type, the upward or downward movement of louvers is carried out by lowering and raising the pivot, respectively. Louvers are made of glass or timber.

17.4.9 Wire gauzed door

This door is used where it is desired to allow free air into the room and avoid the nuisance of insects, mosquitoes, etc. The wire gauzed door consists of vertical stiles and horizontal rails with fine mesh galvanized gauze fixed to the shutters by a bead braded or nailed to the frame. Generally, the frame is provided with two types of shutters. Shutters of ordinary panelled type are provided on the front side and are made to open to the outside of the room, whereas another type of shutter is provided with a wire gauze and is made to open to the inside of the room.

17.4.10 Collapsible steel door

The collapsible steel door neither requires hinges for opening and closing the shutters nor any frame for hanging them. This door is extensively used for the main entrance of residential buildings, shops, garages, etc. where the width of the door is large and the space is sufficient to provide two-leafed hinged shutters. This door being very strong can be used in exposed situations to safeguard against robbers. It may be made of single or double shutters depending upon the size of the opening.

It is fabricated from vertical pieces of rolled steel channels 16–20 mm wide, joined together with the hollows of the channel on the inside, leaving a vertical gap of 12–20 mm between them. Rollers are provided both at their top and bottom or at the top in some case. The doors can be opened or closed by a slight pull or push. The vertical channel pieces are spaced at 10–12 cm centre to centre and are joined to one another by means of hoop iron cross pieces or flats 16–20 mm wide and 5 mm thick, which allow the door to open or fold (Figure 17.9).

17.4.11 Rolling steel shutter door

This door is capable of being rolled up at the top easily and causes no obstruction either in the opening or floor space. It is commonly used for the main entrance of shops, garages, godowns, etc. It is sufficiently strong and offers proper safety to the interior when closed. A rolling steel door consists of a frame, a drum and a shutter of thin steel plates or iron sheets of thickness about 1 mm and width varying up to 6 m. Steel guides are provided on the sides for the movement of shutters. The door is counterbalanced by means of helical springs enclosed in the drum.

17.4.12 Revolving door

This door provides entrance on one side and exit on the other side simultaneously. It keeps the opening automatically closed when it is not in use. These doors are provided where there is constant foot traffic of people entering and leaving the entrance of public buildings. Revolving doors consist of four upright cross

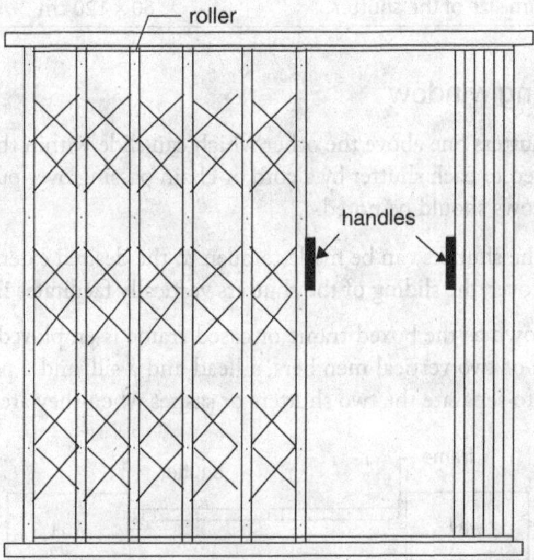

Figure 17.9 Collapsible door

wings, i.e., shutters which are arranged diagonally on the sides of a centrally placed pivot. These shutters revolve about this pivot or vertical axis.

17.4.13 Side sliding door

In this type of doors, the need of hinges for fixing shutters to frames of doors is eliminated. It is commonly used for the entrance of godowns, sheds, shops, garages, etc. The shutter of this type of door consists of one or several panels, which can slide either on one side or both the sides. These doors are operated by sliding the doors on the sides with the help of runners at the top and guides at the bottom.

17.5 TYPES OF WINDOWS

Windows are provided in the wall openings for admission of light, free circulation of air and sunshine. As a general rule, the minimum area of a window or windows should be one-tenth of the floor area of the room and at least half of this area should be made open for ventilation.

17.5.1 Casement window

The simplest type of casement window consists of a square or rectangular window frame of metal or wood, with a casement hinged at one side to the frame to open out. The side-hinged opening part of the window is known as the casement and it consists of glass surrounded and supported by metal or timber strips. The casement is hinged to open out because an outward opening casement can more readily be made to exclude rain and wind than one opening inwards. The usual sizes of the component parts of a casement window are as follows (Figure 17.10):

	Width × thickness
Vertical stiles, top rails and bottom rails	7 × 3.5 cm–9.5 × 3.5 cm
Heads, mullions and transoms	9 × 7 cm–12 × 7 cm
Sash or glazing bars	3.5 × 3.5 cm
Maximum size of the shutter	60 × 120 cm

17.5.2 Double hung window

This window has a pair of shutters one above the other which can slide within the grooves in the frame. Two metallic weights are connected to each shutter by a cord or chain passing over pulleys. The following features regarding double hung windows should be noted.

a. In these windows, the shutters can be made to open to the desired extent and, hence, ventilation can be controlled. Moreover, the sliding of the shutters vertically facilitates in the cleaning of the shutter.

b. A special frame known as the boxed frame or cased frame is employed for sashes sliding vertically. This frame consists of two vertical members, a head and a sill, and a parting bead. The function of the parting bead is to separate the two shutters or sashes when they are opposite to each other.

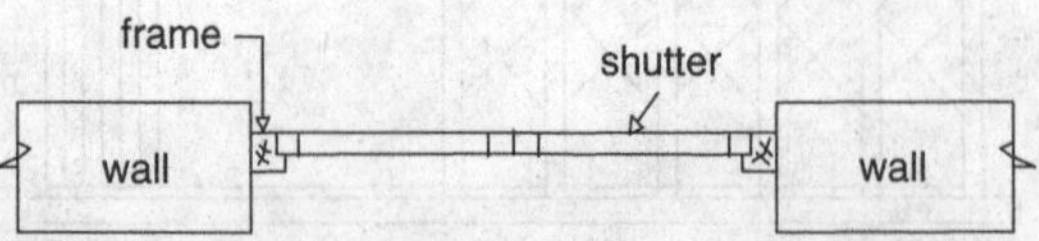

Figure 17.10 Wooden casement window

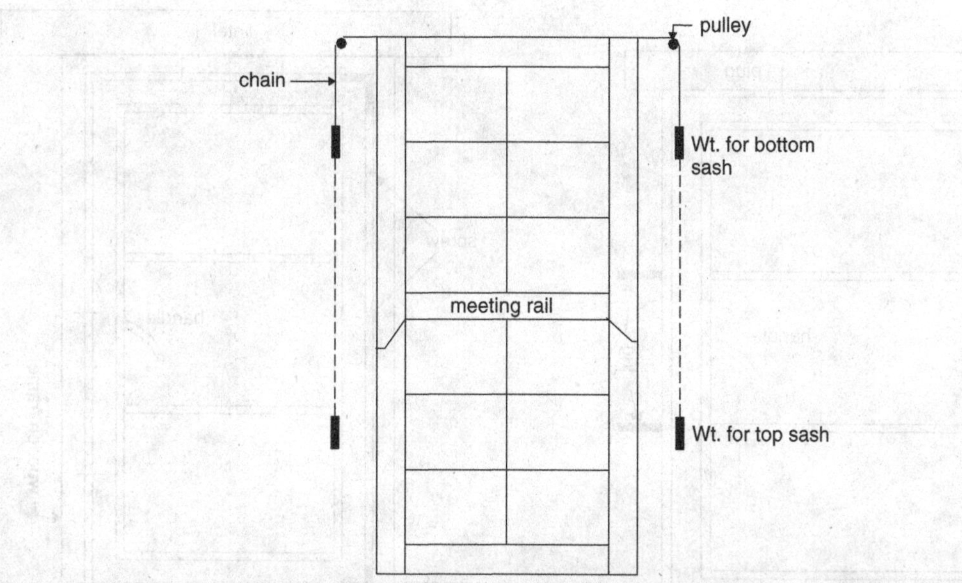

Figure 17.11 Double hung window

c. The shutters are constructed as usual and consist of two stiles, a top rail, a bottom rail and sash bars. The thickness of the shutters is kept to about 5 cm.

d. The metallic counter-weights usually of cast iron are made as heavy as the sash or shutter so that little effort is required to slide the sashes up or down (Figure 17.11).

17.5.3 Sash or glazed window

A sash consists of stiles, rails, transoms and mullions. The glass panes are cut into smaller size than the size of the panels into which they are to be fitted. It is essential to allow slight movement of the sash due to temperature changes. The sash may be rigidly fixed to the frame, hinged at the stiles or may be slided horizontally or may be pivoted to rotate horizontally or vertically. Generally, the sash is hinged to one stile and opens out to keep it watertight.

17.5.4 Louvered window

This window provides free passage of air and sufficient light even when closed. It affords sufficient privacy and also provides protection against excessive daylight and glare inside buildings. It may be fixed or moving type.

17.5.5 Metal window

In the modern age, this type of window is becoming more popular. It is fabricated from light rolled steel sections which form the window frame. The glass panes are fixed into the frame. The glazed shutter may be hung at the top, bottom, side or pivoted to rotate in any direction. The double hung type window may be

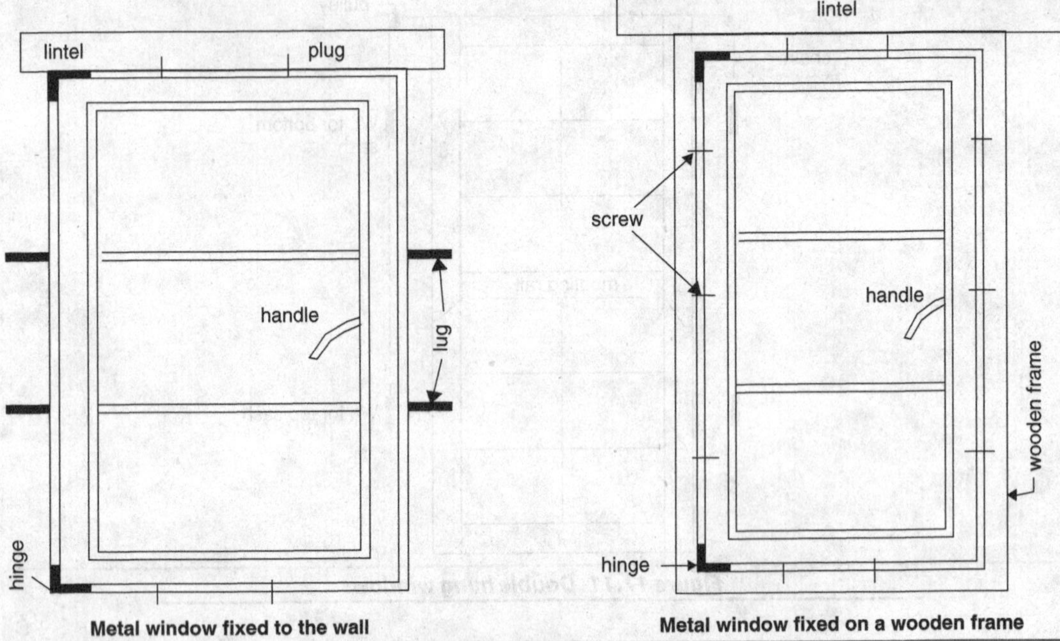

Figure 17.12 Metal window

fabricated with metal frames. The frame is attached to the masonry wall in grooves with cement grout. Sometimes, the timber frame and sill are used to fox up the steel shutter. Steel windows must be properly painted. The hollow metal window is fabricated by annealed steel, bronze, copper, nickel steel or galvanized steel. It is either casement type or double hung type. It is more fire resistant though more costly (Figure 17.12).

17.5.6 Sliding window

This window works exactly on the same principle as sliding doors. It is made of shutters, which move horizontally or vertically on small roller bearings. Suitable openings or cavities are provided in the frame or walls to receive the shutters when the windows are opened out. The windows of this type are commonly used in trains, buses and counter-windows.

17.5.7 Pivoted window

In such type of windows, shutters swing around pivots. It may be horizontally pivoted or vertically pivoted. Such windows are easier to clean and they allow more light to come inside the room. The frame of this type of window is similar to that of a casement window but no rebates are made in the frame (Figure 17.13).

17.5.8 Bay window

It is a window in the building which is projected beyond the walls of the room. It is provided to improve the architectural appearance of the building. Additional space is obtained to allow light and air into the room.

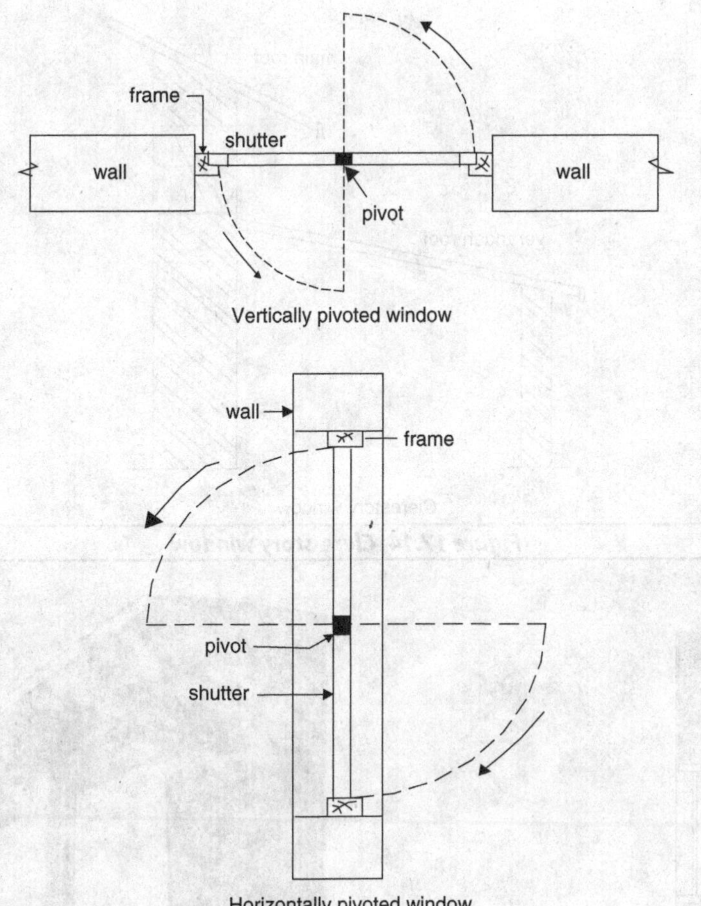

Figure 17.13 Pivoted window

17.5.9 Clere-story window

This window is usually provided near the top of the main roof and is made to open to the adjoining verandah or lean to the roof. This window is made to swing on two horizontal pivots provided in the side stiles. The shutter of the window is opened or closed by means of two cords, one from the top rail and the other from the bottom. It should be noted here that the upper part of the shutter opens to the inside of the room and the lower part opens outside. This is essential to exclude the rainwater (Figure 17.14).

17.5.10 Corner window

As the name implies, this type of window is placed in the corner of a room in a building. This window provides light and ventilation from two directions at right angles to each other and improves the appearance of the building (Figure 17.15).

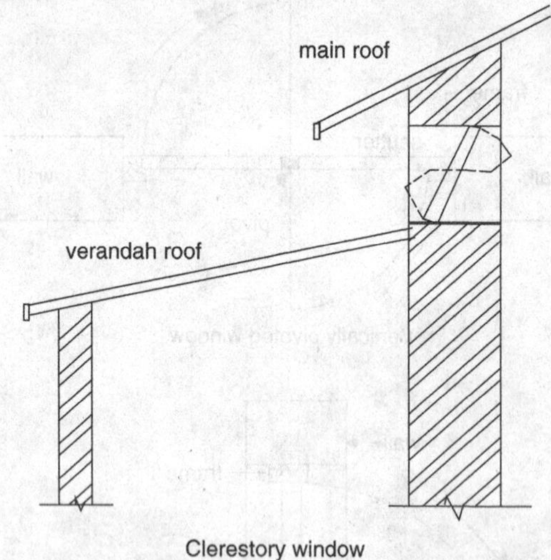

Figure 17.14 Clere-story window

Figure 17.15 Corner window

17.5.11 Dormer window

These are the vertical windows built on the sloping sides of a pitched roof. These windows are provided to admit light and air to the rooms or the enclosed space below the roof slopes. These windows add to the appearance of the building (Figure 17.16).

17.5.12 Lantern light

Sometimes, the light entering from the windows in the walls is inadequate. In this case, some more windows are provided on the flat roofs to admit more light into the room. This type of window is known as lantern. It may be curved, rectangular or square. Generally, glass panels are used to cover the sides of the lantern. Pivoted shutters may be used when it is desired to admit air and light (Figure 17.17).

Figure 17.16 Dormer window

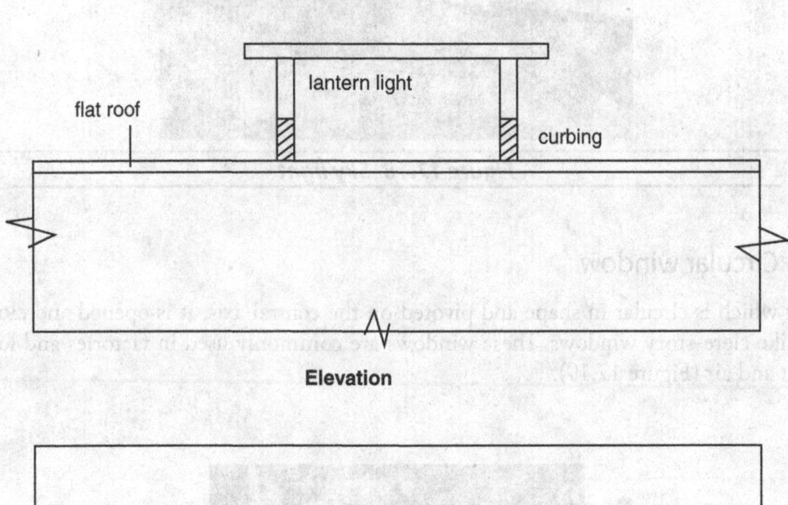

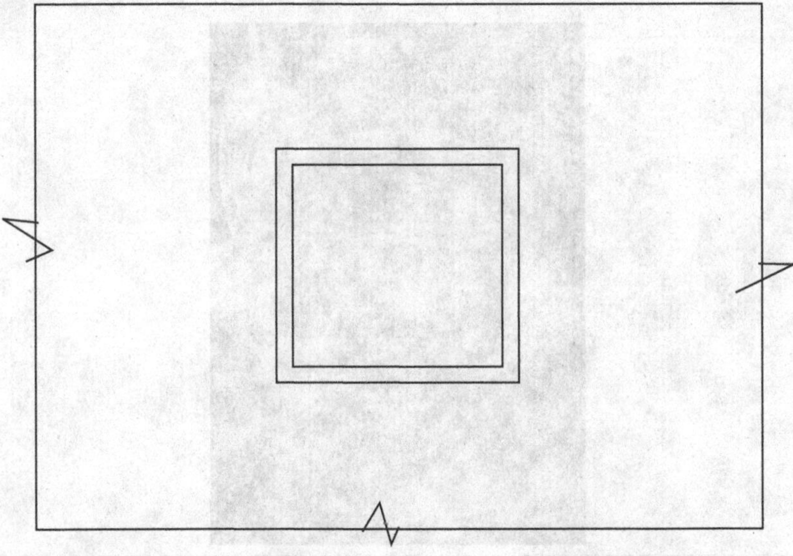

Figure 17.17 Lantern light

17.5.13 Sky light

This type of window is fixed on the sloping surface of an inclined roof. It is meant to admit light into the room and generally is fixed with glass panes to cover it. Reinforced or ribbed glass is preferred. A curb frame is provided with the common rafter to support the window (Figure 17.18).

Figure 17.18 Sky light

17.5.14 Circular window

It is a window which is circular in shape and pivoted on the central axis. It is opened and closed by means of two cords, like clere-story windows. These windows are commonly used in factories and lofty rooms for admitting light and air (Figure 17.19).

Figure 17.19 Circular window

17.6 FIXTURES AND FASTENINGS FOR DOORS AND WINDOWS

The various fixtures used are:

1. Hinges
2. Bolts
3. Locks
4. Handles

The most commonly used type of hinges is the butt hinge whose length varies from 1 to 20 cm. Various other types of hinges like counter flap hinges, spring hinges and parliamentary hinges are also being used.

Barrel bolts and tower bolts are the most commonly used type and their length varies from 10 to 40 cm. Hook and eye is another type which is commonly used for windows. Hasp and staple bolt and Aldrop bolt is used where padlocks are used.

Mortice locks are used for thicker doors and are embedded in the door frame. Otherwise, pad locks are also used.

REVIEW QUESTIONS

1. What are the main functions of doors and windows in a building?
2. What are the functional requirements for doors and windows to perform their basic functions?
3. What are the rules to be observed while locating doors and windows in a building?
4. Briefly discuss the classification of doors according to the operational point of view.
5. Briefly discuss the classification of windows according to the operational point of view.
6. Briefly discuss
   a. Battened and ledged doors
   b. Battened, ledged and braced doors
7. Briefly discuss framed and panelled doors and flush doors.
8. With a neat sketch give a brief description about collapsible doors.
9. Depending upon the material used how are windows classified?
10. What are the general fixtures and fastenings used for doors and windows?

chapter 18

Roof

A roof is the uppermost part of a building whose main function is to enclose the space and to protect the same from the effects of weather elements such as rain, wind, sun, heat and snow. A good roof is just as essential as a safe foundation. As a well-designed foundation secures the building against destruction starting at the bottom, similarly a good roof affords protection for the building itself and what the building contains and prevents deterioration starting from the top. To fulfil this main function efficiently, the roof should satisfy the following functional requirements in its design and construction.

i. ***Strength and stability:*** The roof structure should be strong and stable enough to take up the anticipated loads safely.

ii. ***Weather resistance:*** The roof covering should have adequate resistance to resist the effects of weather elements such as wind, rain, sun and snow.

iii. ***Heat insulation:*** The roofs should provide adequate insulation against heat, particularly in the case of single-storeyed buildings where the roof area may exceed that of walls with a consequent greater heat loss.

iv. ***Sound insulation:*** The roof construction for all buildings should provide adequate degree of insulation against sound from external sources.

v. ***Fire resistance:*** The roof should offer an adequate degree of fire resistance in order to give protection against the spread of fire from any adjacent building and to prevent early collapse of the roof. The form of construction should also be such that the spread of fire from its source to other parts of the building by way of roof cannot occur.

The roofs should be well designed and constructed to meet the requirements of different climates and the covering materials available. From experience it is found that pitched or sloping roofs are very suitable in coastal regions where rainfall is heavy and flat roofs are suitable in plains where rainfall is low and temperatures are high.

The roofs may be classified as follows:

1. Pitched or sloping roofs
2. Flat roofs
3. Shell roofs
4. Domes

18.1 TECHNICAL TERMS

1. ***Shed roof or lean to roof:*** This type of roof slopes in one direction only and is used for smaller spans.
2. ***Gable roof:*** This roof slopes in two directions so that the end formed by the intersection of the slopes is a vertical triangle.

3. ***Hip roof:*** This roof slopes in four directions such that the end formed by intersection of slopes is a sloped triangle.

4. ***Gambrel roof:*** This roof like the gable roof slopes in two directions but there is a break in the slope on each side.

5. ***Mansard roof:*** This roof like the hip roof also slopes in four directions but there is a break in slopes.

6. ***Ridge:*** It is an apex line of a sloping roof.

7. ***Ridge piece or ridge beam or ridge board:*** This is a wooden piece or board, which runs horizontally at the apex (highest point on the roof). The common rafters are fixed to this piece and are supported by it.

8. ***Common rafters or spans:*** These are inclined wooden members supporting the battens or boarding to support roof covering. They run from a ridge to the eaves (edges). They are normally spaced at 30–45 cm centre to centre depending upon the roof covering material.

9. ***Hip:*** It is the line produced when two roof surfaces intersect to form an external angle, which exceeds 180°. Hipped end is a portion of the roof between two hips.

10. ***Jack rafters:*** These are common rafters shorter in length, which run from a hip to the eaves or from a ridge to a valley. A hip or valley is formed by the meeting of jack rafters.

11. ***Valley rafters:*** These are sloping rafters which run diagonally from ridge to the eaves for supporting valley gutters. They receive the ends of the purlins and ends of jack rafters on both sides.

12. ***Valley:*** A valley is the reverse of a hip. It is formed by the intersection of two roof surfaces having an external angle, which is less than 180°.

13. ***Eaves (edges):*** These are the lower edges of the inclined or pitched roof from which the rainwater from the roof surface drops down. Normally, gutters are fixed along the eaves to collect and drain the rainwater.

14. ***Eaves board:*** This is a wooden board fixed to the feet of the common rafters at eaves. The ends of the lower most roof covering material rest upon it. The eaves gutter can also be secured against it. Normally, eaves board is 15–20 cm wide and 20–25 mm thick.

15. ***Barge boards:*** These are wooden planks on boards fixed on the gable end of a roof. They connect the ends of ridges, purlins and wall plates.

16. ***Battens:*** These are thin strips of wood which are fixed on the common rafters or on the top of ceiling boards to support the roofing materials.

17. ***Cleats:*** These are small blocks of wood or steel that are fixed on the principal rafters to support the purlins.

18. ***Purlins:*** These are horizontal wooden or steel members laid on principal rafters on wall to wall to support common rafters of a roof when the span is large.

19. ***Wall plates:*** These are long wooden members, which are embedded from the sides and bottom in masonry on top of walls, almost at the centres of their thickness. This is essential to connect the walls to the roof. The feet of the common rafters are fixed to the wall plates by means of simple notching and nails.

20. ***Truss:*** A roof truss is a framework of triangles designed to support the roof covering or ceiling over rooms. The use of interior columns is avoided.

21. ***Span:*** A span or clear span is the clear horizontal distance between the internal faces of wall or supports. The effective span is the horizontal distance between the centres of walls or supports.

22. ***Rise:*** This is the vertical height measured from the lowest to the highest points. In the case of pitched roof it is the vertical distance between the wall plate and the top of the ridge.

18.2 PITCHED ROOF OR SLOPING ROOF

The following are the different types of pitched roofs.

18.2.1 Lean to roof

This is the simplest type of pitched roof and consists of rafters that slope in one direction only. Generally, it is used to cover the verandah of a building and projects from the main wall of the building. At the upper ends the rafters are fixed by nails to the wooden wall plates, which are placed on the corbel of the main wall. The lower ends of the rafters are notched and nailed to the wooden post plate. The post plate is of timber section, which runs parallel to the wall and is supported on the intermediate columns or posts. Battens are placed and fixed over the rafters and it is finally covered by suitable roof covering materials. It is suitable for spans up to 2.5 m (Figure 18.1).

18.2.2 Couple roof

In this type of roof, each couple or pair of common rafters is made to slope upwards from the opposite walls and they are supported at the upper ends by the ridge piece or ridge board in the middle. The lower ends of the common rafters are fixed to the wall plates embedded in the masonry on the top of the walls. The use of this form of roof is not much favoured as it has a tendency to spread at the feet and thrust out the walls. The couple roof is therefore adopted only for a maximum span of 3.5 m (Figure 18.2).

18.2.3 Couple close roof

This type is similar to a couple roof except that the legs of the common rafters are closed by a horizontal tie known as tie beam. This tie beam is connected at the feet of the common rafters to check their tendency of spreading outwards and hence saves the walls from the danger of overturning. The tie may be a piece of wood or steel rod in tension. The connection between the ties and the feet of rafters is usually obtained by means

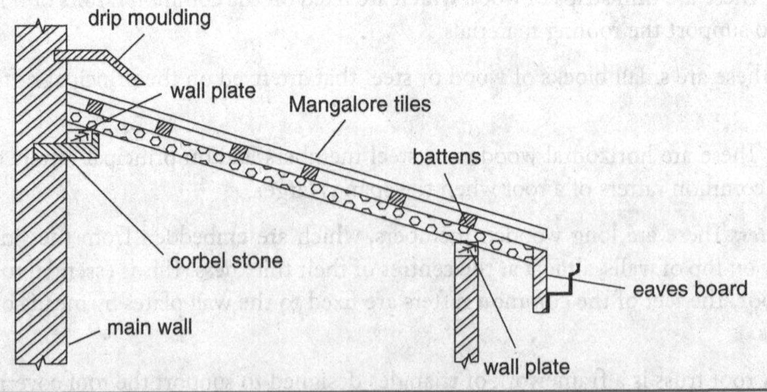

Figure 18.1 Lean to roof

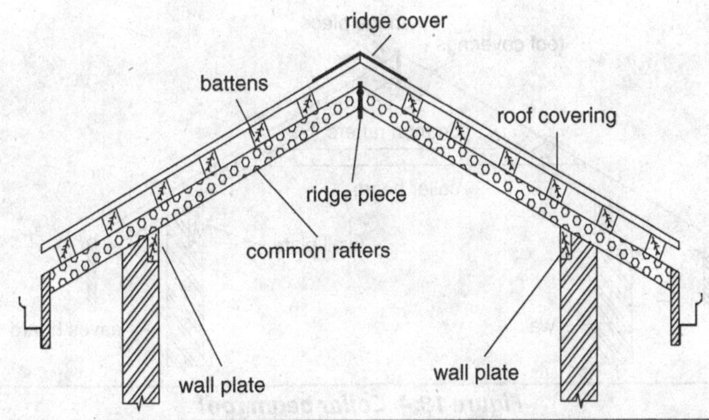

Figure 18.2 Couple roof

of dovetailed halved joint, but for inferior work the ties are just spiked to the rafters. Under normal loading conditions this type of roof can be used for a maximum span of 4.5 m. However, for increased spans or greater loads the rafters have a tendency to sag in the middle. To check this tendency a couple close roof is supported by a central vertical rod known as king rod or king bolt between the ridge piece and the centre of the tie beam (Figure 18.3).

18.2.4 Collar beam roof

It is used for spans between 4 and 5.5 m. A collar of the same width as the rafter is fixed to every pair of rafters and it is attached at a height of half to one-third of the vertical height between the wall and the ridge. The collar is dovetailed with the rafter and the bolts can be used for additional safety. It is desirable to place the collar as low as possible to provide maximum strength to the roof (Figure 18.4).

18.2.5 Collar and tie roof

It is used when the roof spans exceed 5.5 m. It is a combination of collar beam roof and couple close roof. The rafters are supported by purlins and the purlins rest at the ends on walls. A collar and strut are employed

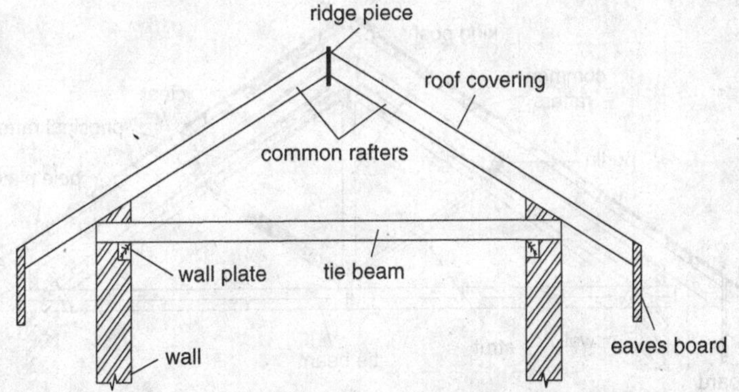

Figure 18.3 Couple close roof

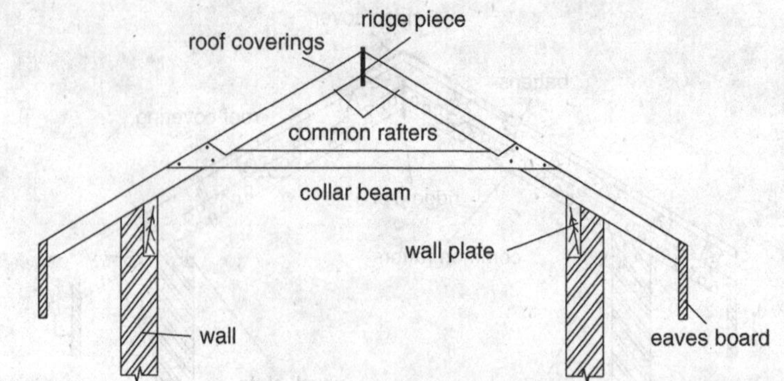

Figure 18.4 Collar beam roof

to support the purlins and rafters. Its use is recommended when purlins may be supported at the ends with reasonable economy.

18.2.6 King post truss

For spans greater than 4.8 m, when no intermediate supporting walls for the purlins are available, framed structures known as trusses are used. The spacing between trusses is guided by the load coming on the roof, material of the truss, span and the location of cross walls.

In a king post truss, the central vertical post called as king post provides a support for the tie beam. The inclined members are known as struts and are used to prevent the principal rafters from bending at the centre. A king post truss can be used economically for spans 5–8 m.

The joint between the king post and the tie beam is an ordinary mortise and tenon joint. An iron stirrup is also provided to strengthen the joint further. For joining principal rafters and the king post, a tenon is cut in the principal rafter and the corresponding mortice into the head of the king post. A bridle joint is provided to connect the principal rafter with the tie beam. Joints between the king post and the strut are also mortice and tenon joints (Figure 18.5).

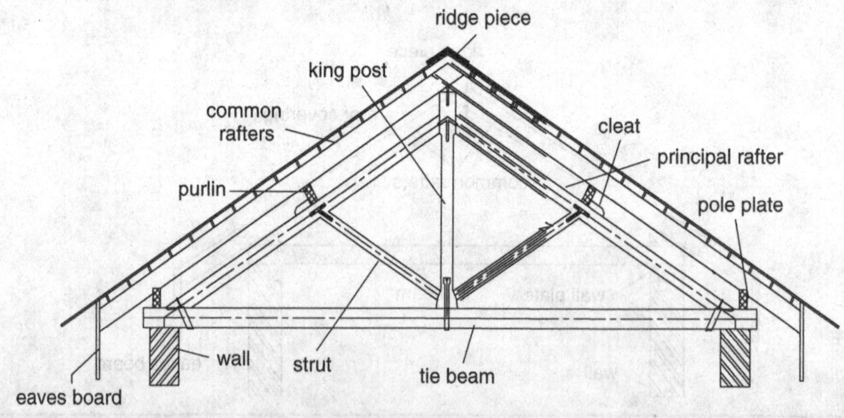

Figure 18.5 King post truss

18.2.7 Queen post truss

It can be used for spans 9–14 m. It varies from the king post truss in having two vertical members known as queen posts. The heads of the queen posts are put apart by a horizontal member known as straining beam. The head of the queen post is made wider to receive the principal rafter and the straining beam. The top end of the principal rafter and the end of the straining beam are tenoned into the widened head of the queen posts. A three-way iron or mild steel strap is fixed to further strengthen the joint. The bottom end of the queen post is tenoned into the tie beam and a steel stirrup strap is fixed by jibs and cotters to make the joint stronger. The tenon of the inclined strut is inserted into the splayed shoulder of the queen post. The other joints in this truss are similar to that of the king post truss (Figure 18.6).

18.2.8 Mansard truss

It is a combination of king post truss and queen post truss. The upper portion has the shape of a king post truss and the lower portion resembles a queen post truss. The truss has two pitches. The upper pitch varies from 30 to 40° and the lower pitch varies from 60 to 70°. This type of truss is economical and in the span an extra room may be provided. This type of truss is now rarely used due to its ugly appearance. The construction of various joints is similar to that of the king post trusses.

18.2.9 Belfast roof trusses

This truss is in the form of a bow and is also called bow string or latticed roof truss. It is made of thin sections of timber. This truss can be used for big spans up to 30 m provided light roof coverings are used. The central rise in this type of truss is usually kept about one-eighth of the span.

18.2.10 Steel trusses

The use of steel trusses has become economical for spans greater than 12 m. Various standard shapes and sizes of rolled steel are available for the fabrication of steel trusses. This type of truss is designed in a manner that members are either in compression or in tension and bending stress is not allowed to develop in them.

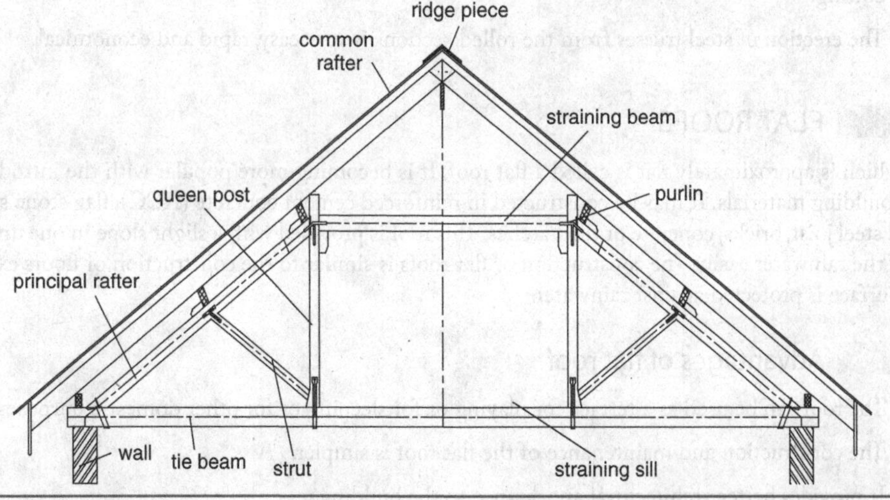

Figure 18.6 Queen post truss

The size and type of the truss depends upon the roof slope, span, centre-to-centre distance of the trusses and the load coming over the roof. T-sections are best suited for use as principal rafters, whereas angle iron or channel section is used as struts. The tension members should preferably be of a flat or round section. The different members of the truss may be fabricated with two or more sections joined together. The members of a truss are joined by rivets or bolts or by welding the plates known as gusset plates. The minimum spacing of the rivets should not be less than 3 times the diameter and the maximum spacing is limited to 15–20 cm in compression and tension members.

The minimum number of rivets to be used at any joint should not be less than two. Gusset plates are designed for the forces coming at the junction but the least thickness should be adopted as 6 mm. The ends of the trusses are placed on bed plates provided on the walls. The bed plate maybe of stone or concrete. The ends of the truss are bolted down with lewis or rag bolts which hold down the truss firmly. The small trusses are pre-fabricated in the workshop on the ground and are then placed in the required position. The bigger trusses are pre-fabricated in smaller parts and then erected in the required position and fixed by gusset plate and riveting or welding.

The relative advantage of steel roof trusses over timber sloping trusses are as follows:

a. Steel sections forming the roof truss are light in weight and can be fabricated in different shapes and sizes. It suits the structural as well as architectural requirements.

b. Steel trusses being made of mild steel sections are free from the attack of white ants and dry rot.

c. Steel trusses are much stronger than timber trusses and they are equally strong in tension and compression.

d. These trusses have a greater resistance against fire and hence are especially suited where fireproof construction is desired.

e. Timber trusses can only be used up to a minimum span of 14 m or so, whereas there is no span restriction in case of steel trusses. Steel trusses are used for structures requiring large spans such as industrial buildings, large sheds, assembly halls, hangers and auditoriums.

f. The various sections forming a steel truss can be easily machined and shaped in the workshop and subsequently packed and transported to the site for assembling. Moreover, there is no wastage in cutting.

g. The erection of steel trusses from the rolled sections is very easy, rapid and economical.

18.3 FLAT ROOFS

A roof which is approximately flat is called a flat roof. It is becoming more popular with the introduction of suitable building materials. It may be constructed in reinforced cement concrete (RCC), flag stone supported on rolled steel joist, bricks, concrete or tiled arches. This roof is provided with a slight slope in one direction to drain off the rainwater easily. The construction of flat roofs is similar to the construction of floors except that the top surface is protected against rainwater.

18.3.1 Advantages of flat roof

a. The roof can be used as a terrace for playing or for sleeping or for other domestic purposes.

b. The construction and maintenance of the flat roof is simpler.

c. It provides better architectural appearance to the building.

d. It is easier to make the flat roof fire resistant.

e. It possesses good insulating properties.

f. It avoids the need for a false ceiling.

g. The construction work of upper floors can be readily taken up in the case of flat roof, whereas in the case of pitched roof the entire roof has to be dismantled before construction.

h. Pitched roof needs much more area of roofing material than flat roofs.

18.3.2 Disadvantages of flat roof

a. A flat roof cannot be used for long spans without using columns and beams.

b. In areas of heavy rainfall, flat roofs are not suitable.

c. The initial cost is more.

d. Due to greater variations in the temperature cracks sometimes develop on the surfaces of the roof, which is difficult to repair.

e. The speed of construction is slower than that of a pitched roof.

f. If proper slope is not provided on the roof to drain off the rainwater, pockets of water are formed on the surface of the roof, which leads to leakage in the roof.

18.3.3 Types of flat roof

The various types of flat roof constructions include the following.

18.3.3.1 Madras terrace roof

Bricks are the major constituent and they are supported on wooden and steel joists.

i. Wooden joints are kept over rolled steel joists with a furring piece in between them.

ii. A course of thoroughly burnt terrace bricks (15 × 8 × 2 cm) is laid on the edge diagonally across the joists in 1:1.5 lime mortar.

iii. After the bricks are completely set, a 10 cm thick layer of brick bat concrete is laid over the course of brick. It is beaten down to 7.5 cm by wooden hand beaters. The beating is continued till the beater fails to make an impression on the roof surface.

iv. Two layers of flat tiles (15 × 10 × 1.2 cm) are laid over the layer of concrete in 1:1.5 lime mortar.

v. Finally, three coats of lime plaster are applied to the surface and it is rubbed to a polished surface. Generally, a slope of 1:36 is provided to the roof by giving slope to the joists and not by increasing the thickness of terracing.

18.3.3.2 Jack arch flat roof

The brick or concrete arches are constructed with rolled steel joists, which are supported on the walls of the rooms. Some inert materials fill up the spandrels between the arches. Over it, a 10 cm thick average lime concrete (LC) terracing is provided to make it waterproof. Some waterproofing material may be used with lime concrete and the LC terracing is beaten thoroughly by wooden beaters to make it more compact.

18.3.3.3 Reinforced concrete slab roof

An average of 10–13 cm thick LC terracing with some waterproofing compound is provided over the RCC slab to make the roof leak proof. The lime concrete is thoroughly beaten by several labourers with wooden beaters for several days to make the terracing more compact and impervious. At the junctions with wall, the lime terracing is taken inside the wall for 10–15 cm depth and the corner is given a smooth and round shape such that water may not accumulate. The lime terracing is provided with a slope (1 in 60 to 1 in 100) to drain off the rainwater easily from the roof to the gutters provided for the purpose.

18.4 SHELL ROOF

Shell roofs are very useful for covering large structures, e.g., assembly halls, recreation centres, libraries, theatres and factories. RCC shell roofs are becoming very popular these days. Very less quantity of materials are required to build up a shell roof as compared to other conventional methods of roofing for the same span. The design of the shell is made as thin as practical requirements will allow, such that the dead load is decreased and the shell acts as a membrane free from large bending stresses. Least quantity of materials is used to the maximum structural advantage.

The following are the common types of shell roof:

1. North light shell roof which is used mostly in factories, workshops and places where good daylight is desired.
2. If good day lighting is not a requirement, long multiple cylindrical shells with feather edge beams may be useful.
3. Double curved shells are structurally more efficient than single curved shells, but it presents more difficulties in preparing the centring for it. Though consumption of materials is less, sometimes the costs of formworks make the shell roofs quite expensive. It proves to be more costly when only a few similar units are to be constructed. Thin shell roofs are economical when many identical units are to be built and the forms can be reused several times. The forms are usually fabricated from timber battens lined with steel sheets or plywood. Sometimes, plastic forms are also used to obtain special surface textures. The materials of formwork and the lining are selected in consideration of the number of reuses in a particular project. Economy may be achieved in two ways for the formwork. Firstly by using moveable formwork when the shell is to be cast in situ. The second way is to use the precast shells.

18.5 DOMES

A dome is a special type of shell roof of semi-spherical or semi-elliptical shape. The modern thin shell dome may be considered as an evolution of a structural form known and used by man from ancient ages.

Dome structures may be divided into two main divisions:

1. Smooth shell domes
2. Ribbed domes

Smooth shell domes may be divided into:

1. Domes with shells of uniform thickness.
2. Domes with shells of uniformly varying thickness.

Smooth shell domes are constructed by brick, stone, concrete or tile. Ribbed domes may be built in steel, concrete or wood. A dome may be constructed with or without a lantern.

The structure of the dome is such that within certain height and diameter ratios very small thickness is required. They are used where architectural treatment is required such as in monumental structures or where roofs have to be constructed on buildings circular in plan or hexagonal in plan.

18.6 SELECTION OF ROOF COVERING MATERIALS

The following factors require due consideration in selecting a roof covering for a building:

1. ***Climatic conditions of the locality:*** The climatic conditions have a marked influence on the performance and durability of roof coverings. Strong winds may damage the roof covering such as slates, tiles and AC sheets by blowing it off, if not properly fixed in position. Extreme temperature changes may cause the sheets to crack and the joints to leak, if not properly protected. Atmospheric effects of fog, salt air, smoke and other gases may result in corrosion of metal roofing if not protected by painting; clay tiles, slates, AC sheets and built-up roof coverings are unaffected by atmospheric action.
2. ***Slope of roof:*** The flatter the slope the greater would be the effect of rains, causing leakage in the roof. However, the steeper the slope the greater would be the effect of wind action, requiring heavier supports and more covering material. Thus, the slope of the roof influences both the strength and the economy of the roof and is decided by considering climatic conditions of the locality and the material of roof covering. Thatch, tiles and slate usually require steeper slopes to prevent water infiltration whereas corrugated sheets and concrete terraces require flatter slopes.
3. ***Nature of the building and of the roof deck:*** A roof covering must be in conformity with the nature of the building as well as of the roof deck. It is the roof deck to which all the types of roof covering are fastened. The decks are supported on principal supporting members such as girders, trusses and rigid frames. Sometimes the deck serves as a principal supporting member in shell roofs.
4. ***Initial cost:*** The initial cost of roofing material is not a definite value and depends upon the time and place and materials availability. For example, a roofing of slates or wood may be cheaper in hilly regions. In the cost analysis of roof covering, it should be noted that the cost of supporting structures and the deck should be accounted in the initial cost because the materials, which are cheap and heavier, may require a strong and costly supporting structure as in the case of slate roofing in hilly regions.
5. ***Maintenance cost:*** While considering the overall economy of the roof covering, the maintenance cost may become as important as the initial cost. Wood shingles and tiles need less maintenance and repairs than thatch roofing. Asbestos sheets and slates may be broken and require replacements occasionally. Galvanized iron sheets require only periodic painting to check its tendency to corrosion.
6. ***Durability:*** The durability or life is an important factor in assessing the economic value of roof coverings. The life of a roof covering is affected by its quality, its suitability for the purpose used, climatic conditions, workmanship in fixing and laying, degree of maintenance and many other factors. Under normal weather conditions, when all roof covering are laid in the best possible way, clay tiles, slates and galvanized iron (GI) sheets are said to have longer life.

7. ***Resistance to fire:*** The roof covering should offer sufficient resistance in the event of fire. From the fire resistance point of view, roof covering of slates, AC sheets, GI sheets, etc. are considered to be quite satisfactory, whereas thatch and shingles are unsatisfactory for this purpose.
8. ***Heat insulation:*** The roof covering should offer adequate insulation against heat so that the inside of rooms can be kept cool and comfortable for living. This is particularly important in tropical countries.
9. ***Weight of roofing material:*** The weight of roof covering affects the design, weight and the cost of roof deck as well as supporting structure and frame work. Heavier roof covering requires stronger supporting structure, which adds to the cost.
10. ***Appearance:*** It is an important factor for residential and other public buildings but is of less significance in the case of industrial buildings. The appearance depends upon the architectural style of the building and the class of occupancy.

18.7 ROOF COVERING MATERIALS

Roof covering material provides protection to the roof and the structure. It prevents rainwater, moisture, heat, dust, etc. from entering into the building from top. The roof covering does not share load in the building. It is rigidly fixed to the roofing structure with various types of fittings and fixtures. The various types of roof covering materials used include the following.

18.7.1 Thatch covering

This form of covering is extensively used in sheds, low-cost houses and village buildings. It is considered suitable for rural areas because it forms the cheapest and the lightest material as a roof covering.

The frame work for supporting the thatch consists of round bamboo rafters spaced at 30 cm and tied with split bamboos or bamboo reapers laid at right angles to the rafters. The thatch is tightly secured to the framework or battens with the help of ropes or twines dipped in tar. Sometimes fire-resisting properties are imparted to the thatch by soaking it in specially prepared fire-resisting solutions that are very costly. For adequate drainage of rainwater the thatch covering should be at least 15 cm thick and laid with a slope of 45°.

18.7.2 Shingles

The use of wood shingles as a roof covering is generally restricted to hilly areas where wood is easily and cheaply available in abundance. Wood shingles are nothing but the sawn or split thin pieces of wood resembling slates or tiles. These sawn shingles, which are obtained from well-seasoned timber, are dipped in creosote to impart preservative qualities. Shingle strips are driven on rafters and shingles are nailed on their top. Shingles are commonly obtained in length varying from 30 to 50 cm and in width varying from 5 to 25 cm.

18.7.3 Tiles

The use of tiles is one of the oldest method of roof covering. The tiles are named according to their shape and pattern and they are manufactured in a similar manner as bricks. The clay tiles are of various types such as flat tiles, pan tiles, pot tiles or half-round country tiles and patent tiles such as Mangalore and Allahabad tiles. Sometimes cement concrete tiles are also used, but is limited on account of high cost and the difficulties in their manufacture. Clay tiles have been widely used as a roof covering material for residential buildings.

18.7.3.1 Advantages of clay tiles as a roof covering

i. Clay tiles, being non-conductors of heat, prevent the building from extreme changes of temperature outside and keep them cool.

ii. These tiles provide quite a durable roof covering when made of well-burnt good materials.

iii. They are quite strong and pleasing in appearance.

iv. If properly selected and laid they have good resistance against fire and moisture penetration.

v. These tiles provide a very economical roof covering with aesthetic values and hence are used for urban and rural houses.

However, these tiles suffer from the limitation of being heavy in weight. The weight of the roof covering is further increased as the rafters are kept closer to reduce the span of timbers and to throw off the rainwater. The average weight of a tile roof is about 75 kg/m^2.

Flat or plain tiles are manufactured in rectangular shapes (size, 25 × 15 cm to 28 × 18 cm) in thickness varying from 9–15 mm. Tiles are not perfectly flat but they have a slight camber usually 5–10 mm. These tiles have two small projecting nibs and two or more nail holes at one end of their surface. These nibs and holes help to fix the tiles on the battens of the roof truss. The tiles should be laid at proper gauge and overlap both at the sides and edges as it is important for their strength, durability and imperviousness.

Curved pan tiles are shorter, less curved, heavier, stronger and more durable than the pot tiles. These tiles are moulded flat first and then given the required curvature. These tiles are about 30–35 cm long and are about 20–25 cm wide.

Pot tiles or half-round country tiles are very commonly used for rural houses as they offer a very cheap roof covering. These tiles are laid in pairs of under tiles (concave upwards) and over tiles (convex upwards) with a proper overlap of at least 8 cm on all the sides. The under tiles are flat with a broad head tapering towards the tail while the over tile has a wider tail and a narrower head which is segmental in section. Country tiles are laid in two layers one over the other and the roof is called doubled-tiled roof. This type of roof requires heavy supporting timbers of greater strength than the usual ones.

Patent tiles are generally rectangular in plan with face corrugations so arranged that the corrugations of tiles fit in or interlock with those of other tiles. These interlocking tiles, which are machine made, provide a lighter roof covering with a decent appearance. In ordinary works the groundwork for these tiles consist of battens only. In superior type of construction the tiles are laid on boarding covered with a protective coat. Boarding is directly nailed to purlins and tiles are on battens nailed on the boarding.

18.7.4 Asbestos cement sheets

Asbestos cement is a material which consists of Portland cement and asbestos fibres (about 15 per cent). Roof covering made of this material is cheap, tough, durable, watertight, fire resisting and light in weight. Asbestos does not require any protective coat. Asbestos cement roof coverings are supplied in flat corrugated and ribbed sheets in various sizes.

18.7.5 Corrugated GI sheets

These are used for the roof coverings of workshops, factories and temporary sheds. GI sheets are available in various sizes. Generally 22 gauge sheets are used. End lap of 15 cm and side lap of two corrugations are provided at the time of fixing the GI sheets at the top of the roof. It is light and simple to fix. The only disadvantage with the use of these sheets for roof covering is that during summer season, the rooms under the roof are heated too much. To protect against the sun, sometimes a layer of ordinary tiles are provided at the top of GI sheeting.

REVIEW QUESTIONS

1. What are the main functions of roofs?
2. To fulfil the main functions, what are the functional requirements to be satisfied by roofs in its design and constructions?
3. How are the roofs generally classified?
4. Define hip roof, gabled roof, ridge, eaves, valley, purlins and rise.
5. Write short notes on
   a. Sloping roof
   b. Couple cross roof
   c. Colour beam roof
6. Explain briefly the king post truss.
7. Explain briefly the queen post truss.
8. What are the advantages of steel roof truss over timber sloping truss?
9. What are the advantages and disadvantages of flat roofs?
10. Explain the different types of flat roofs.
11. What is a dome and what are the main divisions of dome structures?
12. What are the factors to be considered in selecting a roof covering for a building?
13. What are the different materials used for roof covering?

chapter 19

Floors

Floors are provided to divide a building into different levels for creating more accommodation one above the other within a certain limited space. The bottom floor near the ground is known as the ground floor and the other floors above it are termed as upper floors, like first floor and second floor. If there is any accommodation constructed below the natural ground level, it is known as basement and the floor provided in it is known as the basement floor.

A floor may consist of two main components:

a. A sub-floor that provides proper support to the floor covering and the superimposed loads carried on it.

b. A floor covering which provides a smooth, clean, impervious and durable surface.

19.1 FACTORS AFFECTING SELECTION OF FLOORINGS

Each type of floor has its own merits and there is not even a single type which can be suitably provided under all circumstances, and more so when floors have to serve different purposes in different types of building.

1. ***Initial cost:*** The cost of construction is very important in the selection of a type of floor. A floor covering of marbles, granite, special clay tiles, etc. is considered to be very expensive, whereas a flooring of cork, slate, vinyl tile, etc. is moderately expensive. The floors made of concrete and brick offer the cheapest type of floor construction. It should be ensured during the comparison of cost for different floors that the cost of both covering and sub-floor has been accounted for.

2. ***Appearance:*** Flooring should produce the desired colour effect and architectural beauty in conformity with its use in the building. Generally, flooring of terrazzo, tiles, marble and cement mortar provides a good appearance whereas the asphalt covering gives an ugly appearance.

3. ***Cleanliness:*** A floor should be non-absorbent and capable of being easily and effectively cleaned. All joints in flooring should be such as to offer a watertight surface. Moreover, greasy and oily substances should neither spoil the appearance nor have a destroying effect on the flooring materials.

4. ***Durability:*** The flooring material should offer sufficient resistance to wear and tear, temperature, chemical action, etc. so as to provide long life to the floors. From the durability point of view, flooring of marble, terrazzo, tiles and concrete is considered to be of the best type. Flooring of other materials such as linoleum, rubber, cork, bricks, wood blocks, etc. can be used where heavy floor traffic is not anticipated.

5. ***Damp-resistance:*** All the floors, especially ground floors, should offer sufficient resistance against dampness in buildings to ensure a healthy environment. Normally, floors of clay tiles, terrazzo, concrete, bricks, etc. are preferred for use where the floors are subjected to dampness.

6. ***Sound insulation:*** According to modern building concepts, a floor should neither create noise when used nor transmit noise. Sometimes, it is required that any movement on the top floors should not

disturb the persons working on the other floors. Suitable flooring is provided which is somewhat noiseless when travelled over.

7. ***Thermal insulation:*** It should be possible for a building to maintain constant temperature or heat inside the building irrespective of the temperature changes outside. Thermal insulation is needed to reduce the demand of heating in winter and refrigeration in summer. It is important in the case of wooden floors where heat losses are considerable and in solid floors with heating pipes or cables where the heat losses at the edges of the floor slab can be higher. Floors of wood, cork, etc. are best suited for this purpose.
8. ***Smoothness:*** The floor covering should be of superior type as to exhibit a smooth and even surface. However, at the same time, it should not be too slippery which will otherwise endanger safe movements over it, particularly by old people and children.
9. ***Hardness:*** It is desirable to use good quality floor coverings, which do not give rise to any form of indentation marks, imprints, etc. when used for either supporting the loads or moving the loads over them. Normally, the hard surfaces rendered by concrete, marble, stone, etc. do not show any impressions, whereas the coverings like asphalt, cork, plastic, etc. do form marks on the surfaces when used in traffic.
10. ***Maintenance:*** It is always desired that the maintenance cost should be as low as possible. Generally, a covering of tiles, marble, terrazzo or concrete requires less maintenance cost as compared to the floors of wood blocks, cork, etc. It should, however, be noted that the repairing of a concrete surface is more difficult than the floorings of tiles, marbles, etc.

19.2 TYPES OF FLOOR

The various types of floor commonly used are as follows.

19.2.1 Basement or ground timber floor

Timber floors are constructed on ground floors, generally in theatres. Several sleeper walls or dwarf walls of half brick thickness or full brick thickness are constructed at an interval of 1.5 m to support the timber floor. Wall plates are placed on walls and sleeper walls to support the joists supporting the floors. The joists are provided at a distance of about 30 cm and the timber planks are closely fitted over the joists to provide the floor. The arrangement for proper air circulation is made in the floor, otherwise timber will be attacked by dry rot. The following precautions are recommended:

a. Well-seasoned timber should be used in the construction of such floors.
b. Plain cement concrete 1:2:4 of 10 cm thickness is to be provided over the soil beneath the timber floor.
c. The empty space between the floor and the concrete base should be filled up with sand.
d. The damp proofing courses are to be placed in the external walls and at the top of the sleeper walls.

19.2.2 Single joist timber floor

This type of floor is used for residential buildings where spans are comparatively small and the loads are lighter. The wooden joists are placed at about 30 cm centre to centre, spanning the rooms in the shorter direction. Wooden planks are laid over these joists. The timber joists are supported on wall plates. Corbels may be required to support the joists if the width of the wall is not sufficient. Joists must be strong enough to withstand the loads and at the same time they should not deflect too much. If the length of the joist is more than 3.5 m, then struts are provided in the joists to check side buckling. The wooden planks are about 4 cm thick and 10–15 cm wide.

19.2.2.1 *Advantages*

i. Single joist timber floors are simple to construct.
ii. They require less initial cost.
iii. Distribution of loads on the wall is more uniform as the joists are spaced closely.

19.2.2.2 *Disadvantages*

i. The joists may sag and, hence, cracks will develop in the ceilings.
ii. They are not soundproof.
iii. Deep joists are required for larger spans that increase the weight and construction cost of the floor.
iv. The loads are transmitted to the openings such as windows or door lintels because of evenly spaced joists.

19.2.3 Double joist timber floor

This type of floor is stronger than the single joist timber floor. They are used for longer spans of 3.6–7.5 m and prevent the travel of sound waves to a great extent. Intermediate supports called binders are placed for bridging the joists. Binders are spaced at a centre-to-centre distance of about 2 m. The ends of binders are kept on wooden or stone blocks and they should not be embedded in the masonry wall. The ceilings may be fixed to the bottom of the binders by fixing a ceiling joist to the binders. Lathing is fixed to the ceiling joist.

19.2.3.1 *Advantages*

i. The loads are transmitted to the wall at certain specified points and, hence, door and window openings may be avoided.
ii. This is a more rigid type of flooring and, hence, there is less chance of developing cracks in the plastered ceiling.
iii. It is more soundproof.
iv. The use of additional binders near the walls can eliminate the need of wall plates.

19.2.3.2 *Disadvantages*

i. More labour is required
ii. The depth of the floor is considerably increased and, thus, the head room is reduced

19.2.4 Framed timber floor

This type of timber floor is used for spans of more than 7.5 m. Girders are placed between the walls and the binders are put on the girders and the bridging joists rest on the binders. The spacing between girders depends on the type and size of the girder and the size of the binders. The ends of girders are put on stone or concrete templates in the wall. The ceilings are fixed directly to the binders or ceiling joists may be employed.

19.2.5 Filler joist floor

Small sections of rolled steel joists are encased in the concrete. The joist is supported on walls or on steel beams. The joists are placed at a centre-to-centre distance of 60–90 cm and act as reinforcement in the concrete. The rolled steel joists and beams should be completely encased in the concrete.

19.2.6 Jack arch floor

Bricks or concrete may be used for the construction of jack arch floor. The arches are provided between the lower flanges of rolled steel joists at a centre distance of not more than 1.5 m. The rise of the arch is generally one-twelfth of the span. Mild steel ties are provided in the end spans to take up the tension developed due to the arch action of the floor. The diameter of tie rods is 18–25 mm and their spacing is 1.8–2.4 m. In addition, they are rigidly fixed on the walls, the side filling is done by lime concrete and the desired floor covering is provided. Plain ceiling is not obtained in this case and it may be considered as a shortcoming of such construction.

19.2.7 Double flagstone floor

For spans less than 4 m only rolled steel joists are placed, but if the span exceeds 4 m, a framework of rolled steel beams and joists is formed. Flagstones of about 40 mm thickness and proper width are placed on the lower flanges and upper flanges. The joints of flagstones in the top layer are given a fine finish for improved appearance. The empty space is filled up with selected earth.

19.2.8 RCC floor

Reinforced cement concrete (RCC) slab is being more commonly used in the construction of modern buildings. For small spans and comparatively lighter loads, a simple RCC slab is suitable. If the ratio of the length and width of a room is more than 1.5, the slab is designed to span along the shorter direction. The main reinforcement is provided along this shorter dimension of the room. The thickness of the slab is guided by the superimposed loads, span and type of concrete used. The end of the slab rests on the wall. When the building is constructed in reinforced concrete frames, it is essential to construct the slab monolithic with the supporting beams. For larger spans and greater loads, RCC beams and slab construction are adopted in the construction of buildings. The slab acts as a flange of the beam and is cast monolithic with the beams. In this case, the size of the beam is greatly reduced. Over the RCC floor suitable covering is laid to get the desired finish.

19.2.9 Flat slab floor

It is directly supported on the columns without providing any intermediate beams. This type of construction is adopted when the use of beams is forbidden.

19.2.9.1 Advantages

i. More clear head room is available for use.

ii. Even for quite heavy loads, thinner slabs are required.

iii. No projecting of beams is to be seen and, therefore, the need of false ceiling is eliminated.

iv. It is convenient to make lighting arrangements.

v. The framework and construction of flat slabs are simpler.

Flat slabs are commonly used in commercial buildings, factories and warehouses, etc., but they are not economical for lighter loads.

19.2.10 Hollow tiled ribbed floor

To reduce the weight of a solid floor structure, a hollow tiled ribbed floor is constructed. In this type of construction, hollow blocks of clay or concrete are used. These hollow bricks or tiles are placed at about 10 cm apart. In this space of 10 cm, mild steel bars of 8–12 mm diameter are placed. The surfaces of the hollow tiles are kept rough to develop a better bond with the surrounding concrete. A minimum cover of 8 cm is provided at the top of the tiles. The empty spaces are filled up with concrete. These floors are fireproof, soundproof, damp proof, light and economical. A properly designed floor of this type can carry considerably heavy loads.

19.3 FLOOR COVERING

Floor coverings are provided to improve the appearance, cleanliness, noiselessness and damp proofing. Various types of materials are used and different treatments are done. The following types of floor coverings are generally employed.

19.3.1 Mud flooring

This is mainly used in unimportant buildings, particularly in the villages. They are cheap, hard, fairly impervious and easy in construction and maintenance. Thermal insulation properties are also very high. To prepare this a 25-cm thick layer of selected moist earth is spread over a prepared bed and rammed well to get a consolidated layer of 15 cm thickness.

19.3.2 Brick floor covering

It is employed for cheap constructions such as godowns, sheds, stores and barracks and where good bricks are available. Over well-compacted and levelled ground a layer of lean cement concrete mix (1:6:18) of 10 cm thickness is laid. Over this bedding, bricks are placed in proper bonds on their edges. They are joined with cement or lime mortar. Sometimes, the joints are pointed to obtain a better appearance. The only drawback of brick floor covering is that it absorbs water.

19.3.2.1 Merits

i. It offers a durable and sufficiently hard floor surface.

ii. It provides a non-slippery and fire-resistant surface.

iii. It is cheaper in initial cost as compared to cement concrete, mosaic and terrazzo flooring.

iv. It is easy to maintain.

19.3.2.2 Demerit

i. This type of flooring acts as an absorbent

19.3.3 Stone floor covering

Square or rectangular slabs of stones are used as the floor covering. Generally, 20–40 mm thick stone slabs of size 30 cm × 30 cm, 45 cm × 45 cm, 60 cm × 60 cm, 45 cm × 60 cm, etc. are used. The stone should be hard, durable, tough and of good quality. The earthen base is levelled, compacted and watered. On this surface a layer of 10–15 cm thick concrete is laid and properly rammed. Over this concrete bed the stone slabs are fixed with a thin layer of mortar. Before fixing the stone slabs in position, they are dressed on all the edges and the joints are finished with cement. The stone surface may be rough or polished. A rough surface is provided in rough works like godowns, sheds, stores, etc. and a polished surface is provided in superior type of works. A slope of 1:40 should be provided in such type of floor covering for proper drainage.

19.3.4 Concrete floor covering

The concrete flooring consists of two layers:

a. A base course or the subgrade and

b. A wearing course

The concrete flooring consists of a topping of cement concrete 2.5–4 cm thick laid on a 10–15 cm thick base of either lime or cement concrete. The actual construction operation consists of:

a. Ground preparation

b. Formation of base course

c. Laying of topping concrete

d. Laying of wearing coat

e. Grinding and polishing and

f. Curing

19.3.4.1 Merits

i. It is non-absorbent and, hence, offers sufficient resistance to dampness. This is used for water-retaining floors as well as stores.

ii. It possesses high durability and, hence, is employed for floors in kitchens, bathrooms, schools, hospitals, etc.

iii. It provides a smooth, hard, even and pleasing surface.

iv. It can be easily cleaned and overall has proved economical due to less maintenance cost.

v. Concrete being a non-combustible material offers a fire-resistant floor required for fire-hazardous buildings.

19.3.4.2 Demerits

i. Defects once developed in concrete floors, whether due to poor workmanship or materials, cannot be easily rectified.

ii. The concrete flooring cannot be satisfactorily repaired by patchwork.

iii. It does not possess very satisfactory insulation properties against sound and heat.

19.3.5 Tiled floor covering

Clay tiles of different sizes, shapes, thickness and colours are prepared and they are used as floor coverings. They are placed in position on a concrete base with a thin layer of mortar. When these tiles are to be fixed on timber floors, special beds of emulsified asphalt and Portland cement are used.

19.3.5.1 Merits

i. It provides a non-absorbent, decorative and durable floor surface.

ii. It permits quick installation or laying of floors.

iii. It is easily repaired in patches.

19.3.5.2 Demerits

i. It is generally costly in initial cost as well as in maintenance cost.

ii. On becoming wet, it provides a slippery surface.

Vitrified tiles

Vitrified tiles have zero water absorption property. They resemble granite but offer a great variety in terms of finish, colour and design options. Polished vitrified tiles such as mirror stone, granamite and marbogranite are cheaper than marble and granite.

Ceramic tiles

Ceramic tiles are non-slippery and are used in wet areas like bathrooms and kitchens. They are available in a variety of interesting shapes, wide range of colours and textures. They are used in living rooms also. Ceramic tiles are usually embedded in mortar. Special tile adhesives and tile grouts are also available which allow easy laying and render the tiled area useable within 24 hours.

Laying of tiles

Use a waterproof, floor tile adhesive which allows slight flexibility when set. Follow the manufacturer's instruction and use a notched or plain trowel, as directed, to spread the adhesive on the floor over a manageable area for laying approximately 10 tiles.

Use a layer of adhesive on the back of the tile and press into the desired position. It is very important to lay the first tile correctly, as its position will determine the position of all the other tiles in the room. Use a batten nailed to the floor to give a straight edge to guide the positioning of the tiles. Remember to use plastic spacers or a thick card to regulate the distance between the tiles. These areas will be grouted when the floor is complete and must be equally spaced for neat, accurate results.

Use a spirit level to check the horizontal level and a straight edge to continually check the position of the tiles on the floor. Continue across the room and work towards the door. Leave the room for 24 hours. Then remove the battens and cut the border tiles and fix in a similar way. Remove the plastic spacers or thick card and grout the tiles.

Grout is available in a variety of colours, but the standard colours are white, grey or brown. However, most floor tiles are grouted with a mortar mix. Use a plastic scraper or a rubber blade to push the grout between the gaps in the tiles. Make sure all the spaces are evenly filled and then wipe the grout off the tile surface before it dries. Use a blunt edge of a stick or tool carefully to smooth the surface of the grout in the gaps, but do not 'dig down' into the grout. Remove any excess grout before it dries. Allow the floor to dry completely before polishing the surface of the tiles with a dry cloth.

19.3.6 Wooden floor covering

This type of floor covering is the oldest type, but nowadays it is used for some special-purpose floors such as theatres and hospitals. It possesses natural beauty and has enough resistance to wearing. Wooden floor covering may be carried out in one of the following three types:

a. Strip floor covering: This is made up of narrow and thin strips of timber which are joined to each other by tongue and groove joints.
b. Planked floor covering: In this type of construction, wider planks are employed and these are joined by tongue and groove joints.
c. Wood block floor covering: It consists of wooden blocks which are laid in suitable designs over a concrete base. The thickness of a block is 20–40 mm and its size varies from 20 × 8 to 30 × 8 cm. The blocks are properly joined together with the ends of the grains exposed.

19.3.7 Mosaic floor covering

This type of floor covering is commonly used in operation theatres, temples, bathrooms, etc. A concrete base is constructed for laying the floor covering. Over this base lime or cement mortar is placed to a depth of about 2 cm and it is levelled up. A layer of cementing material about 3 mm in thickness is spread. The cementing material consists of two parts of slaked lime, one part of powdered marble and one part of pozzolana. After 4 hours of laying this cementing material, a mixture of coloured cement and chips are laid. This surface is left for 24 hours and then it is rubbed with pumice stone to get a smooth and polished surface. The polished surface is left for about 2 weeks before use.

19.3.8 Rubber floor covering

It consists of pure rubber mixed with cotton fibre, granulated cork or asbestos fibre and the desired colouring pigments. A small amount of sulphur is also added. Its thickness varies from 3 to 10 mm and it is available in many designs and patterns. It is available in the form of tiles or sheets and can be directly laid over the floor by the vulcanizing process. It is mostly used in hospitals, radio stations, etc. The flooring is elastic, attractive, comparatively warm and soft.

19.3.9 Linoleum floor covering

It is the fabricated form of a mixture of resins, linseed oil, gums, pigments, wood flour, cork dust and other filler materials. It is available in the market in rolls of width about 2–4 m. The thickness varies from 2 to 6 mm. These tiles are also manufactured in various sizes, shapes and patterns. This floor covering is laid over an effective damp-proof course. It is cheap, durable, attractive, comfortable and moderately warm. It can be cleaned easily.

19.3.9.1 Merits

i. It provides an attractive, resilient durable and cheap surface.
ii. It offers a surface that can be easily washed and cleaned.

iii. Being moderately warm with cushioning effect, it provides comfortable living and working conditions.
iv. It offers adequate insulation against noise and heat.

19.3.9.2 Demerits

i. It is subjected to rotting when kept wet for sufficient time and its use is not recommended for basements.
ii. It does not offer resistance against fire, being combustible in nature.
iii. This covering when applied over a wooden base may get torn under excessive sub-floor movements.

19.3.10 Glass floor covering

It is used when it is desired to admit light to the floor below. Structural glass is available in the form of slabs or tiles. They are fitted within frames of different types. The members of the frame are closely spaced such that the glass floor covering can take up the superimposed loads without breaking. This type of floor covering is not commonly used.

19.3.11 Magnesite floor covering

It is known as composite flooring or jointless flooring. It is composed of a dry mixture of magnesium oxide, a pigment and inert filler materials, e.g., wood flour, asbestos or sawdust. Liquid magnesium chloride is mixed to this powder and a plastic material is obtained in situ. This plastic material is laid over the floor and the surface is levelled with a trowel. It can be directly laid over stone, concrete or wooden floor base. It is cheap and is used as floor covering for office buildings, schools, factories, etc.

19.3.12 Plastic floor covering

Thermoplastic tiles can be economically used as floor covering on the concrete floor base. It is generally not laid on a wooden floor base as the preparation of the wooden surface for receiving the tiles is very costly. Plastic floor covering has been used with success in all types of buildings like offices, hospitals, shops, schools and residential buildings.

19.3.13 Terrazzo floor covering

Terrazzo is a mixture of cement and marble chips and the surface is polished with carborundum stone to obtain a smooth finish at the top. The base for this type of floor covering is concrete and is laid in the ordinary way. On the 3 cm concrete (1:3) base, a thin layer of sand is sprinkled evenly and it is covered by tarred paper. A layer of rich mortar is spread over it and then terrazzo mixture is placed over it evenly. Marble chips of 3–6 mm are mixed with white or coloured cement in the proportion 1:2 or 1:3 to get the terrazzo mixture. Dividing strips of metal, 20 gauges thick, are inserted into the mortar base to form the desired pattern and in these small bays the terrazzo mixture is laid alternatively. The terrazzo is levelled in position by a trowel. If required some additional chips are also added at the surfaces so that about 70 per cent of the surface area is covered by the marble chips.

When the terrazzo has hardened, the surface is rubbed by coarse and fine carborundum stones, respectively, to get a smooth finished surface. It is kept wet with water while rubbing. The surface is cleaned with water and soap solution and then ax polish is applied to the surface. This type of floor covering is very costly and is used to obtain a clean, attractive and durable surface in public buildings, hospitals bathrooms, etc.

19.3.14 Marble and granite flooring

Naturally available stones like marble and granite are used as flooring materials. They are available in the form of flat slabs and can be laid above the prepared concrete base. Marble slab is to be polished with carborundum stone, whereas granite does not require any polishing. Their hardness, durability and aesthetic appearance have increased its demand as a flooring material.

REVIEW QUESTIONS

1. What is the main function of a floor and what are its main components?
2. What are the factors affecting selection of floors?
3. Briefly discuss the different types of floors.
4. What are the advantages of flat slab floors?
5. What is the necessity of floor coverings?
6. Briefly explain the different types of floor coverings.
7. Briefly discuss RCC flooring.
8. Write short notes on
   a. Wooden floor covering
   b. Mosaic floor covering
   c. Glass floor covering
   d. Granite flooring

chapter 20

Stairs

A few technical terms generally used for the design of stairs are defined below:

1. ***Baluster:*** It is a vertical member supporting the handrail. The combined framework of handrail and baluster is known as balustrade.
2. ***Flight:*** It is a series of steps without any platform or landing or break in their direction.
3. ***Tread:*** It is an upper horizontal part of a step on which the foot is placed while ascending or descending a stairway.
4. ***Step:*** This is a portion of a stair which comprises the tread and riser. This permits ascending or descending from one floor to another.
5. ***Riser:*** This is a vertical member between two treads. This provides support to the tread.
6. ***Rise:*** This is the vertical distance between the upper faces of any two consecutive steps.
7. ***Flier:*** It is a straight step having a parallel width of tread.
8. ***Landing:*** This is a platform provided between two flights.
9. ***Nosing:*** This is the outer projecting edge of a tread. This is generally made round to give an appearance that is more pleasing and makes the stair easy to negotiate.
10. ***Going:*** This is the width of the tread between two successive risers. In other words it is the horizontal distance between the faces of two consecutive risers.
11. ***Winders:*** They are tapering steps used for changing the direction of a stair.
12. ***String or Stringer:*** This is a sloping member which supports the steps in a stair.
13. ***Newel Post:*** This is the vertical post placed at the top and bottom ends of flights supporting the handrails.
14. ***Run:*** This is total length of stairs in a horizontal plane, including landings.
15. ***Soffit:*** This is the underside of a stair.
16. ***Header:*** This is a horizontal structural member supporting stair stringers or landings.
17. ***Carriage:*** This is a rough timber supporting the steps of wooden stairs.
18. ***Staircase:*** It is the space or enclosure or room which contains the complete stairway.

20.1 REQUIREMENTS OF A GOOD STAIR

1. Stairs should be so located that they can be easily accessible from the different rooms of the building.
2. It should have provision for adequate light and proper ventilation.

3. It should have sufficient stair width to accommodate a number of persons in peak hours or emergencies. Generally, for interior stairs the clear width may be required to be at least 50 cm in one and two family dwellings, 90 cm in hotels, motels, apartment building and industrial buildings and 1.1 m for other types of occupancy.
4. The number of steps in a flight should generally be restricted to a maximum of 12 and minimum of 3.
5. Ample head room should be provided not only to prevent tall people from injuring their head, but also to give a feeling of spaciousness. Vertical clearance should never be less than 2.15 m.
6. Sizes of risers and treads should generally be proportioned from the comfort point of view. Treads should be 25–32.5 cm wide, exclusive of nosing. Treads less than 25 cm width should have a nosing of about 25 cm. The most comfortable height of the riser is 17.5–18.5 cm. Generally the following formulae should be used:
   a. Product of riser and tread must be between 400 and 410.
   b. Riser plus tread must equal 42.5–43.5 cm.
   c. Sum of the tread and twice the riser must lie between 60 and 64 cm.
7. Stair width depends upon the purpose and importance of the building. In the case of residential buildings it should be kept as 1 m.
8. The number of stairways required should be controlled on the maximum floor area contributory to a stairway. The number of persons that may be served by stairs per floor per 55 cm unit of stair width should be 15 for such buildings as hospitals and nursing homes, 30 for other institutional and residential buildings, 45 for storage buildings, 60 for mercantile, business, educational, industrial buildings, theatre and restaurants, 80 for churches, concert halls and museums, and 320 for stadium and amusement structures.
9. The minimum width of landing should be equal to the width of the stairs.
10. The maximum and minimum pitch should be 40° and 25° respectively in any stairs.
11. Winders should be provided at the lower end of the flight, only when it is essential. Generally, the use of winders in a staircase should be avoided.
12. In open-well stairs in order to avoid the danger of accidents balustrade must be provided.
13. The live loads to be considered on stairs have been stipulated by IS:875-1964. The stairs and landings should be designed for a live load of 300 kg/m^2 in buildings where there are no possibilities of overcrowding. In the case of public buildings and warehouse, where overcrowding is likely, the live load may be taken as 500 kg/m^2.

20.2 TYPES OF STAIRS

Generally, stairs are of the following types (Figure 20.1):

1. Straight stairs
2. Quarter turn stairs
3. Half turn stairs
4. Three quarter turn stairs

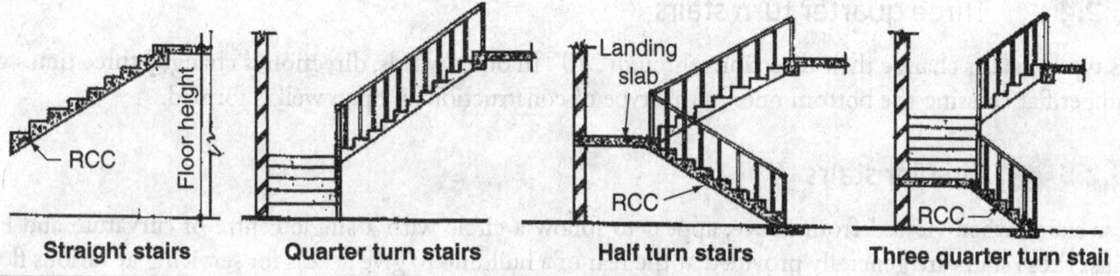

Figure 20.1 Different types of stairs

5. Circular stairs
6. Spiral stairs
7. Curved stairs
8. Geometrical stairs
9. Bifurcated stairs

20.2.1 Straight stairs

These are the stairs along which there is no change in direction on any flight between two successive floors. The straight stairs can be:

a. Straight run with a single flight between floors
b. Straight run with a series of flights without change in direction
c. Parallel stairs
d. Angle stairs
e. Scissors stairs

Straight stairs can have a change in direction at an intermediate landing. In case of parallel stairs, the stairs require a complete reversal of direction. In case of angle stairs, the successive flights are at an angle to each other. Scissor stairs are comprised of a pair of straight runs in opposite directions and are placed on opposite sides of a fire-resistive wall.

20.2.2 Quarter turn stairs

They are provided when the direction of flight is to be changed by 90°. The change in direction can be effected by either introducing a quarter-space landing or by providing winders at the junctions.

20.2.3 Half turn stairs

These stairs change their directions through 180°. It can be either 'dog-legged' or 'open newel type'. In case of dog-legged stairs, the flights are in opposite directions and no space is provided between the flights in plan. On the other hand, in open newel stairs, there is a well or opening between the flights and it may be used to accommodate a lift. These stairs are used at places where sufficient space is available.

20.2.4 Three quarter turn stairs

This type of stairs change their directions through 270°. In other words, direction is changed three times with its upper flat crossing the bottom one. In this type of construction an open well is formed.

20.2.5 Circular stairs

These stairs, when viewed from above, appear to follow a circle with a single centre of curvature and large radius. These stairs are generally provided at the rear of a building to give access for servicing at various floors. All the steps radiate from a newel post in the form of winders. These stairs can be constructed in stone, cast iron or RCC.

20.2.6 Spiral stairs

These stairs are similar to circular stairs except that the radius of curvature is small and the stairs may be supported by a centre post. Overall diameter of such stairs may range from 1 to 2.5 m.

20.2.7 Curved stairs

These stairs, when viewed from above, appear to follow a curve with two or more centres of curvature, such as an ellipse.

20.2.8 Geometrical stairs

These stairs have no newel post and are of any geometrical shape. The change in direction in these stairs is achieved through winders. The stairs require more skill for its construction and are weaker than open newel stairs. In these stairs, the open well between the forward and the backward flights is curved.

20.2.9 Bifurcated stairs

These stairs are so arranged that there is a wide flight at the start which is subdivided into narrow flights at the mid-landing. The two narrow flights start from either side of the mid-landing. Generally, these stairs are more suitable for public buildings.

20.3 STAIRS OF DIFFERENT MATERIALS

The following materials are used in the construction of stairs:

20.3.1 Stone stairs

The stones employed in the construction of stairs are hard, durable, weather resistant and fire resistant. The stone steps can be either

a. Cantilevered from the wall

b. Built into the wall at both ends

20.3.2 Brick stairs

They are rarely used except as entrance steps. A brick stair may be built of solid masonry work. The surface of the stair may be given any suitable type of floor finish.

20.3.3 Wooden stairs

Wooden stairs are light in weight and they are generally used for residential buildings. The greatest limitation of wooden stairs is that it catches fire easily, and in case of a fire, the occupants of upper floors will find no way of escape. If good quality of timber of proper thickness is used, it resists fire to a great extent and the occupants will have enough time to escape.

20.3.4 RCC stairs

These stairs are commonly used in all types of constructions. They can resist wear and fire better than any other material and can be casted to the desired shape. The advantages of the RCC stair are as follows:

a. It can resist fire in a better way.

b. Movement over it produces less noise.

c. It can be cast.

REVIEW QUESTIONS

1. Define
   a. Rise
   b. Tread
   c. Landing
   d. Flights of a staircase
2. What are the requirements of good steps?
3. Briefly discuss the different types of stairs.
4. Explain different types of stairs depending upon the different materials used.

chapter 21

Plastering

Plastering is the process of covering rough walls and uneven surfaces in the construction of houses and other structures with a plastic material, called plaster, which is a mixture of lime or cement concrete and sand along with the required quantity of water.

21.1 REQUIREMENTS OF GOOD PLASTER

1. It should adhere to the background and should remain adhered during all climatic changes.
2. It should be cheap and economical.
3. It should be hard and durable.
4. It should be possible to apply it during all weather conditions.
5. It should effectively check the entry or penetration of moisture from the surfaces.
6. It should possess good workability.

21.1.1 Objective of plastering

a. To provide an even, smooth, regular, clean and durable finished surface with improved appearance.

b. To conceal defective workmanship.

c. To preserve and protect the surface.

d. To provide a base for the decorative finish.

e. To cover up the use of inferior quality and porous materials of the masonry work.

21.2 METHODS OF PLASTERING

The plaster may be applied in one or more coats, but the thickness of a single coat should not exceed 12 mm. In the case of inferior or cheaper type of construction, the plaster may usually be one coat. For ordinary type of construction, the plaster is usually applied in two coats, whereas for superior type of works it is applied in three coats. The final setting coat should not be applied until the previous coat is almost dry. The previous surface should be scratched or roughened before applying the next coat of plaster. In plastering, the plaster mix is either applied by throwing it with great force against the walls or by pressing it on the surface.

Table 21.1 Plaster in Three Coats with Lime Mortar

	Name of coat	Thickness	Remarks
First coat	Rendering coat	12 mm	This is left for a period of 2 days to set and is not allowed to dry.
Second coat	Floating coat	6–9 mm	This coat is applied with trowels and rubbed with a straight edge. The water is sprinkled on the surface and the surface is well rubbed with floats to make it an even surface.
Third coat	Setting coat or finishing coat	3 mm	This coat is applied after 5 days. Neeru or sagol is used to prepare a smooth surface. After giving a rest of 24 hours to the plastered surface, the work should be watered well for a fortnight or so.

The methods of applying the following common types of plasters are as follows.

21.2.1 Lime plastering

Lime plastering is the process of covering the surfaces by lime plaster or mortar in various proportions depending upon the nature of work and the number of coats to be applied (Table 21.1).

21.2.1.1 Preparation of a surface for plastering

When a surface is to be plastered, the surface is prepared in the following manner:

i. All the mortar joints of the wall to be plastered are left rough and projecting, so as to give a key or hold to the plaster. All the joints and surfaces are well cleaned with a wire brush and ensured that they are free from oil, grease, etc. If the surface is smooth or the wall to be plastered is an old one, then the mortar joints are raked out at least to a depth of 12 mm to give bond to the plaster.

ii. Projections more than 12 mm over the surface are knocked off so as to obtain a uniform surface of wall and also to reduce the consumption of plaster.

iii. Similarly, all the cavities and holes inside the surface are properly filled up in advance.

iv. All woodwork to be plastered is roughened.

v. Finally, the mortar joints and surfaces of the wall are well washed, wetted with water and kept for at least 6 hours before plastering.

21.2.1.2 Groundwork of plaster

In order to obtain a true surface and uniform thickness, vertical strips called screeds and bands are formed on the wall surface by fixing dots. These dots are first of all applied horizontally and vertically at a distance of about 2 m, covering the entire wall surface. These dots are fixed by laying a small quantity of plaster on the surface in the forms of squares (15 cm × 15 cm) having thickness of about 10 mm. The verticality of dots, one over the other, is checked by means of a plumb bob. After fixing dots, the vertical strips of plaster known as screeds are formed in between the dots. These screeds serve as the gauges for maintaining even thickness of the plaster being applied. Mortar is then applied on the wall between groundworks of screeds so prepared.

21.2.1.3 Application of plaster coats

i. *First coat (or rough course of plastering or rendering coat):* It is usual to provide an average thickness of first coat of plaster equal to 12 mm over brick or ashlar masonry and 20 mm on rubble masonry, the larger thickness on rubble masonry being due to the roughness of its surface and the necessity of providing a minimum covering of 6 mm of mortar to rubble. The first coat (rendering coat) of plaster is generally applied by dashing against the wall surface between the screeds mentioned above. It is then sprinkled frequently with sufficient quantity of water and rubbed well by means of floats.

If a second coat, called floating coat, is to be applied, the surface of the first coat is left exposed to air for a period of 2 days to set but not to dry. After this period, the surface of the first coat is swept clear of any dust or loose particles, sprinkled with water and well beaten with thin strips of bamboo or cane. The surface of the first coat is kept wet till the second coat is applied.

ii. *Second coat or floating coat:* The second coat is applied after preparing the surface of the first coat as mentioned above. The second coat is spread out uniformly with trowels. It is pressed and rubbed with a wooden straight edge, to obtain the desired surface. It is finally finished by slightly sprinkling water over the plastered surface and rubbing it with the floats. The thickness of the second coat is usually between 6 and 9 mm.

iii. *Third coat or final coat or finishing coat:* This coat is applied after 5 days of the second coat. This coat consists of a cream of white or fat lime (called neeru or plaster's putty) and fine white sand in the ratio of 1:2 laid in thickness of 3 mm with straight plane and is rubbed with a straight edge. The surface is well rubbed with a wooden float and then finally finished with a trowel to obtain the desired surface. A polishing stone is used to obtain a fine polished surface.

21.2.2 Cement plastering

Cement plastering is an ideal plastering for external renderings. It is specially suited for damp conditions such as bathrooms, reservoirs, water tanks, floors, copings, etc. where non-absorbent surfaces are desired. Cement plaster is usually applied in a single coat. However, in certain cases when the thickness of the plaster is more than 15 mm or it is desired to have a finer finish, the plaster is applied in two coats (Table 21.2).

The process of cement plastering in two coats is as follows:

a. *First coat or rough coat:* Usually, the average thickness of the first coat of plaster is 12 mm on brick masonry or ashlar masonry and 23 mm on rubble masonry. In the case of concrete masonry, this thickness varies from 9 to 15 mm depending upon the nature of work. For the first coat, cement plaster with mix proportions as 1:3 or 4 (1 cement:3 or 4 sand) is generally used. The first coat of plaster is placed between the spaces or bays formed by the screeds on the wall surface. This plaster is applied with a mason's trowel. The surface is then levelled by means of flat wooden floats and wooden straight edges and finally finished by polishing with a trowel. If a second coat or fine coat is to be applied, the surface of the first coat is not polished, but roughened with a scratching tool to form a key to the second coat of plaster.

Table 21.2 Plaster in Three Coats with Cement Mortar

	Name of coat	Thickness	Remarks
First coat	Rendering coat	9-10 mm	This is left for a period of 3-4 days to harden. Its surface is kept rough.
Second coat	Floating coat	6–9 mm	The purpose of this coat of plaster is to bring the work to an even surface.
Third coat	Setting coat or finishing coat	3 mm	This coat is similar to the second coat of a two coat plaster.

b. *Second coat or fine coat:* Before applying the second coat, the first coat is left to set for at least 7 days and is roughened to form a proper key with the second coat. The second coat, consisting of pure Portland cement mixed with sufficient quantity of water, is applied after 6 hours. This second coat is laid in a thin layer of 3 mm maximum thickness over the rough and moist surface of the first coat. Finally, this coat is well trowelled and rubbed smooth. Each coat should be kept damp continuously by curing for at least 5–10 days.

21.2.3 Mud plastering

This is the cheapest type of plastering, generally employed in the construction of village houses, temporary sheds and structures of temporary importance. Besides being cheap it provides insulation against heat and keeps the house cool for comfortable living. Mud plastering (in two coats) is then carried out as below:

a. The mud plaster, consisting of well-tempered clay, chopped straw and cow dung, is prepared in a homogenous mass as described earlier.

b. The preparation of the wall surface and groundwork for plastering is done exactly in the same manner as that for lime or cement plastering.

c. The first coat is then applied in a thickness of 12 mm by dashing the plaster or by placing it in the spaces formed by the screeds. The surface is then finished by means of a straight edge and a wooden float.

d. After 24 hours of setting (but not drying) of the first coat, the second coat is applied in a thickness of 6 mm.

e. No curing is done in this case but the surface is given a wash of fine white earth, cow dung and cement in a mix proportion of 3:2:1 respectively.

REVIEW QUESTIONS

1. Why is plastering required for walls?
2. What are the requirements of a good plaster?
3. What are the objectives of plastering?
4. Briefly discuss the different types of plastering.
5. Briefly discuss the process of cement plastering in two coats.
6. How is cement plaster prepared?

chapter 22

Painting

Paints are coatings of fluid materials which are applied as a final finish to surfaces like walls, ceiling, wood and metal works.

Painting is done to protect the surface from the effects of weathering, to prevent wood from decay and metal from corrosion, to provide a decorative finish and to obtain a clean, hygienic and healthy living atmosphere.

22.1 TYPES OF PAINTS

1. **Enamel paints:** These paints are available in numerous shades. They mainly consist of white lead or zinc white, resinous matter and petroleum spirit. Their formation into hard, impervious, decay-resistant enamel-like surface soon after application protects it from being affected by acids, alkalies, fumes and gas, hot and cold water, etc. They can be used for internal as well as external purposes.

2. **Cement paints:** These include a variety of paints in which cement is the main constituent responsible for the hardness and durability of the painted surface. They are available in dry, powder form. They are waterproof. It is desirable to provide cement paint on a rough surface rather than a smooth surface because its adhesion power is more on rough surface than on smooth surface. They prove to be economical as compared to oil paints. They are suitable for painting fresh plasters having high alkalinity because cement paints are not likely to be attacked by the alkalinity of the masonry surface. It is not necessary to remove the existing paint for the application of new paint.

3. **Oil paints:** They are generally applied in three different layers with varying composition. These are termed as primes, undercoats and finishing coats. The dampness of the wall affects the life of the oil paint, hence it must not be applied during damp weather.

4. **Cellulose paints:** They are prepared from nitro cotton, celluloid sheets, photographic films, etc. The cellulose paints harden by evaporation of thinning agents. The surface painted with cellulose can be washed and cleaned easily. They are little costlier than other paints.

5. **Aluminium paints:** Finely ground aluminium is suspended in either quick-drying spirit varnish or slow-drying oil varnish as per requirement. As the spirit or oil evaporates, a thin film of aluminium is formed on the surface. This paint forms a better protective surface over steel and iron. They are impervious to moisture and possess high electrical resistance. They have a good appearance and are visible in darkness.

6. **Emulsion paints:** These paints contain polyvinyl acetate, synthetic resins, etc. They are easy to apply and are retained for a long period and can be cleaned easily with water. For a rough plastered surface, a thin coat of cement paint may be first applied to smoothen the surface. It is necessary to have a sound surface to receive the paint.

7. **Anticorrosive paints:** These consist of oil and a strong drier. The pigments such as chromium oxide, lead or zinc chrome is taken and after mixing it with a small quantity of very fine sand it is added to the paint. These are cheap and last for a long duration. They are black in colour.

8. **Synthetic rubber paints:** These paints are prepared from resins. They have an excellent chemical resistant property. They can be applied to surfaces that may not be completely dry. They offer good resistance to water and are not affected by heavy rains. They dry very quickly. They are not affected by weather and sunlight and are quite easy to apply.

22.2 USES OF PAINT

1. It protects the surface from weathering effects of the atmosphere and the actions by other liquids, fumes and gases.
2. It prevents decay and corrosion in metal.
3. It is used to give good appearance to the surface.
4. It provides a smooth surface.

22.2.1 Characteristics of a good paint

A good paint should possess high spreading power and should be durable, tough and resistant to wear on drying. It should work smoothly and freely and should not crack, fade or change colour. Its surface should become dry in 9 hours and hard enough to take another coat in 24 hours. We should be able to spread it into a very thin layer and it should provide a smooth and pleasing appearance.

22.3 PAINTING OF DIFFERENT SURFACES

22.3.1 New woodwork

Normally, four coats of paint are required for new woodwork. The process of painting is as follows:

a. The surface of the woodwork is prepared to receive the paint for satisfactory working. It is necessary that the woodwork is sufficiently seasoned and it does not contain more than 15 per cent moisture at the time of painting. The surface of the woodwork is thoroughly cleaned and the heads of nails are punched to a depth of 3 mm below the surface.
b. The surface of the woodwork is then knotted.
c. The process of cleaning the surface using sand paper is then carried out.
d. The subsequent coats of paint, namely undercoat and finishing coats, are then applied on the surface. Extreme care should be taken to see that the finishing coat presents a smooth and even surface and that no brush marks are seen on the finished surface.

22.3.2 Repainting old woodwork

If the paint on the old woodwork has cracked or has developed blisters, it is to be removed. If the surface has become greasy, it should be cleaned by rubbing down with sandpaper or fine pumice stone. The old paint can also be removed by applying any one of the following three paint solvents.

a. A solution containing 200 g of caustic soda in a litre of water is prepared and used to wash the surface. The paint dissolves and the surface becomes clean.
b. A mixture consisting of one part of soft soda and two parts of potash is prepared and one part of quick lime is added afterwards. This mixture is applied on the surface in a hot state and allowed to stay for about 24 hours. The surface is then washed with hot water.

c. A mixture consisting of equal parts of washing soda and quicklime is brought to a paste form by adding required quantity of water. It is applied on the surface and kept for about an hour. The surface is then washed with water.

After removing the old paint from the surface, painting is carried out in 2 or 3 coats of an oil paint.

22.3.3 Painting new iron work and steel work

The surface of iron or steel to receive the paint should be free from rust, grease, dirt, etc. Suitable equipments such as wire brushes and scrapers are used to remove all loose scales, marks, etc. from the surface. Water with caustic soda or lime is used to remove grease. The cleaned surface is provided with a film of phosphoric acid. This film protects the surface from rust and it also facilitates the adhesion of paint. The coats of paint are then applied. The paint suitable to iron and steel surfaces should be selected for each coat. The priming coat or first coat is made by dissolving fine red lead powder in boiled linseed oil. After this coat has dried, two or more coats of the desired paint are applied over the priming coat, with the help of a brush or spray gun The finishing coat should present a smooth finish and precaution should be taken to avoid the presence of brush marks on the final painted surface.

22.4 WHITE WASHING

The process of white washing process can be done through carried out under the following operations:

22.4.1 Preparation of white wash

The white wash is prepared from fresh burnt shell or pure stone lime mixed with water. Shell lime is preferred to pure lime as it is whiter and slakes more perfectly to a smoother paste.

To prepare white wash, fresh lime is slaked at the site of work and is dissolved in a tub with sufficient quantity of water. After slaking, it is allowed to remain in the tub of water for 2 days and then stirred up with a pole until it attains the consistency of thin cream. This mixture is then strained or screened through a clean coarse cloth. Clean gum dissolved in hot water is then added at the rate of 2 kg/m^3 of lime to the white-wash water. The solution so formed is called as white wash. To prevent the glaring effect due to whitewash, sometimes copper sulphate at the rate of 4 kg/m^3 of thin cream is added. In order to have better adhesive properties, alum or common salt may be added in the same proportion as gum.

22.4.2 Preparation of surface

Before applying white wash to a new wall surface, it is essential that the surface should be cleaned, brushed and made free from loose materials and any other foreign matter. If the surface to be coated is oversmooth, then the coats will not stick to it. In such cases, the surface should be rubbed with sand paper to ensure proper adhesion of white wash.

In case of re-white washing, all loose materials and scales should be scrapped off. The old loose white wash is removed by rubbing with sand paper. All holes on the wall, irregularities of surface and minor repairs are corrected in advance by filling with lime putty.

All greasy spots should be given a coat of a mixture of rice water and sand so that the finishing wash may stick to the surface. If old white wash is discoloured by smoke or other reasons as in kitchens, factories, restaurants, etc., then in such cases the surfaces should be given a wash of a mixture of wood ashes and wastes or yellow earth, before the application of white wash. Cement plastered walls should be washed with a weak solution of soap and dried before applying white wash.

22.4.3 Application of white wash

The white wash is applied to a specified number of coats with a brush. Usually, three coats are required for new work and scrapped surfaces, while one or two coats are considered sufficient for old work. For each coat, one stroke is given from the top downwards and the other from the bottom upwards over the first stroke, and similarly one stroke from the right and another from the left over the first brush before it dries. Each coat should be allowed to dry before applying the next coat. The finished dry surface should not show any signs of cracking or peeling and should not come off readily on fingers when rubbed.

22.5 COLOUR WASHING

A colour wash is usually prepared by adding the necessary colouring pigments in suitable quantities to the screened white wash or liquid mixture of white wash. It should be ensured that the colouring pigment is not affected by the presence of lime. The colour wash is applied in exactly the similar manner as white wash. The colour is constantly stirred with a stick during its use. The quantity of colour wash, which is just enough for the day's work, should be prepared at a time in the morning.

1. Normally, the ceilings are white washed and the walls are colour washed.
2. Before applying colour wash on new surfaces or scrapped surfaces, a coat of white wash should be applied. This coat will act as a priming coat and incidentally result in economy also.
3. In the case of old work, a coat of colour wash is first applied over the patches or repaired work and rectified surface spots. Two or more coats of colour are then applied on the entire surfaces till it presents the desired finish.
4. A satisfactory finish should not give out the powder when rubbed with fingers.

22.6 DISTEMPERING

The main object of applying distemper to the plastered surfaces is to create a smooth surface. The distempers are available in the market under different trade names. They are cheaper than paints and varnishes and they present a neat appearance. They are available in a variety of colours.

22.6.1 Properties of distempers

a. On drying, the film of distemper shrinks. Hence, it leads to cracking and flaking, if the surface to distemper is weak.

b. The coating of distemper is usually thick and they are more brittle than other types of water paints.

c. The film developed by distemper is porous in character and it allows water vapour to pass through it. Hence, it permits new walls to dry out without damaging the distemper film.

d. They are generally light in colour and they provide a good reflective coating.

e. They are less durable than oil paints.

f. They are treated as the water paints and they are easy to apply.

g. They can be applied on brickwork, cement plastered surface, insulating board, etc.

h. They exhibit poor workability.

i. They prove to be unsatisfactory in damp locations such as kitchens and bathrooms.

22.6.2 Process of distempering

The application of distemper is carried out in the following manner:

22.6.2.1 *Preparation of surface*

The surface to receive the distemper is thoroughly rubbed and cleaned. The important points to be noted are:

i. The new plastered surfaces are to be kept exposed for a period of 2 months or so to dry out before the distemper is applied on them. The presence of dampness on the surface results in failure of the distemper coating.
ii. The surface to receive the distemper should be free from any efflorescence patches. The patches are to be wiped out by a clean cloth.
iii. The irregularities of the surface such as cracks and holes are to be filled by lime putty or gypsum and are allowed to become hard before distemper is applied on the surface.
iv. If distemper is to be applied on the existing distempered surface, the old distemper should be removed by profuse watering.

22.6.2.2 *Priming coat*

After preparing the surface to receive the coats of distemper, a priming coat is applied and it is allowed to become dry. For readymade distempers, the priming coat should be composed of materials as recommended by the makers of distempers. For locally made distempers, the milk can be used for priming coat. One litre milk will cover about 10 m^2 of the surface.

22.6.2.3 *Coats of distemper*

The first coat of distemper is then applied on the surface. It should be of a light tint and applied with great care. The second coat of distemper is applied after the first coat has dried and become hard.

i. The distempering should be done in dry weather to achieve better results.
ii. The oil-bound distemper or washable distemper adheres well to oil-painted walls, woods, corrugated iron, etc. But a priming coat of pure milk should be applied before distempering is done on such surfaces.
iii. The application of distemper by a spraying pistol is superior to that by brushes. Spraying affords a smooth and durable film of distemper.

22.7 TILING OF WALLS

Nowadays walls of the bathrooms, kitchens, work area, etc. are given a covering of glazed tiles in order to provide improved cleanliness and aesthetic appearance. Tiles are available in various forms, shapes, sizes and colours.

Before fixing the tiles, the verticality of all the corners should be checked and all the concealed plumbing and electric works should be completed and checked for their leak proofness. The plastered surface should then be given a final bedding of cement paste 1:2 with very fine sand. Tiles should be kept soaked in clean fresh water for 30 minutes before mixing. Cement mix should evenly be applied on the backside of the tiles and fixed in position by tapping into the correct position. The joint should be kept minimum (1.5 mm). Then the corners should be stroked with a mallet to check whether there is any hollow sound, which indicates poor workmanship. The joints should then be cleaned and a finish of white or coloured cement should be given. Curing should be continued for at least 15 days.

REVIEW QUESTIONS

1. Why is painting done on wooden surfaces?
2. What are the types of paints mainly used?
3. What are the main uses of paints?
4. What are the main characteristics of a good paint?
5. How is painting done on wooden surfaces?
6. What is white washing and how is white wash prepared?
7. What are the properties of distempers?
8. Briefly discuss the process of distempering.
9. What do you mean by tiling of walls?

chapter 23

Damp Proofing

23.1 SOURCES OF DAMPNESS

Dampness in buildings is generally due to one or more of the following reasons:

1. Faulty design of structure
2. Faulty construction or poor workmanship
3. Use of poor materials in construction

These causes give rise to an easy access to moisture to enter the building from different points, such as rising of moisture from the ground and rain penetration through walls, roofs and floors. The moisture entering the building from the foundation and roofs travels in different directions further under the effects of capillary action and gravity, respectively. The entry of water and its movements, in different parts of the building, are positively due to one or more of the causes listed above. The various sources that create dampness in buildings are as follows.

23.1.1 Rising of moisture from the ground

The sub-soil or ground on which the building is constructed may be made of soil that easily gives access to water to create dampness in buildings, through the foundations. Generally, foundation dampness is caused when the building structures are constructed on low-lying waterlogged areas, where a sub-soil of clay or peat is commonly found, through which dampness would easily rise under capillary action unless properly treated. This dampness further finds its way to the floors, walls, etc. through the plinth.

23.1.2 Action of rainwater

Whenever the faces of walls are not suitably protected from the exposure to heavy showers of rains, they become the sources of dampness in a structure. Similarly, the poor mortar joints in walls and cracked roofs also allow dampness to enter a building structure. Sometimes, due to faulty eave board the rainwater may percolate through roof coverings.

23.1.3 Rain penetration from tops of walls

All parapet walls and compound walls of buildings, which have not been protected from rain penetration by using damp-proof course or by such measures on their exposed tops, are subjected to dampness. This dampness in buildings is of serious nature and may result in unhealthy living conditions or even in structurally unsafe conditions.

23.1.4 Condensation due to atmospheric moisture

Whenever the warm air in the atmosphere is cooled, it gives rise to the process of condensation. On account of the condensation, the moisture is deposited on the whole area of walls, floors and ceilings. However, this source of dampness is prevalent only in certain places in India, where very cold climates exist.

23.1.5 Miscellaneous sources or causes

The various other sources or causes, which may be responsible for dampness in buildings, are mentioned below:

a. ***Poor drainage of the site:*** The structure if located on low-lying sites causes waterlogged conditions when impervious soil is present underneath the foundations. Therefore, such structures that are not well drained cause dampness in buildings through the foundation.

b. ***Imperfect orientation:*** Whenever the orientation of the building is not proper or geographical conditions are such that the walls of the building get less of direct sunrays and more of heavy showers of rain, then such walls become liable to dampness.

c. ***Constructional dampness:*** If more water has been introduced during construction or due to poor workmanship, the walls are observed to remain in a damp condition for sufficient time.

d. ***Dampness due to defective construction:*** The dampness in buildings is also caused due to poor workmanship or methods of construction, namely inadequate roof slopes, defective rainwater pipe connections, defective joints in the roofs, improper connections of walls, etc.

23.2 EFFECTS OF DAMPNESS

1. A damp building creates unhealthy living and working conditions for the occupants.
2. Presence of damp conditions causes efflorescence on building surfaces, which ultimately may result in the disintegration of bricks, stones, tiles, etc., and hence in the reduction of strength.
3. It may result in softening and crumbling of plaster.
4. It may cause bleaching and flaking of the paint, which result in the formation of coloured patches on the wall surfaces and ceilings.
5. It may result in corrosion of metals used in the construction of buildings.
6. The materials used as floor coverings, such as tiles, are damaged because they lose adhesion with the floor bases.
7. Timber, when in contact with damp conditions, gets deteriorated due to the effects of warping, buckling and rolling.
8. All electrical fittings get deteriorated, causing leakage of electric current with the potential danger of a short circuit.
9. Dampness promotes the growth of termites and, hence, creates unhygienic conditions in buildings.
10. Dampness, when accompanied by warmth and darkness, breeds the germs of tuberculosis, neuralgia, acute and chronic rheumatism, etc., which sometimes result in fatal diseases.

23.3 TECHNIQUES AND METHODS OF DAMP PREVENTION

The following precautions should be taken to prevent the dampness in buildings before applying the various techniques and methods described later:

1. The site should be located on a high ground and well-drained soil to safeguard against foundation dampness. It should be ensured that the water level is at least 3 m below the surface of the ground or

lowest point even in the wet season. For better drainage, the ground surface surrounding the building should also slope away from the house or structure.

2. All the exposed walls should be of sufficient thickness to safeguard against rain penetration. If walls are of bricks, they should be made of at least 30 cm thickness.
3. Bricks of superior quality, which are free from defects such as cracks, flaws and lump of limestones, should be used. They should not absorb water more than one-eighth of their own weight when soaked in water for 24 hours.
4. Good quality cement mortar (1 cement:3 sand) should be used to produce a definite pattern and perfect bond in building units throughout the construction work. This is essential to prevent the formation of cavities and occurrence of differential settlement, due to inadequate bonding of units.
5. Cornices and string courses should be provided. Window sills, coping of plinth and string courses should be sloped on top and throated on the underside to throw the rainwater away from the walls.
6. All the exposed surfaces like tops of walls and compound walls should be covered with waterproofing cement plaster.
7. Hollow walls (i.e., cavity walls) are more reliable than solid walls in preventing dampness and, hence, the cavity wall construction should be adopted wherever possible.

23.3.1 Prevention of dampness

The various techniques and methods, generally adopted to prevent the defects of dampness, are as follows:

a. Use of damp-proofing courses (DPC) or damp-proofing membranes.
b. Waterproof or damp-proof surface treatments.
c. Integral damp-proofing treatments.
d. Cavity walls or hollow walls.
e. Guniting or shot concrete, or shotcrete.
f. Pressure grouting or cementation.

23.3.1.1 *Use of damp-proofing courses*

These are the layers or membranes of water repellent materials, such as bituminous felts, mastic asphalt, plastic sheets, cement concrete, mortar, metal sheets, slates and stones, which are interposed in the building structures at all locations wherever water entry is anticipated or suspected. These damp-proof courses of suitable materials should be provided at appropriate locations for their effective use. Basically, DPC is provided to prevent the water from rising from the sub-soil or ground and getting into the different parts of the building. The best location or position for DPC, in case of buildings without basements, lies at the plinth level or, in case of structures without plinth, it should be laid at least 1.5 cm above the ground level. These damp-proof courses may be provided horizontally or vertically in floors, walls, etc. In the case of basements, laying of DPC is known as 'tanking'.

23.3.1.2 *Waterproof (or damp proof) surface treatments*

The surface treatment consists of filling up the pores of the material exposed to moisture by providing a thin film of water repellent material over the surface. The surface treatments can be either external or internal; the external treatment is effective in preventing dampness whereas internal only reduces it to a certain extent.

Many surface treatments like pointing, plastering, painting and distempering are given to the exposed surfaces and also to the internal surfaces. The most commonly used treatment to protect walls against dampness is lime cement plaster of (1 cement:1 lime:6 sand) mix proportions. A thin film of waterproofing material can be applied to the surface of concrete after it is laid. Some of the materials generally employed as waterproofing agents in surface treatments are sodium or potassium silicates, aluminium or zinc sulphate, barium hydroxide and magnesium sulphate in alternate applications, soft soap arid alum also in alternate applications, lime and linseed oil, coal tar, bitumen, waxes and fats, resins and gums, etc.

23.3.1.3 Integral damp-proofing treatment

The integral treatment consists of adding certain compounds to the concrete or mortar during the process of mixing, which when used in construction act as barriers to moisture penetration under different principles. Compounds like chalk, talc and fuller's earth have a mechanical action principle, i.e., they fill the pores present in the concrete or mortar and make them denser and waterproof. The compounds like alkaline, silicates, aluminium sulphate and calcium chlorides work on a chemical action principle, i.e., they react chemically and fill in the pores to act water resistant. Similarly, some compounds like soaps, petroleum oils and fatty acid compounds such as stearates of calcium and sodium ammonium work on a repulsion principle, i.e., they are used as admixtures in concrete to react with it and become water repellent.

23.3.1.4 Cavity walls (or hollow walls)

A cavity wall consists of two parallel walls or leaves or skins of masonry separated by a continuous air space or cavity. Cavity walls consist of three main parts, namely,

i. The outer wall or leaf (10 cm thick) which is the exterior part of the wall.

ii. The cavity or air space of 5–8 cm.

iii. The inner wall or leaf (minimum 10 cm) which is the interior part of the wall.

The two leaves, forming a cavity in between, may be of equal thickness or the thickness of the inner leaf may be increased to take the greater proportion of the imposed loads transmitted by the floor and roof. The provision of continuous cavity in the wall efficiently prevents the transmission or percolation of dampness from the outer wall or leaf to the inner wall or leaf. Based on the climatic conditions in India, i.e., hot dry (hot humid), this cavity type of construction is most desirable as it offers many advantages such as better living and comfort conditions, construction economy and preservation of the building against dampness.

The cavity wall construction offers the following advantages over solid wall construction:

i. As there is no contact between the outer and inner walls of a cavity wall except at the wall ties, which are of impervious material, the possibility of moisture penetration is reduced to a minimum. It has been established that a cavity wall having 10 cm thick internal and external leaves with 5 cm cavity or air space in between is better and more reliable than a solid wall of 20 cm thickness, in respect of damp penetration.

ii. As air in the cavity is a non-conductor of heat, it prevents the transmission of heat through the walls and maintains better consistency of temperature inside the building. In this regard, it has been established that cavity walls provide an improvement of 25 per cent in heat insulation over the solid walls of the same cross section less the cavity thickness. Therefore, cavity wall construction is best suited for a tropical country like India.

iii. The cavity walls also offer good insulation against sound.

iv. The cavity tends to reduce the nuisance of efflorescence.

v. This type of construction also offers many other benefits such as economy, better comfort and hygienic conditions in buildings.

23.3.1.5 Guniting (or shot concrete)

This consists in forming an impervious layer using a rich cement mortar (1 cement:3 sand or fine aggregate mix) for waterproofing over the exposed concrete surface or over the pipes, cisterns, etc., for resisting the water pressure.

Gunite is a mixture of cement and sand or well-graded fine aggregate, the usual proportions being 1:3 or 1:4 (i.e., 1 cement:(3 or 4) sand or fine aggregate). A machine known as cement gun, having a nozzle for spraying the mixture and a drum of compressed air for forcing this mixture under desired pressure, is used for this purpose. Any surface, which is to be treated, is first thoroughly cleaned of any dirt, grease or loose particles and then fully wetted. The mixture of cement and sand or aggregates is then shot under a pressure of 2-3 kg/cm^2 by holding the toe nozzle of the cement gun at a distance of 75–90 cm from the surface of the wall. The necessary quantity of water is added by means of a regulating valve soon after the mixture comes out from the cement gun. Thus, the mixture of desired consistency and thickness can be sprayed or deposited to get an impervious layer. This impervious surface should be watered for at least 10 days.

By this technique, an impervious layer of high compressive strength (560–700 kg/cm^2 after 28 days) is obtained and, hencc, this is also very useful for reconditioning or repairing old concrete works and brick or masonry works, which have deteriorated either due to climatic effects or inferior workmanship.

23.3.1.6 Pressure grouting (or cementation)

Cementation is the process or technique of forcing the cement grout (i.e., mixture of cement, sand and water) under pressure into the cracks, voids or fissures present in the structural component or ground. That is, all the components of a structure in general and foundations in particular, which are liable to moisture penetration, are consolidated and, hence, made water resistant by this cementation process.

In this process, holes are drilled at selected points in the structure and the cement grout, of sufficiently thin consistency, is forced under pressure to ensure complete penetration into cracks or voids. This makes the structure watertight and restores its stability and strength to some extent.

Similarly, when the structure is resting on hard but loose textured ground, this process can increase its strength. For this, pipes are driven into the ground, cores within the pipes are removed by means of earth auger and finally the grout is forced (pumped) into the ground to fill the voids, loose pockets, etc.

This technique is also used for repairing structures, consolidating ground to improve bearing capacity, forming water cut-offs to prevent seepage, etc.

23.4 MATERIALS USED FOR DAMP PROOFING (DPC)

There are various materials, which are used as damp-proof courses depending upon the location, economy and degree of damp proofing desired. However, while selecting a particular damp-proofing material, the following requirements of an ideal damp-proofing material should be remembered.

1. The material for DPC should be impervious and durable, i.e., the material should be effective during the useful life of the building.
2. The material should be capable of resisting both dead loads and superimposed loads without being disintegrated.

3. The material should remain steady in its position, without any movements, so that the walls overlying the DPC do not develop any cracks.
4. For DPC above the ground level, with wall thickness up to 40 cm, any material listed below for DPC can be used.
5. For DPC to be laid over larger areas, such as floors and roofs, and thicker walls, a DPC material which provides lesser number of joints should be used, such as mastic asphalt, bitumen sheeting and plastic sheeting.
6. The material for parapet walls and in other situations where differential thermal movements are expected due to exposure should be of flexible material, like mastic asphalt, bituminous felt and metal sheets.
7. In water-retaining structures or situations, a jointless DPC should be provided to take care of the risk of leakage.
8. In cavity or hollow walls, the cavity over the door or window openings should be bridged by flexible materials, like bitumen sheet, strips of lead and copper.

The materials generally used for DPC are listed below.

1. **Flexible materials:** Hot bitumen, bituminous felts, bituminous sheet, plastic sheet (polythene sheets), metal sheets of lead, copper, etc.
2. **Semi-rigid materials:** Mastic asphalt or a combination of materials or layers.
3. **Rigid materials:** Use of I-class bricks, stones, slates, etc.

23.4.1 Hot bitumen or hot asphalt

This is a flexible material, which is first heated and then spread over the bedding of concrete or mortar (i.e., over walls). This should not be applied in thickness less than 3 mm. Bitumen or asphalt forms an excellent damp-proof course, as it offers an impervious, indestructible and tough surface.

23.4.2 Bituminous felts (6 mm thick sheet or asphaltic felt)

This is also a flexible material, which is available in rolls of normal wall widths. For placing this in position, first a layer of cement mortar is laid on the brickwork and then DPC is bedded on it. An overlap of 10 cm in case of joints and full width overlap in case of angles and crossings should be provided. Bitumen felt is capable of accommodating slight movements but cannot withstand heavy loads.

23.4.3 Sheets of lead, copper and aluminium (metal sheets)

These are used as membranes for damp proofing and are of flexible type.

23.4.3.1 Sheets of lead

The thickness of sheet should be such that the weight of sheet is not less than 20 kg/m^2. These are spread on the walls and overlapped at the joints. The sheets of lead should be embedded in lime mortar and not in cement mortar (because cement chemically reacts with lead and destroys it). The surfaces of lead should be protected by a coating of bitumen against corrosion. DPC formed by lead sheet provides an impervious and highly resistant surface against lateral movements.

23.4.3.2 *Sheets of copper (minimum 3 mm thick)*

Like lead sheets, these are spread, lapped and jointed. They are embedded in lime or cement mortar. This is another excellent DPC material which possesses characteristics such as high durability, good resistance to dampness and high resistance against sliding action.

23.4.3.3 *Sheets of aluminium*

These can also be used for DPC but not as good as lead or copper sheets. These sheets should be protected with a layer of bitumen.

23.4.4 Combination of sheets and bituminous felts

A lead foil is sandwiched between the sheets of asphalt or bituminous felt. This combined sheet is known as 'lead core' and DPC of this core possesses the characteristics of easy laying, durability, efficiency and economy.

23.4.5 Mastic asphalt

This is obtained by heating the asphalt with sand and mineral fillers. This is a semi-rigid material and forms an excellent impervious layer for damp proofing, i.e., DPC. However, it requires special care in its laying. Good mastic asphalt has many characteristics, such as high durability, excellent water-proofing quality and reasonable elasticity. This can withstand only very slight distortion and is liable to lateral movements under heavy pressures or very hot climates.

23.4.6 Bricks

Good dense bricks, which absorb water less than 4 per cent of their weight, are suitable as a DPC material at places where damp is not excessive. The joints are left open. They are widely used for providing or inserting DPC membrane in an existing wall.

23.4.7 Stones

Generally, dense and sound stones, such as granite, trap and slates, are laid in cement mortar in two courses to form an effective DPC. The stones are used for the full width of the wall. While laying the stones, care should be taken in breaking the continuity of vertical joints.

23.4.8 Cement concrete layers

A cement concrete layer, having mix proportions 1:2:4 (1 cement:2 sand:4 aggregates) with waterproofing agents, is used as DPC at the plinth level. It is effective in stopping the water rising due to capillary action but allows the water to pass through the cracks, etc., and hence is suitable as DPC material where dampness is not in excess. Concrete layer is used as a horizontal main DPC in thickness varying from 4 to 15 cm followed by two coats of hot bitumen paint.

23.4.9 Mortar

It is used in two ways, either (i) as a bedding layer for taking up other types of DPC or (ii) as a waterproofing plaster. For bedding layer, the mortar is prepared by mixing cement and sand in proportions 1:3 and adding slight lime to increase the workability. For waterproofing plasterwork, the mortar is prepared by mixing either of the following:

a. 1 cement:2 sand pulverized alum and soap water,

b. Cement:sand in proportions 1:3, with patented waterproofing material like pudio, dampro and sika. After applying this plaster in 2–4 cm thickness, it is painted with two coats of hot bitumen.

REVIEW QUESTIONS

1. What are the sources of dampness?
2. What are the effects of dampness?
3. What are the various techniques and methods adopted to prevent dampness?
4. Explain:
   a. Damp proofing
   b. Integral damp-proofing treatment
   c. Cavity walls
   d. Pressure grouting
5. What are the materials used for damp proofing?

chapter 24

Building Services

24.1 AIR CONDITIONING

24.1.1 Principles of air conditioning

Air conditioning is the process of treating air so as to control simultaneously its temperature, humidity, purity and distribution to meet the requirements of the conditioned space, such as comfort and health of human beings and other needs of the situation.

24.1.2 Purposes of air conditioning

a. To improve the quality of the products in industrial processes such as artificial silk and cotton cloth. In the case of other industries, it helps in providing comfortable working conditions for the workers, resulting in the increase of production.

b. In commercial premises such as theatres, offices, banks, shops and restaurants, air conditioning is done to improve the working atmosphere and maintain comfort within these concerns.

c. To give comfort to the residents of private buildings. The air-conditioning system in this case serves a small number of persons.

d. In travel by air, railway, road and water, air conditioning imparts facility and comfort by conditioning the quality of air in aeroplanes, railway coaches, road cars, buses, ships, etc.

The principle involved in air conditioning is:

a. Sucking the air through a filtering media.

b. Cooling it (in summer) or heating it (in winter).

c. Dehumidifying, if it is to be cooled, or humidifying, if it is to be heated.

d. Forcing it into the rooms for proper circulation.

e. The used air is collected through an exhaust and mixed with some outside fresh air and sucked again through the filtering medium, thus completing the cycle.

24.1.3 Principle of comfort air conditioning

Comfort feeling is a good indication of healthy atmosphere and it depends upon the temperature, air motion or air velocity and humidity changes for different seasons of the year.

The principle of air conditioning should involve proper control of temperature, humidity and air velocity

a. ***Temperature control:*** Comfortable zone is the temperature range suitable for majority of the people. The comfortable zones are different for summer and winter due to the clothing worn in these two seasons. The effective temperature zone for summer is 20–23°C and for winter is 18–22°C. A temperature of 21–25°C is required for comfort conditions regardless of the outside temperature.

b. ***Air velocity control:*** Air velocity control is also an important factor. The increase in velocity results in the decrease of inside effective temperature below the outside temperature. Therefore, the velocity of air is generally taken as 6–9 m/sec, which is considered as relatively still air.

c. ***Humidity control:*** Dry air imparts great strain for the human body. Due to this reason moisture is added to the heated air (i.e., humidification) in case of winter air conditioning and moisture is extracted from the cooled air (i.e., dehumidification) in case of summer air conditioning. An average value of relative humidity between 40 per cent and 60 per cent is considered desirable. During the summer season 40–50 per cent is comfortable and for winter 50–60 per cent is suggested.

24.1.4 Systems of air conditioning

Depending on the location of air-conditioning equipment, the system of air conditioning is classified as follows:

a. ***Central system:*** In this system, all the equipments pertaining to air conditioning are installed at one focal or central point and then the conditioned air is distributed to all the rooms or enclosures by ducts. This type of system requires less space for installation and the maintenance is also easy. It proves to be economical. Due to the presence of ducts, it requires large space.

b. ***Self-contained or unit system:*** In this system, special portable attractive cabinets which fit in with the decoration of modern rooms are placed inside the room near the ceiling or window. They are self-contained in every respect and conditioned air is formed inside the unit itself. The conditioned air is then directly thrown into the room without the help of any ducts.

c. ***Semi-contained or unitary central system:*** In this system, every room is provided with an air-conditioning unit and the room unit obtains its supply from the central system. Such a system results in the smaller size of ducts. Another form of this system is adopted in which conditioned air may be supplied from a central unit but the heating or cooling may be carried out in the room itself.

d. ***Combined system:*** A combined system may consist of (i) central and self-contained system (ii) central and semi-contained system and (iii) self-contained and semi-contained system

The choice of a particular system of air conditioning depends upon several factors such as the size of structure, method of heating, volume and type of air conditioning unit, period of year for which air conditioning is required and number of rooms to be served.

24.2 FIRE PROTECTION

It should be the objective of every engineer and architect, while planning and designing the buildings, that the structures offer sufficient resistance against fire so as to afford protection to the occupants in the event of fire. This objective is achieved by adequate planning, use of fire-resisting materials and construction techniques and by providing quick and safe means of escape in the building.

The building should be so planned or oriented that the elements of construction or building components can withstand the fire for a given time depending upon the size and use of building and the various compartments should be isolated so as to minimize the spread of fire. Suitable separation is necessary to prevent fire, gases and smoke from spreading rapidly through corridors, staircases, lift shafts, etc. Adequate separation from adjacent buildings should also be planned.

All the structural elements, such as floors, walls, columns and beams, should be made of fire-resisting materials so that life, goods or contents and activities within the building can be protected.

The construction of structural elements, namely walls, floors, columns, lintels, arches, etc., should be made in such a way that they should function at least for the time, which may be sufficient for the occupants to escape safely in times of fire. Escape elements like stairways and staircases, corridors, lobbies and entrances should also be constructed out of fire-resistant materials and be well separated from the rest of the building.

Adequate means of escape are provided for the occupants, to leave the building quickly and safely, in times of outbreak of fire. This objective is attained by providing an exit from within a building by way of definite escape ways, corridors and stairs to a street or an open space or roof of an adjoining building from where access to escape may be found. The desired degree of fire resistance largely depends upon the use of the buildings. In India, the types of building construction and fire zones in a city are classified on the basis of fire-resistance and fire-hazard characteristics, respectively.

24.2.1 Fire-resisting materials

24.2.1.1 Timber

Timber, though itself a combustible material, offers sufficient resistance to fire when used in adequate sizes. Timber also possesses the properties of self-insulation and slow burning. Timber on exposure to fire first gets charred; this charring provides a protective coating to the inner portions of the timber and prevents it from rapid combustion, even if subjected to a temperature up to 500°C. At still higher temperatures, under continued exposure, it is dehydrated giving rise to combustible volatile gases, which readily catch fire. Additional fire resistance is achieved by impregnating timber with large quantities of fire-retarding chemicals, like ammonium phosphate and sulphate, borax and boric acid and zinc chloride, because these chemicals retard the rate of temperature rise during fire. To make a timber structure more fire resistant the following points should be given due consideration:

i. Instead of using a number of smaller sections for joists and floor beams, thicker sections at a wider spacing should be used.

ii. The number of corners and the area of exposed surface should be reduced to a minimum. All sharp edges should be rounded.

iii. Timber should not be treated with oil paints or varnishes, which are liable to catch fire. Instead of this, timber ceilings and partitions should be treated with asbestos or ferrous oxide paints if needed.

iv. In a multi-floor timber structure, there should be a minimum number of floor openings and no through opening in multi-floor levels should be provided. A through opening spreads the fire in the vertical direction and behaves like a chimney and induces draught.

v. Adequate fire steps or barriers should be provided in the floors and walls.

24.2.1.2 Stone

The use of stone in a fire-resisting construction should be restricted to a minimum as this material cannot resist the effects of sudden cooling. After becoming hot, if it is cooled, it breaks into pieces. Granite, when subjected to excessive heat, crumbles to sand or cracks and turns to pieces with a series of explosions and disintegration. The use of limestone is not at all desirable as it gets crumbled and ruined (turns to quick lime) under the effects of fire. The compact sandstone has better resistance against fire than limestone as it can stand the exposure to moderate fire without serious cracks.

24.2.1.3 Bricks

First-class bricks are practically fire proof as they can withstand the exposure of fire for a considerable length of time. Being poor conductors of heat the bricks can withstand high temperatures up to 1300°C without causing serious effects. Firebricks are best for use in fire-resisting construction. The degree of fire resistance of bricks depends upon factors like size of bricks, composition of brick clay and method of construction. Though brick has its own structural limitations for use in buildings, brick masonry has been proved to be most suitable for safeguarding the structure against fire hazards.

24.2.1.4 Terra-cotta

Like bricks, it is also a clay product which possesses better fire-resisting qualities than bricks. Being costlier, its use is restricted in the construction of fire-resisting floors.

24.2.1.5 Steel

Steel, although an incombustible material, has a very low fire-resistance value. With increase in temperature, it gets softened and, hence, there is reduction in resistance to the effects of tension and compression. At about 600°C, its yield stress is reduced to only one-third of its value at normal temperatures. When the members made of steel come in contact with water used for extinguishing the fire, they tend to contract, twist or distort and thus the stability of the entire structure is endangered. It has been observed in practise that unprotected steel beams sag and steel columns buckle, resulting in collapse of the structure. It is, therefore, necessary in the fire-resisting characteristics of a structure to protect all the structural steel members with some covering of insulating material. This can be achieved by covering the steel members completely with materials like bricks, burnt clay blocks, terra-cotta or concrete.

24.2.1.6 Wrought iron and cast iron

Wrought iron behaves almost in a similar way as steel when subjected to fire except that it has lesser elasticity and lower strength in compression and tension as compared to steel. Cast iron is rarely used from the fire-resisting point of view in construction, as on sudden cooling it gets contracted and breaks into pieces or fragments. For using cast iron in fire-resistive construction, it should be protected by a suitable covering of bricks, concrete, etc.

24.2.1.7 Aluminium

In some advanced countries, aluminium is being used for reinforcement purposes in multi-storeyed structures because of its light weight and anti-corrosion properties. However, it has a very poor performance as a fire-resisting material and its use (as alloy) should be restricted to those structures which have low fire risks. It is a good conductor of heat and possesses enough tensile strength.

24.2.1.8 Concrete

In general, it is a bad conductor of heat and possesses good fire-resisting characteristics. The actual degree of fire resistance of this material depends upon the nature of aggregates used and its density. In the case of RCC and prestressed construction, it also depends upon the position of steel in concrete. It is found that ordinary concrete, when exposed to fire, gets dehydrated and results in shrinkage cracks. (This happens because on heating aggregates in concrete expand whereas cement shrinks and these two opposite actions lead to the development of cracks.) Coarse aggregates like foamed slag, blast furnace slag, crushed brick, crushed limestone and cinder are best suited for concrete from the viewpoint of fire resistance. Aggregates like flint, gravel and granite possess poor fire-resisting characteristics. It has been observed that in the event of average fire the concrete surface gets disintegrated for a depth of about 25 cm because of the fact that the mortar in concrete gets dehydrated by the fire. Hence, in the case of reinforced concrete fire-resistive construction, a cover of sufficient thickness should be provided (cracks generally originate from the reinforcement). RCC structures are considered superior to steel-framed structures since less steel is used and that too are well protected by the mass concrete.

24.2.1.9 Glass

Because of its low thermal conductivity, this material undergoes very small volume changes during expansion or contraction and, hence, is considered as a good fire-resisting material. However, sudden and extreme changes in temperature result in fracture or cracks. When glass is reinforced with steel wire netting, e.g., in wire glass, its fire resistance is considerably increased and its tendency to fracture with sudden changes in temperature gets minimized. Reinforced glass has a higher melting point and, hence, is commonly used for making fire-resisting doors, skylight, windows, etc. in construction work.

24.2.1.10 Asbestos cement

This material, which is formed by combining fibrous mineral with Portland cement, has a great fire-resistive value. Asbestos cement products are largely used for the construction of fire-resistive partitions, roofs, etc. Being poor conductors of heat and incombustible material, the structural members blended with asbestos cement offer great resistance to cracking, swelling or disintegration when subjected to fire.

24.2.2 Causes of fires and their prevention

Fire may not occur under any one of the following conditions:

a. Absence of a component necessary for combustion.
b. Improper ratio of combustible material to oxygen for the formation of a combustion system.
c. Heat source available is insufficient to ignite a combustion system.
d. Heat source is not available for a sufficient time to ignite a combustion system

It is possible to establish the different causes of fire which occur as a result of various optimum combinations of combustion systems and heat sources necessary for a fire to start. Fire prevention and limitation of fire spread are achieved by taking action with respect to the combustion systems and heat sources so as to avoid conditions under which fires can originate. Proper design and planning of buildings are important in fire safety.

24.2.3 Fire protection of buildings

All the structural components of a building should be constructed in such a way and of such materials that they withstand, as an integral member of the structure, for the period desired according to the type of construction, in the event of fire (i.e., 1–4 hours for type 4 buildings to type 1 buildings).

The load-bearing walls or columns of masonry should be thicker in section to resist fire for a longer time. Bricks are more preferred to stones and all walls should be plastered with fire-resistive mortar. All steel members should be embedded in dense concrete or some other fire-resistant material. There should be sufficient cover for all the embedded steel material (a minimum of 5 cm). RCC floors and jack arch floor with steel joists embedded in concrete are more preferred. The wall openings provided which act as escape passages should be properly protected, otherwise they help in the on spread of fire. So, all doors and windows are to be provided with fire-proof shutters and frames. Thick wooden members and steel shutters offer resistance to a certain extent. In addition, to the internal stair, fire escapes should be provided which in the form of external stairs. These should be straight flight type with a minimum width of 75 cm and 20 cm rise and 15 cm tread. Non-combustible hand rails should be provided for a minimum height of 100 cm. Ramps with gradient not more than 1 in 10 can also be provided. Alarm systems and fire extinguishing systems are also to be provided. Generally, one fire hydrant per 4,000–10,000 m^2 area is provided depending on the density of population and the importance of the region.

24.3 VENTILATION

24.3.1 Necessity of ventilation

a. To prevent an undue concentration of body odours, fumes, dust and other industrial by products.

b. To prevent an undue concentration of bacteria-carrying particles.

c. To remove products of combustion, and, in some cases, to remove body heat and the heat liberated by the operation of electrical and mechanical equipment.

d. To create air movement, so as to remove the vitiated air or replace it by fresh air.

e. To create healthy living conditions by preventing the undue accumulation of carbon dioxide and moisture and depletion of the oxygen content of the air. For comfortable working conditions, the content of carbon dioxide should be limited to about 0.6 per cent volume (in air).

f. To maintain conditions suitable to the contents of the space.

g. To prevent flammable concentration of gas vapour or dust in case of industrial buildings.

24.3.2 Functional requirements of a ventilation system

A ventilation system should meet the following functional requirements.

a. Rate of supply of fresh air

b. Air movements or air changes

c. Temperature of air

d. Humidity

e. Purity of air

24.3.2.1 Rate of supply of fresh air

The quantity of fresh air to be supplied to a room depends upon the use of the building to which it is subjected. The rate of supply of fresh air is decided by considering several factors such as the number of occupants, type of work and age of occupants.

Type of building	Minimum rate of fresh air supply to buildings (m^3 per person per hour)
Assembly halls, canteens, shops, restaurants	28
Factories and workshops	
i) work rooms	23
Residential buildings	
ii) living rooms	3 air changes per hour
iii) kitchens	6 air changes per hour
iv) bathrooms	6 air changes per hour
v) halls and passages	1 air changes per hour
Hospitals	
i) wards	3 air changes per hour
ii) theatres	10 air changes per hour

24.3.2.2 Air movements (or air changes)

At workplaces, air has to be moved or changed to cause proper ventilation of the space. The minimum and maximum rates of air change per hour are 1 and 60 respectively. If the rate of air change is less than one per hour, then it will not create any effect on the ventilation system. While, on the other hand, if the rate of air change is more than 60 per hour, it may lead to discomfort due to high velocities of air. For effective working of the ventilation system, 5-6 air changes per hour are considered alright. Moreover, the air movements should be uniform and should not allow the formation of pockets of stagnant air at any spot in the room.

In naturally ventilated buildings, cross ventilation is provided to secure air movement, whereas in the case of mechanically ventilated buildings air movement is obtained by either increasing the rate of fresh air supply or by recirculation of a part of air in water. The air movement or rate of air change will depend upon the velocity of incoming fresh air, disposition of inlets, type of activity in the premises, number of occupants, etc. The air movement should be varied both in velocity and direction and this can be achieved by means of fans.

24.3.2.3 Temperature of air

It is desirable that the incoming air for ventilation should be cool in summer and be warm in winter before it enters the room. Whenever the velocity of the incoming air is high, its temperature should not be lower than the room temperature. The usual temperature difference between inside and outside should not be more than 8°C.

The effective temperature should, therefore, be maintained with regard to the comfortable conditions for different seasons of the year. This effective temperature indicates a most suitable temperature to the majority of people, considering the comfort of human body under the probable conditions of humidity and air motion. The value of effective temperature depends upon the type of activity, geographical conditions, amount of heat loss from the body, age of occupants, etc.

24.3.2.4 Humidity

A relative humidity within the range of 30–70 per cent at a working temperature of 21°C is considered desirable and, therefore, should be maintained. When work is required to be done at a higher temperature, low humidity and greater air movements are necessary for removing a greater portion of heat from the body. The value of relative humidity can be obtained by comparing dry bulb and wet bulb temperatures.

Any water vapour within a given space or volume is known as humidity and its ratio with the water vapour it had contained had it been saturated is known as relative humidity. The relative humidity depends only upon the vapour pressure of water vapour and dry bulb temperature. The relative humidity of saturated air is 100 per cent.

24.3.2.5 Purity of air

The purity of air plays a significant role in the comfort of people affected by a ventilation system. Hence, it is essential that the ventilating air should be free from any impurities. To get pure ventilated air, the entry point of the ventilation system should not be situated in the neighbourhood of chimneys, latrines, urinals, etc.

24.3.3 Systems of ventilation

The systems of ventilation are basically divided into the following two categories:

a. Natural ventilation or aeration

b. Mechanical ventilation or artificial ventilation

24.3.3.1 Natural ventilation

In this system of ventilation, the outside air is supplied into a building through windows, doors, ventilators and other openings due to the wind outside and convention effects arising from temperature or vapour pressure differences or both between the inside and the outside of the building.

This system of ventilation may be developed where precise control over the air conditions and the rate of air changes are not required. Natural ventilation is usually considered suitable for houses and flats (i.e., small buildings) and it cannot be adopted for big offices, assembly halls, theatres, auditoriums, large factory workshops, etc. This system is very economical and the desired ventilation can be achieved by providing sufficient windows and other openings which open to the external air. An opening area equal to not less than one-twentieth of the floor area of the room should be provided in view of proper ventilation. The top of this opening area should be not more than 45 cm below the ceiling.

The rate of ventilation by natural means through doors, windows and other openings depends upon the following effects.

Wind effect (or wind action)

In this, ventilation is affected by the direction and velocity of wind outside and the sizes and position of the openings. Wind creates pressure differences and when it blows against a building a positive pressure is created on the windward side and leeward side. Suction will occur on the other side and the wind will blow from the windward side to the other side if there is an opening.

When wind blows at right angles to one of the rectangular faces of the building or an exposed site, a positive pressure is produced on the windward face and a negative pressure on the leeward face. If the wind direction is at 45° to one of the faces, positive pressure will be produced on the two windward faces and negative pressure on the two leeward faces.

The rate, at which the air change or airflow will occur, depends upon the pressure difference between the inside and the outside. The greater the wind speed the greater will be the pressure difference and sometimes the air changes can occur quickly.

Stack effect

In this, the ventilation rate is affected by the convection effects arising from temperature or vapour pressure difference or both, between the inside and outside of the room, and the difference of height between the outlet and inlet openings. If the air temperature inside is higher than that of outside, the warmer air tries to rise and pass through the opening in the upper part of the building.

At the same time, the incoming cooler air from outside through the opening at the lower elevation replaces it. The rate of air flow, in addition to the temperature or pressure difference and height difference, also depends upon the ratio between the areas of the two openings.

General considerations and rules for natural ventilation

The following considerations and rules should be followed for promoting natural ventilation in buildings:

- Inlet openings in the buildings should be well distributed and should be located on the windward side at a low level. Outlet openings should be located on the leeward side near the ceiling in the side walls and in the roofs.
- Inlet and outlet openings should preferably be of equal size for greatest air flow, but when the outlet is in the form of a roof opening the inlet should be larger in size.
- Where the wind direction is variable, openings should be provided in all walls with suitable means of closing them.
- Inlet openings should not be obstructed by adjoining buildings, trees, signboards, partitions or other obstructions in the path of air flow.
- Increased height of the room gives better ventilation due to stack effect.
- The long narrow rooms should be ventilated by providing suitable openings in short sides.
- The rate of air change in the room mainly depends on the design of the opening location of the inlet and outlet and the difference in temperature between the inside and outside air. Generally, the outside air is cooler than the inside air. Hence, the cooler air enters from the bottom and after becoming hot during its stay in the room leaves from the top. It would, therefore, be advantageous to provide ventilators as close to the ceilings as possible.
- The efficiency of roof ventilators depends on their location, wind direction and the height of the building.
- It is found that the ventilation through windows can be improved by using them in combination with a radiator, deflector and exhaust duct.
- For cross ventilation, the position of outlets should be just opposite to inlets. The openings over the doors of back walls create good conditions for cross ventilation.
- Windows of living rooms should either open directly to an open space or the open space created in buildings by providing adequate courtyards.
- If the room is to be used for burning gas or fuel, enough quantity of air should be supplied by natural ventilation for meeting the demands of burning as well as for ventilation of the room.

24.3.3.2 Mechanical or artificial ventilation

In this system of ventilation, outside air is supplied into a building either by positive ventilation or by infiltration by reduction of pressure inside due to exhaust of air, or by a combination of positive ventilation and exhaust of air. The supply of outside air by means of a mechanical device such as a fan is termed as 'positive ventilation', whereas the removal of air and its disposal outside by such a device is termed as 'exhaust of air'. For positive ventilation, centrally located supply fans of centrifugal type, and for exhaust of air, wall- or roof-located exhaust fans of propeller type are normally used. So, this ventilation involves the use of some mechanical arrangement for providing enough ventilation to the room.

Mechanical ventilation is recommended in all the cases where a satisfactory standard of ventilation in respect of air quantity, quality or controllability cannot be obtained by natural means. A mechanical system is capable of meeting the requirements of air quantity and quality (of air) regarding humidity, temperature, etc and produces the comfortable conditions at all times during the year. Though this system is comparatively costly, it results in the considerable increase in the efficiency of the persons under the command of this system. This system is adopted for big offices, banks, assembly halls, auditoriums, theatres, large factories, workshops, places of entertainment, etc. This system may be regarded as generally desirable in all rooms occupied by more than 50 persons, where the space per occupant is less than 3 cu.m.

The following methods of mechanical or artificial ventilation are in common use.

i. Extract or exhaust systems
ii. Supply or plenum systems
iii. Combination of exhaust and supply systems or balanced systems
iv. Air conditioning

Extract or exhaust systems

In this system, a partial vacuum is created in the inside of the room by exhausting or removing the vitiated inside air by means of propeller type fans. The extraction of air from inside permits the fresh air to flow from outside to inside and thus it becomes possible to provide fresh air to flow from outside into the room through doors and windows.

These fans for exhaust are installed at suitable places in the outside walls or roofs and they are further connected to different rooms through a system of duct-work.

These exhaust systems are best confined to situations where it is essential to create an air flow towards the ventilated rooms, such as in kitchens, lavatories and industrial plants. This system is useful for removing smoke, odours, fumes, dust, etc from the above-mentioned rooms. In this system, the ducts are placed near the place of formation of smoke, fumes, odours, dust, etc.

Supply or plenum systems

As the name implies, in this system the space is filled with air by means of fans, but no special provision is made to remove it. In plenum ventilation, the air inlet is selected in the side of the building where the air is purest. In this opening, screens or filters may be fixed and fine stream of water may be impugned in the path of the incoming air. The disinfection of incoming air is achieved by adding ozone at the point of inlet. Thus, by this system of mechanical ventilation, it is possible to control the quality, humidity and temperature of incoming air.

Ventilation by plenum process may be downward or upward. In downward ventilation, the incoming air is allowed to enter at the ceiling height and is taken out through outlets situated at the floor level. In upward ventilation, the fresh air is allowed to enter at the floor level and the outlet is provided at the ceiling height.

These ventilation systems are costly and are used for factories, big offices, theatres, etc. and also for supplying air to the air-conditioned buildings.

Combination of exhaust and supply systems or balanced systems

This balanced system is a combination of the above-said two systems and makes use of fans to supply and extract air (i.e., input fans and exhaust fans). This system enables full control over the air movement and conditions to be obtained and should be used where accurate performance is desired. In most buildings, it is desirable to extract only 75 per cent of the quantity of air supplied so that positive pressure is maintained within the rooms. This is essential to prevent the entry of hot air when the doors are opened and also to prevent the infiltration of dust and air-borne contaminants. Moreover, the recirculation of air is possible in this system.

Air conditioning

This is the most effective system of artificial ventilation, in which provision is kept for humidifying or dehumidifying, heating or cooling, filtrations, etc. of the air to meet the possible requirements.

24.4 LIFTS AND ESCALATORS

24.4.1 Elevators or lifts

Elevators are used in buildings having more than four storeys. They are used for providing vertical transportation of passengers or freight. They can be either electric traction elevators or hydraulic elevators. Electric traction elevators are used exclusively in tall buildings. Hydraulic elevators are generally used for low-rise freight service which rises up to about six storeys. Hydraulic elevators may also be used for low-rise passenger service.

The different components of an electric traction elevator are the car or cab, hoist wire ropes, driving machine control equipment, counterweight, hoistway rails, penthouse and pit. The car is the load carrying element of the elevator and a cage of light metal supported on a structural frame, to the top of which the wire ropes are attached. The ropes raise and lower the car in the shaft. They pass over a grooved, motor-driven sheave and are fastened to the counter weights. The paths of both the counter weights and the car are controlled by separate sets of T-shaped guide rails. The control and operating machinery may be located in a penthouse above the shaft or in the basement. Safety springs or buffers are placed in the pit, to bring the car or counterweight to a safe stop. For elevators serving more than three floors, means should be provided for venting smoke and hot gases from the hoist ways to the outer air in case of fire. Vents may be located in the enclosure just below the uppermost floor, with direct openings to the outside or with non-combustible duct connections to the outside. Vent area should be at least 35 per cent of the hoistway cross-sectional area.

24.4.1.1 Design considerations

The key considerations which affect elevator system design are:

i. Number of floors to be served
ii. Floor-to-floor distance
iii. Population of each floor
iv. Location of building
v. Specialist services within the building
vi. Type of building occupancy
vii. Maximum peak demand in passenger per 5-minute period

24.4.1.2 Design parameters

There are numerous parameters which can be used to judge elevator system performance. The principal one is based on quality of service and quantity of service.

The 'quality of service (or interval)' is related fundamentally to the time interval a passenger has to wait. It can also be said as the expected interval (in seconds) between the arrivals of elevators in the main floor. For a large building, the quality of service can be categorized as

i. average interval 20–25 seconds – excellent

ii. average interval 35–40 seconds – fair

iii. average interval 45 seconds – poor

The 'quantity of service (or handling capacity)' of a system is expressed in the elevator industry design terms as a function of expected building population. A large building with single tenancy usually provides heavier peak flows than those with multiple tenancy.

The following handling capacity should be used as a basis for design to meet up morning peak. Single tenancy – 15–25 per cent of the total building population entering in a 5-minute period. Multiple tenancy – 10–15 per cent of the total building population entering in a 5-minute period.

24.4.1.3 Location of elevators

The most efficient method of locating elevators to serve an individual building is to group them together. A group has a lower average interval between car arrivals than a single elevator.

Groups should be located:

i. For easy access to and from the main building entrance.

ii. Centrally for general ease of passenger journey.

If a building has areas which give long distance to the central group elevator, then it may be efficient to provide an additional elevator for local areas.

24.4.2 Ramps

They are sloping surfaces used to provide an easy connection between the floors. They are especially useful when large number of people or vehicles have to be moved from floor to floor. They are usually provided at places such as garages, railway stations, stadiums, town hall, office buildings and exhibition halls. Sometimes, they are provided in special-purpose buildings such as schools for physically handicapped children. They should be constructed with a non-slippery surface.

Ramps are generally given a slope of 15 per cent. But a slope of 10 per cent is usually preferred. The space required for ramps is more. The ramp need not be straight for the whole distance. It can be curved, zigzagged or spiralled. Ramps and landings should be designed for a live load of at least 21.2 kg/cm^2. Minimum width of pedestrian ramps is 75 cm for heights between landings not exceeding 3.6 m. Landings should be at least as wide as the ramps. Powered ramps, or moving walks, carrying standing passengers may operate on slopes up to 8° at speeds up to 60 m/min and/or slopes up to 15° at speeds up to 47 m/min.

24.4.3 Escalators

These are powered stairs. They are used when it is necessary to move large number of people from floor to floor. These stairs have continuous operation without the need of operators. These escalators are in the form of an inclined bridge spanning between the floors. The components of an escalator are a steel trussed framework, hand rails and an endless belt with steps. At the upper ends of an escalator are a pair of motor-driven sprocket wheels and worm gear driving machine. At the lower end is a matching pair of sprocket wheels. Two precision-made roller chains travel over the sprockets pulling the endless belt of steps. Escalators are reversible in direction. They are generally operated at a speed of 30 or 40 m/min. Slope of stairs is standardized at 30°. For a given speed of travel, the width of steps determines the capacity of the powered stairs.

Escalators should be installed where traffic is heaviest and where it is convenient for passengers. In the design of a new building, adequate space should be allotted for powered stairs. Structural framing should be made adequately to support them.

Escalators are generally installed in pairs. One of them is used for carrying upgoing traffic and the other for traffic moving down. The arrangement of escalators in each storey can be either parallel or criss cross. Criss-cross arrangement is more compact. It reduces walking distance between stairs at various floors to a minimum. This is why criss-cross arrangement is preferred over parallel arrangement.

Number of floors	Speed
4-5	0.5–0.75 m/s
6–12	0.75–1.5 m/s
13–20	above 1.5 m/s

REVIEW QUESTIONS

1. What are the purposes of air conditioning a building?
2. What are the principles of comfort air conditioning?
3. What are the systems of air conditioning?
4. What is the necessity of fire protection of a building?
5. Briefly discuss fire-resisting materials.
6. What are the main causes of fire and how are they prevented?
7. What are the necessities of ventilation?
8. What are the functional requirements of a ventilation system?
9. What are the general considerations and rules for natural ventilation?
10. What are the functional requirements of a lift in a building?
11. What is the difference between a lift and an escalator?

chapter 25

Building Maintenance

A safe and trouble-free structure is the dream of every engineer and inhabitant. Leaks and cracks are the major causes of worry as far as medium-sized and small buildings are concerned. An engineer or supervisor should be able to identify the type of distress or damage and the cause of the defect, so that he is able to suggest the repair measures accordingly.

25.1 DETERIORATION OF CONCRETE

Concrete is the most important building material and the worldwide consumption of concrete today is 8.8 billion tons per year. Concrete, although a very durable material, can show loss of strength and deterioration in course of time because of various reasons.

25.1.1 Design deficiency

Faulty design is one of the major causes of structural failure. This may include underestimation of loads, faulty mix design, poor detailing, faulty analysis, poor foundation, etc.

25.1.2 Material deficiency

Use of poor quality materials, physical and chemical properties of concrete, increased water–cement ratio, etc. are also some of the major reasons.

25.1.3 Construction deficiency

Poor workmanship is one of the most important causes of distress in buildings. Inadequate control in batching, mixing, transporting, placing, compacting, finishing and curing of concrete leads to various problems including honeycombing, segregations, poor strength, etc. Very often sufficient cover is not provided for the reinforcement. Bad-quality formwork and poor preparation of joints also lead to distress.

25.1.4 Environmental deficiency

Attack of chemicals and acids, proximity to marine environment, thermal variations, entrapped air, leaching because of the presence of hydrated compounds, etc. can also end up in problems to the structure.

25.1.5 Man-made deficiencies

Reckless modifications and alterations, installation of additional load, poor or no maintenance, etc. make the performance of the building poor.

Deterioration of concrete basically starts from leaks and cracks. Because of any of the reasons said above, concrete can become porous and leakage can happen. Because of poor workmanship, improper mixing, inadequate compaction or poor quality formwork, etc. segregation of material can occur and honeycombs can be formed. The quality of the joints is also a major cause of leakage.

25.1.6 Cracks in concrete

The cracks in concrete vary in width from 0.1 to 1.0 mm. Concrete starts cracking, even at the plastic state and when hardened concrete is exposed to weather, because of thermal and shrinkage strains. The cracks in concrete are caused primarily because of the following reasons.

a. Temperature gradient including frost action
b. Humidity gradient (drying shrinkage)
c. Structural overloading, cyclic or impact loading
d. Rapid drying conditions (plastic shrinkage)
e. Inadequate structural design and detailing
f. Chemical causes including corrosion of reinforcement

The shrinkage strain causes tensile stress in concrete, and when this exceeds the tensile strength of concrete, cracks begin to happen. However, due to the viscoelastic behaviour (creep) of concrete, some stress is relieved and it is the residual stress, after the relaxation due to creep, that is responsible for cracking.

Cracks on concrete surfaces seriously affect the durability of concrete and 0.15 mm is often recommended as the maximum allowable crack width.

25.2 DETERIORATION IN MASONRY WORKS

Cracks can occur in concrete as well as masonry works and can cause distress of the structure. Cracks can happen because of high ambient temperature, low humidity, improper curing, poor quality and improper combination of materials and above all poor workmanship.

Vertical cracks in brickwork can happen because of foundation movement, thermal movement, overloading, drying shrinkage, differential expansion between various members, shrinkage, etc. These cracks usually occur at the corners, below the window openings, around staircase openings, at the junctions between old structures and new extensions, at the junctions of reinforced cement concrete (RCC) columns or lintels and masonry works, etc.

Horizontal cracks usually occur below the slabs, at the lintel level or at the sill level, on the mortar joints, at the corners, etc. The main reasons for horizontal cracks are the horizontal movement due to shrinkage, lifting of the slabs, thermal gradient, drying shrinkage, differential expansion, chemical attack, ageing, etc.

Diagonal cracks can happen at the corners, especially along the cross walls even up to the foundations, over the RCC lintels, etc. Drying shrinkage, differential strain, moisture movement in the soil, etc. are some of the reasons that cause diagonal cracks.

25.3 PREVENTION OF CRACKS AND LEAKS

Prevention is always better than cure. So in addition to proper design, planning and execution, proper control is necessary at every stage of construction.

1. Porous materials, which are likely to shrink on drying, should be avoided. Bricks should be exposed for 2-3 weeks after burning and soaked well before being used in the structure.
2. Fine sand, clay and silt should not be used.
3. In mass concreting, heat of hydration should be controlled.
4. Use of very rich cement mortar, which has high drying shrinkage, should be avoided.

5. Shrinkage cracks can be reduced to a great extent by keeping the water–cement ratio minimum.
6. Large spans of slabs and windows should be avoided.
7. The foundation strata should be well prepared so that differential settlement of foundations is avoided. Proper type of foundation suitable for the type of soil and its bearing capacity should be adopted.
8. It must be ensured that there is no presence of organic matter, brickbats, debris, etc. in the soil.
9. Curing of masonry and concrete should be proper.
10. In RCC-framed structures, panel walls should be constructed from top downwards.
11. Joints for movements should be provided as per standards especially at the change of directions.
12. Plastering on panels and partitions should be delayed as far as possible whereas plastering on concrete is to be done as early as possible.
13. Sufficient bearing length should be provided for lintels to avoid cracks due to stress concentration.
14. Concrete in the marine atmosphere has to be made dense with sufficient cover thickness to avoid corrosion of steel.
15. Differential loading on the foundation should be avoided during construction.
16. Roots of trees removed for construction should be made to dry out before construction is taken up.
17. In concrete utmost care should be taken to avoid entrapped air and segregation of materials. In order to attain these, it should be ensured that the concrete mix possesses good workability and cohesiveness.
18. Before tampering or modifying a concrete or masonry structure (for extensions, modifications or even for providing embedded service conduit for electrical, water and drainage lines) all the aspects should be taken care of.
19. Blended cement, minerals and chemical additives, etc. have a significant role to play in durability and so their use should be encouraged. Water proofing agents can be mixed with roofing concrete.

It is essentially because of ignorance, carelessness, greed and negligence that the failures occur. So utmost care must be there in dealing with all the factors mentioned above.

25.4 REPAIR OF CRACKS AND LEAKS

The defective or cracked portion of the concrete or plaster has to be removed. On doing this care should be taken that the cracks are entirely removed from the surface and a well-defined cavity is formed without affecting the good concrete or plaster. The size of the cut depends upon the size, depth and extent of repair. Before doing this, all the structural load carrying members like columns, beams and slabs, must be supported, propped or isolated so that the load on these members is reduced or temporarily transferred to other good structural members. Care should also be taken during repairs that the weakened members do not lose their strength further due to chipping and other removal procedures.

Rust from the steel must be completely removed by tapping lightly, wire brushing and cleaning. The entire surface should be cleaned from dust and dirt by sand blasting or any other effective method and washed thoroughly. The excess water is removed from the cavity and all the cracks and joints are to be properly inspected and sealed with mortar and cured for 7 days.

The reinforcing bars are to be given a protective coating and the concrete surface is given a coating with a thin layer of cement grout before placing the patching material.

The cavities are then filled with good quality mortar or concrete as required. Repair material should be filled in layers of about 10 mm thick. For columns, beams and slabs, guniting is resorted to with either ordinary cement concrete or polymer modified cement concrete. The surface is properly cured for 7 days.

The surface is then allowed to dry and finishing coats of mortar are applied in two coats. Curing is done for the surface also for 7 days.

Cracks can also be repaired by the use of Ferro cement. This is done by peeling off the affected plastering and replastering the surface after placing a sheet of steel wire mesh of required size inside the plaster in position.

Plugging and sealing of cracks can be done using cement grout or various brands of epoxy compounds available in the market.

It is important that during repair the safety of the inhabitants and other people in and around be given due consideration.

25.5 COST-EFFECTIVE CONSTRUCTION

Building is a product involving time, material, labour and other resources under various constraints. Today's construction scenario presents steeply escalating material costs and time overrun leading to project cost escalation, in spite of the earnest efforts to keep the situation within limits.

The total costs of a building can be defined as the sum of the costs that develop at the planning state, the cost of construction and maintenance after construction. The cost of any building can be reduced by giving proper attention to these three aspects. Cost and quality are two important and interdependent aspects of any product. Saving money along with providing better value is a concept that attracts all. Efficient design and construction management techniques employed in a systematic way are essential to assure cost and quality control. It is not sufficient to merely reduce cost. All the requirements of function, space and quality and strength have to be met too.

With high urban land values and increased cost of construction due to the ever-increasing cost of building materials and labour, people are on the lookout for cost efficiency in building.

Cost efficiency can be achieved by efficient layout design, choice of appropriate materials and appropriate building technology. In any project involving construction of building units, the major share of expenditure is the construction work. Land subdivision and development is an essential requirement for any scheme. Layout determines the environment as well as distribution network for services. Thus, optimization of layout will lead to overall economy in land development works.

25.5.1 Breakup cost for housing

The breakup cost of housing in developing countries is generally as follows:

Cost of materials	–	73%
Cement	–	18%
Iron and Steel	–	10%
Bricks	–	17%
Timber	–	13%
Sand	–	7%
Aggregates	–	8%
Labour	–	27%
Masons	–	12%
Carpenters	–	10%
Unskilled labour	–	5%

25.5.2 Planning considerations for layout planning

a. Layout plan should be prepared after thorough investigations of site conditions, considerations of a number of factors which determine project content, cost levels and levels of facilities to be provided.

b. An analysis of site development costs should be made in terms of density, construction types, topography, grading and local requirements with regard to zoning and utility services.

c. The design of an enclosure of the required volume should be achieved by using minimum amount of materials and labour consistent with the structural stability and it should provide adequate control of the external climate with the lowest maintenance cost.

d. With regard to orientation, buildings in general should be oriented on the east–west axis:

   i. The long elevations should face north and south to reduce heat gain.

   ii. Buildings should be planned around small courtyards as thermal storage is required for most of the year.

   iii. Breeze penetration must be provided.

   iv. Rooms may be single banked with windows in the north and south walls for cross-ventilation.

   v. If rooms are double banked, the plan should allow temporary cross-ventilation through large interconnecting doors.

25.5.3 Means and methods to achieve cost effectiveness in construction

25.5.3.1 Planning aspect

For proper cost-effective construction, planning should be implemented to achieve:

i. Efficient arrangement of spaces which are cost conscious in their utilization of materials and space.

ii. The design of spaces which are function oriented, while at the same time adaptive, so as to accommodate change which may be in the form of increase in size of the spaces or their functional requirements effectively.

iii. A simultaneous synthesis of the infrastructural requirements with respect to both their quantities and their most economic arrangement should be undertaken and their fusion into the design scheme is to be ensured.

25.5.4 Arrangement of spaces

This includes both the arrangement of spaces within a dwelling unit as well as disposition of spaces/zones for any planned development. Design of these spaces should take in cost effectiveness in the use of materials. Materials are to be selected based on their local availability as well as the cost of their conversion into possible building materials.

Spaces should be function specific and space saving. These are achieved through an understanding of the functional requirements they are to accommodate.

25.5.4.1 Adaptability

It is desirable that individual dwelling units have adaptive properties. These are achieved through flexible open plans so that future changes may be accommodated through minimum expenditures. The same holds true for settlements also.

25.5.4.2 *Modular co-ordination*

Modular units ensure that the planning is not random and renders units with simple additive qualities that facilitate easier and more economic adaptation to change, besides ensuring economization through mass production of building units.

25.5.4.3 *Incorporating infrastructure*

The economization of infrastructure is a major priority in the planning process, as a careful layout can reduce costs manifold. Therefore, planning should ensure minimization of infrastructural lines, with the adoption of local know-how and technology with regard to matters such as waste and sewage disposal. A careful analysis of the types of wastes ensures a more eco-friendly development. However, the common tendency of planning along a grid is to be avoided as this hampers growth of the settlement and thus increases costs in the long run.

25.5.5 Cost-effective methods and techniques

Building construction involves two factors:

a. Materials for construction and
b. Techniques employed in construction

25.5.5.1 *Choice of the materials*

While choosing the materials for construction, emphasis should be given to the prevention of environmental degradation, energy conservation and ecological balance. The choice of materials depends a lot on the availability in a particular location.

An efficient type design for a given plinth area is evaluated on the basis of

i. Length of the load-bearing wall
ii. Percentage of wall area by floor area
iii. Area meant for circulation
iv. Wall space provided for storage and arrangement of furniture
v. Adequate lighting and ventilation for the house
vi. Grouping of wet cores
vii. Provision for future expansion of the house

Foundation

Foundation is one of the most important components. There can be no compromise regarding the foundation, as on an average about 20 per cent of the overall cost of a building goes for the foundation.

Superstructure

The life, durability and strength of the superstructure depend mainly on the materials that are used for the construction. About 25 per cent of the overall construction cost should be earmarked for the superstructure. As far as the rural areas are concerned people are well contented with burnt brick masonry using lime or cement mortar.

Roof

The roof consumes about 20 per cent of the overall cost. Periodical replacement, the susceptibility of fire hazards and certain other factors advocate against the use of thatched roof. Though tiled roofs are familiar in many areas, the high cost, thermal discomfort and the non-availability of well-seasoned palmyra or other wooden rafters are major disadvantages.

Asbestos roof substituted the tiled roof to a certain extent, but is not popular due to health hazards, cost and thermal discomfort. Madras Terrace is now obsolete because of the non-availability of skilled labour and due to heavy deadweight.

Reinforced concrete (RC) slab is a very good substitute for flat roof in such conditions.

25.6 ANTI-TERMITE TREATMENTS IN BUILDINGS

The termite proofing treatment should invariably be given in all types of buildings during the construction stage. It is because of the fact that during the post-construction period, it is extremely difficult and costly to control termite growth. Care should be taken to ensure that no bridge is formed between any part of the building and untreated soil. In order to reclaim land by utilizing debris or filling material, great care should be exercised to ensure that the debris is termite free. As far as possible a metal strip or suitable joint filler may be used to make the floor joints free from termite attack. To check termite movement from ground the foundations should be either made of concrete or any other solid material. Also, care should be taken to ensure that the building site is free from dead wood, old tree stumps, etc. The superstructure should be treated with suitable preservatives to make it termite proof. All the wooden members like door frames, staircases, etc. should be set on flooring. They should not be through flooring to prevent ground–soil contact.

25.6.1 Termite proofing methods

Generally, the following two methods are adopted:

a. Chemical treatment of soil
b. Physical structural barriers

25.6.1.1 *Chemical treatment*

In order to provide for an effective control of termites, the soil insecticides are thoroughly mixed and evenly spread in soil. Several patented insecticides like DDT, BHC and PCP have been generally used. However, chemicals like Aldrin 0.5 per cent, chlordane 1.0 per cent, Dieldrin 0.5 per cent and Heptachlor 0.5 per cent by weight in oil solutions or as an emulsion in water are found to be more successful. All these chemicals are chlorinated hydrocarbons and are insoluble in water. These chemicals are not leached away by water and these have proved to be quite effective as a chemical barrier between the building and ground. These are used in damaged portions of masonry and woodwork by injecting them under pressure in drilled holes.

25.6.1.2 *Physical structural barriers*

Continuous physical structural barriers in the form of a concrete layer or metal layer may be provided at the plinth level. These cement concrete layers should be 50–75 mm in thickness and should preferably be kept projecting about 50–75 mm internally and externally. Metal barriers comprising of non-corrodible metal sheets of copper or galvanized iron having a thickness of 0.80 mm when provided have in certain cases got damaged. Thus, these barriers have not proved to be very effective.

The above post-construction treatments are not low cost and simple. Hence, they should be used only where skill is available.

REVIEW QUESTIONS

1. What are the main reasons for the deterioration of concrete?
2. Why do cracks develop in concrete?
3. How does masonry work get deteriorated?
4. How can cracks and leaks be prevented?
5. How are cracks and leaks repaired?
6. What is cost-effective construction?
7. Briefly discuss the breakup cost of a building.
8. Explain the cost-effective methods and techniques in building construction.
9. How is anti-termite treating done in buildings?
10. Explain termite proofing methods.

Part-III

Basic Surveying

chapter 26

Chain Surveying and Modernization in Land Surveying

26.1 CHAIN SURVEYING

A chain is a device for taking linear measurements in the field. Chain surveying is accomplished by taking linear measurements only.

26.1.1 Basic principle of chain surveying

The principles followed in chain surveying are:

a. The whole of the area to be surveyed is divided into a skeleton of framework consisting of a number of well connected networks of 'well conditioned' triangles.

b. The details are to be located with respect to the sides of the triangles or any other subsidiary lines running between the sides by taking lateral measurements called 'offsets'.

26.1.2 Skeleton framework for chain surveying

To decide on the best planning of the skeleton framework for chain surveying a field, one should base his judgment on the minimum labour to produce a map with the required details and minimum distortion. The following points should be taken care of:

a. Through reconnaissance it should be ascertained that due to the presence of thick hedges, trees or any other obstacles, there should not be any obstruction to the laying of chain along the framework, taking the necessary measurements and the intervisibility along the lines of the framework.

b. Preferably a long line should run through the centre of the area to form a base line on which well-conditioned triangles should be based. The base line should prevent the relative rotation of the triangles due to error in the measurements when they are plotted in the map. This arrangement for the framework will limit the distortion in the map to a minimum.

c. The lines should run closer to the details, which are to be described with respect to them by taking lateral measurements called 'offsets'.

d. To avoid long offsets additional lines should run between the sides of the triangles of the framework.

e. There should also be some other additional lines for checking the accuracy of the measurements.

f. Care should be taken so that no portion of the framework goes beyond the boundary of the area to be surveyed.

26.1.3 Terms used in chain surveying

a. *Main station:* Main station is a point in chain survey where the two sides of a traverse or triangles meet. These stations command the boundaries of the survey and are designated by capital letters such as A, B and C.

b. *Tie stations or subsidiary stations:* Tie station is a station on a survey line joining main stations. These are helpful in locating the interior details of the area to be surveyed and are designated by small letters such as a, b and c.

c. *Main survey line:* The chain line joining two main survey stations is called a main survey line. In Figure 26.1, AC represents a main survey line.

d. *Tie line or subsidiary line:* A chain line joining two tie stations is called a tie line, such as jk. It is also called an auxiliary line. These are provided to locate the interior details, which are far away from the main lines.

e. *Base lines:* It is the longest main survey line on a fairly level ground and passes through the centre of the area. It is the most important line as the direction of all other survey lines are fixed with respect to this line.

f. *Check lines:* Check line or proof line is a line which is provided to check the accuracy of the fieldwork. The measured length of the check line and the computed one (scaled off the plan) must be the same.

g. *Offset:* It is the distance of the object from the survey line. It may be perpendicular or oblique.

h. *Chainage:* It is the distance of a well-defined point from the starting point. In chain surveying it is normally referred to as the distance of the foot of the offset from the starting point on the chain line.

26.1.4 Field work in chain survey

Suppose a plan is required for a small area as shown in Figure 26.1. The surveyor should first of all examine the ground to ascertain as to how the work can be arranged in the best possible manner. This is known as reconnaissance survey. In this process, the surveyor selects suitable ground points to be used as stations like A, B, C, etc.

Stations are arranged so that the entire area may be controlled from these points and all the main survey lines AB, BC, CD, etc. run near to the boundaries. The survey lines should not be many and lie over flat level grounds as far as possible. The triangles formed by survey lines should be well conditioned.

The main survey lines are measured with a chain and offsets are taken to the boundaries. Offsets are taken wherever there is a bend or any special feature in the boundary. In cases, where the boundary forms a smooth curve offsets are taken at the end of each chain. Offsets should be short, particularly for locating important details.

The lengths and positions of offsets being known, the boundaries can be plotted to their shapes. The other details which are deep inside the area, such as a well as shown in Figure 26.1, can be located by selecting tie stations, drawing tie lines and taking offsets to the ground features.

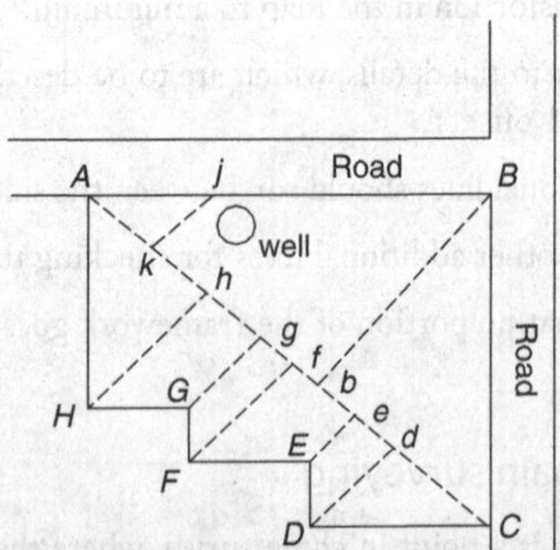

Figure 26.1 The main survey line and tie lines in chain surveying

26.1.5 A survey chain

Gunter, revenue, engineer and metric chains are the various types of chains, which are normally used for surveying. The chains are mostly divided into 100 links. While gunter's chain is 66 ft long, the revenue chain is 33 ft long and the engineer's chain is 100 ft long. Metric chains are either 30 m or 20 m in length.

The different parts of a chain are shown in Figure 26.2. The part marked 1 is a brass handle which contains a semicircular groove at its end for accommodating arrows during chaining. Parts marked 2 and 3 are the collar and eye bolt. 2 and 3 form a swivel joint with the brass handle. Part 4 is a circular ring which connects the end link marked 5 with the eye bolt. The links are connected with each other with three circular rings marked 6.

All links except the end links are of equal lengths. The link length measures the distance between the centres of the central rings of the joints on either side of the link. The end link forms the complete link length with the brass handle. The end link length measures the distance from the end of the handle to the centre of the central ring of the joint next to the end link. The length, of course, is to be measured while the chain is held straight.

The swivel joint allows rotation of the handle preventing the deformation due to twist in the end link. The three-ring joints provide enough flexibility to prevent the deformation of the links by bending.

For the ease of readability of the chain, different brass tag marks or tally marks are provided in it at definite length intervals, identical tags being placed at identical distances from the either end of the chain. This provides an opportunity to read the length on the chain at any of its points from either of its ends (marked 8 in Figure 26.2).

All links are made out of SWG No.8–No.12 wires of galvanized mild steel, in accordance with IS:1492-1970, having a diameter of 4 mm usually. The material is sufficiently soft and ductile to withstand hammering when the bent links are to be rectified. It may be noted that a chain with 100 links will have 99 joints between the links and two joints between the handles and the end links. There are 4 × 101 = 404 pairs of wearing surfaces at those joints.

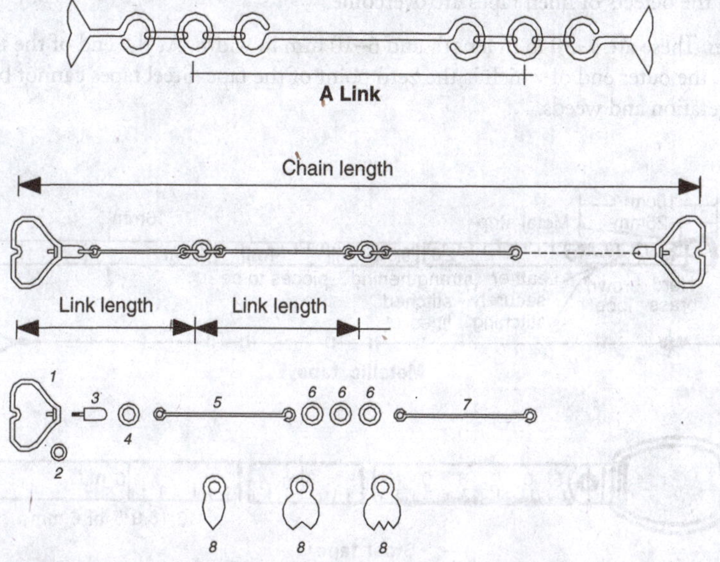

Figure 26.2 Components of a chain

26.1.6 Folding and unfolding chain

Unless great care is taken in folding and unfolding a chain, the links of the chain may bend and the rings of the joints may be deformed, making the length of the chain different from its standard length. This will introduce error in the measurements when employed in survey. Moreover, the chain cannot be boxed without folding it in a proper way.

Before putting it to use the chain bunch is taken out and held in one hand. The rest of the bunch is to be spread on the ground. One of the handles is to be released now and the chain is to be dragged when it is unfolded.

After its use, the chain is to be folded before it is put into its box. The central ring at the middle of the chain or the tally mark there and both the handles of the chain are put together at one end and the central rings quarter length, that is between 25th and 26th links and at 3 quarter length, that is between 75th and 76th links are put together at the other end. The chain laid and folded in this way on the ground will have bunches of 4 links lying side by side along the length. Each bunch of four is to be held and folded over to the next bunch alternatively to complete the folding of the chain.

26.1.7 Tapes

Tapes are available in a variety of materials, lengths and weights. The different types of tape in general use are explained below (Figure 26.3).

a. *Cloth or linen tape:* These are closely woven linen or synthetic material and are varnished to resist the moisture. These are available in 10–30 m in length and 12 to 15 mm in width. The disadvantages of such tapes are that it is affected by moisture and gets shrunk, its length gets altered by stretching and it is likely to twist and does not remain straight in strong winds.

b. *Metallic tapes:* It is a linen tape with brass or copper wires woven into it longitudinally to reduce stretching. As it is varnished, the wires are not visible. These are available in 20–30 m length. It is an accurate measurement device and is commonly used for measuring offsets. As it is reinforced with wires, all the defects of linen tapes are overcome.

c. *Steel tapes:* These are 1–50 m in length and 6–10 mm in width. At the end of the tape a brass ring is attached, the outer end of which is the zero point of the tape. Steel tapes cannot be used in grounds with vegetation and weeds.

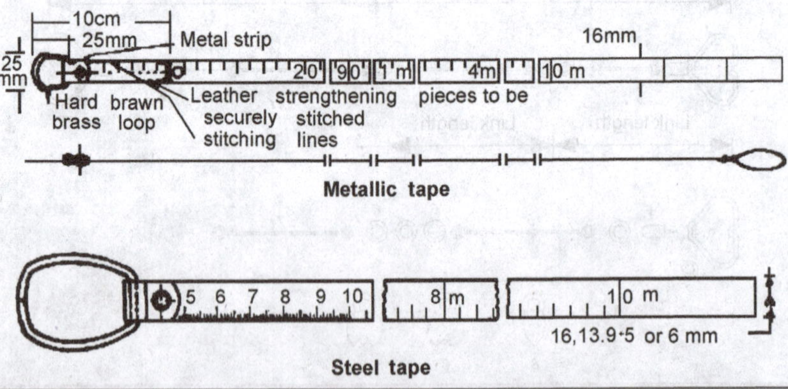

Figure 26.3 Types of tapes

d. *Invar tape:* This is made of an alloy of nickel (36 per cent) and steel having a very low coefficient of thermal expansion (0.122×10^{-6}/°C). These are available in lengths of 30, 50 and 100 m and in a width of 6 mm. The advantages and disadvantages of an invar tape are as follows.

26.1.7.1 Advantages

i. Highly precise.

ii. It is less effective by temperature changes when compared to other tapes.

26.1.7.2 Disadvantages

i. It requires much attention in handling.

ii. It is soft and so deforms easily.

PVC coated fibre glass tapes are most common nowadays, as they are more durable and weatherproof.

26.1.8 Marking station points on the ground in chain surveying

On a normal ground, the station points are to be marked by driving wooden pegs into the ground. Made of hard wood the wooden pegs are 25 mm × 25 mm in cross section and 150 mm in length. When driven, at least 50 mm should be above the ground.

26.1.8.1 Accessories used

i. *Ranging rod:* 2-3 m in length, with well-seasoned timber made into a rod of 30 mm diameter, alternatively painted red and white or black and white over a length interval of 200 mm. The ranging rod is provided with an iron shoe at its bottom and carries a 250 mm × 250 mm flag coloured red and white or yellow and white to make it conspicuous from a distance. It is used to find out the points lying on a line, i.e., in ranging a line. It is also used to mark the station points when the wooden pegs driven for the purpose are not visible from a working distance (Figure 26.4).

Figure 26.4 Ranging rods

ii. *Ranging poles:* These are ranging rods with longer lengths varying from 4 to 6 m. These is employed to range long lines on undulatory grounds.

iii. *Offset rods:* It is almost the same as the ranging rod with the only difference that here the flag at the top is replaced by a stout ring or a hook for pulling or pushing the chain. It is employed to take short offsets in ordinary works.

iv. *Arrows:* Accompanying each chain are 10 arrows. They are also called marking chains or pins and are used to mark the end of each chain during the process of chaining (Figure 26.5).

v. *Pegs:* Wooden pegs are used to mark the positions of the stations. They are made of hard timber and are tapered at one end. They are usually 2.5 cm^2 and 15 cm long. But in soft grounds, pegs 40–60 cm long and 4-5 cm^2 are suitable (Figure 26.6).

vi. *Cross staff:* The cross staff is used for finding the foot of the perpendicular from a given point of a line and for setting out a right angle at a given point of a line. There are three forms of cross staff, namely the open cross staff, the French and the adjustable cross staff (Figure 26.7).

- *Open cross staff:* The simplest form of the cross staff is the open cross staff. It consists of the head and the leg. The head is simply a wooden block octagonal or round around 15 cm side or diameter and 4 cm deep.

  To find the foot of the perpendicular from a given point to a given chain line, i.e., to take the offset, the cross staff is planted or held vertically on the chain line where the offset is likely to occur and turned until one pair of opposite slits is directed to a ranging rod at the forward end of the chain line. Looking through the other pair of slits, it is seen if the point to which the offset

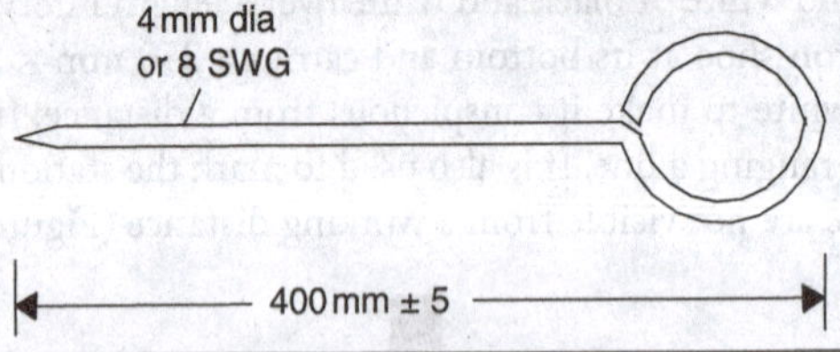

Figure 26.5 Arrow

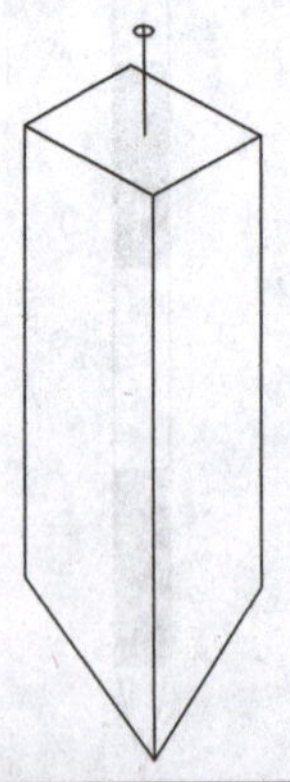

Figure 26.6 Peg

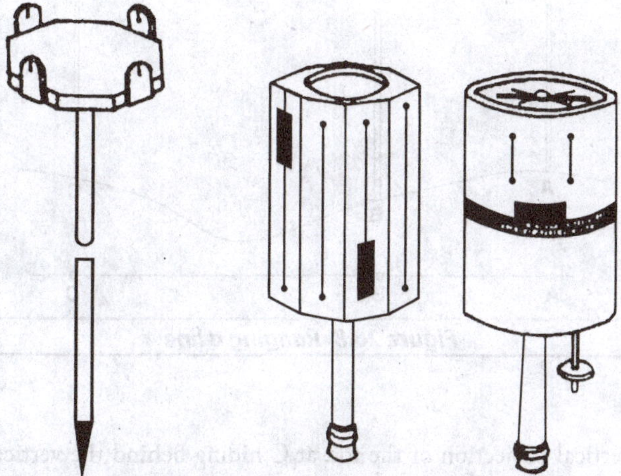

Figure 26.7 Different types of cross staff

is to be taken is bisected. If not, the cross staff is moved forward or backward on the chain line until the line of sight through the pair of slits at right angles to the chain line bisects the point.

- *French cross staff:* It consists of an octagonal brass tube with slits on all eight sides. The base carries a socket so that it may be fitted on the pointed staff when the instrument is to be used. The sights being too close, it is inferior to the open type.
- *Adjustable cross staff:* This may be used for setting out angles of any magnitude. However, the results are only approximate owing to the closeness of the sights.

26.1.8.2 Ranging a line – direct and indirect

Ranging a line means establishing intermediate points on the line. This may be accomplished with visual observations and ranging rods, looking through an instrument called 'line ranger' or looking through theodolite in precision works for ranging long lines. The operation may be classified as follows:

i. Direct ranging

ii. Indirect ranging

i. *Direct Ranging:* In direct ranging, the surveyor can observe and direct the person holding the ranging rod to put the ranging rod at any intermediate point on the line being ranged out.

ii. *Indirect Ranging:* Here, the surveyor cannot find out any intermediate points on the line on his own and has to depend upon the statement of a person at the end of the line and also on the statement of the person carrying the ranging rod.

26.1.8.3 Ranging with ranging rods

In Figure 26.8, A and B are two station points. The line AC is to be ranged. At A and C two ranging rods are to be driven vertically. The verticality of the rods is to be assured with the help of a plumb bob. If the rods at A and C are truly vertical, the surveyor standing about 2 m behind A and looking along the

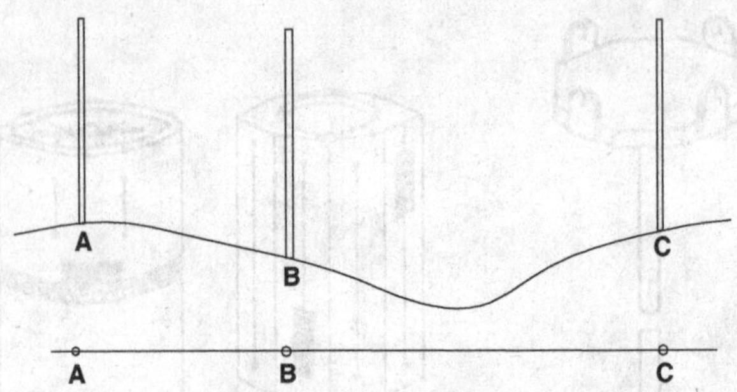

Figure 26.8 Ranging a line

line AC will find the vertical projection of the rod at C hiding behind the vertical projection of the rod at A, as shown in the figure. If the surveyor stands very near to A he may miss the vision for comparison. The surveyor will now send a man with a ranging rod to stand on the line AC at a distance less than one chain length from A. The surveyor should advise the man through hand signals to put a rod vertically on the line AC till he finds with his one eye closed and looking along AC the vertical projections of both the rods hiding behind that of the rod at A. The point B will then lie on AC. This way the other points may be ranged out.

26.1.8.4 Code of signals in ranging

In the field nothing should be communicated through loud speaking because it will lead to exhaustion. The following hand signals are therefore to be followed during the ranging of a line

i. The rapid sweeping movement of the right hand should mean long movement to the right. The slow sweeping movement of the right hand should mean very short movement to the right.

ii. Similarly, the rapid sweeping movement of the left hand should mean long movement to the left and the slow sweeping movement of the left hand should mean short movement to the left.

iii. The right or the left arm extended should mean to continue moving to the right or left accordingly.

iv. The right arm up and moved to the right should mean to make the rod vertical by moving the top of the rod to the right. Similarly the left arm up and moved to the left means to make the rod vertical by moving the top of the rod to the left.

v. Both the hands raised above the head and then brought down should indicate that the position of the rod is on the line AC.

vi. Both arms projected forward horizontally and then brought down should conclude that the rod is to be fixed at that position.

26.2 INTRODUCTION TO RECENT ADVANCES IN SURVEYING

26.2.1 Electronic distance meter (EDM)

EDM is an electronic equipment which is used to measure distances easily.

For measuring distances between two points, the instrument is set up at one point and a pole with a prism is held vertically at the other point. The EDM instrument transmits an infrared beam which is reflected back to the unit by the reflection prism used as the target at the other point. Using the time required by the ray to come back, the distance travelled by the ray is calculated and displayed by the EDM.

The most advanced form of EDM is the laser meter which can be used to measure distances upto 30 m without using any target and upto 100 m using the target. So for small distance measurements even an assistant is not required. If you measure the length, breadth and height together using this instrument, it can be used to compute the area and volume automatically.

26.2.2 Electronic total station

A total station is a combination of electronic theodolite, an electronic distance-measuring device (EDM) and a microprocessor with memory unit. With this device, one can determine angles and distances from the instrument to the points to be surveyed. With the aid of trigonometry, the angles and distances may be used to calculate the actual positions (x, y and z or northing, easting and elevation) of surveyed points in absolute terms.

The measurements are taken with the help of a telescope, which is mounted on a tripod and levelled and focussed on to the prisms held as targets at the points to be surveyed. The instrument is attached with an alphanumeric key board and LCD display. This works with the help of a rechargeable compact battery. The instrument has a built-in automatic atmospheric sensor that measures the atmospheric pressure and temperature in real time and automatically applies the necessary corrections.

The total station can be used to make linear measurements to an accuracy of 0.1 mm and angle measurements to 1" accuracy. All the measurements are done with speed and accuracy and are recorded with the help of a memory card (usually a PCMCIA card or so). The range of the equipment varies from 2,000 to 3,000 m using single prism and from 3,000 to 3,600 m using multiple prisms. All the calculations can be carried out automatically and the data can be transferred to any computer using communication software and a cable. Any required drawings such as contours can then be prepared using any standard software. The instrument is very compact and light in weight and can be used for various other purposes such as area calculation, traverse, road mapping and three-dimensional cross sectioning (Figures 26.9–26.12).

Figure 26.9 An electronic total station

Courtesy: M/s. Sokkia, Japan.

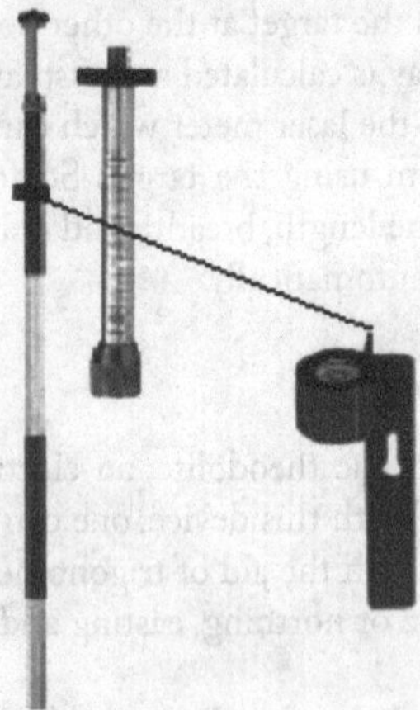

Figure 26.10 Typical standard rod (w/bubble level)

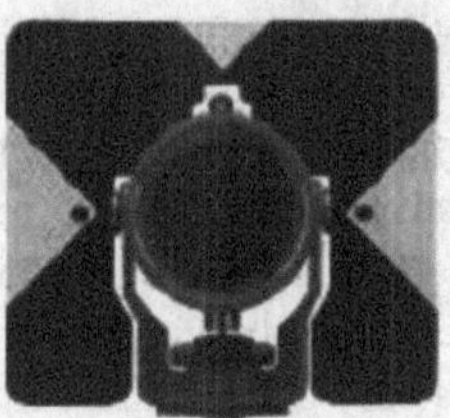

Figure 26.11 Typical standard prism and sight

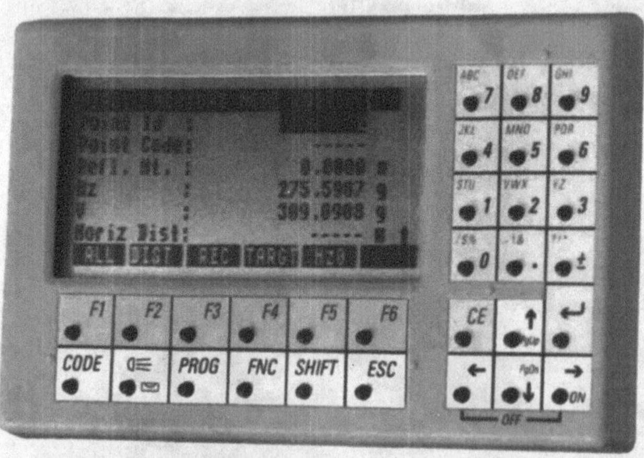

Figure 26.12 Display (Leica)

26.2.2.1 *Advantages of total station*

i. Quick setting of the instrument on the tripod using laser plummet.

ii. On board area computation program will compute the area of the field and the area computation is more accurate.

iii. Graphical view of plots and land for quick visualization.

iv. As soon as the field jobs are finished, the map of the area with dimension is ready after data transfer.

v. Enormous plotting and area computation at any user required scale.

vi. Integration of database (Exporting Map to GIS packages).

vii. Automation of old maps.

26.2.3 Global positioning system (GPS)

GPS is a satellite navigation system designed to provide instantaneous position, velocity and time information almost anywhere on the globe at any time and in any weather (Figure 26.13).

Present usage of GPS for positioning includes personal navigation (hiking, boating, hunting, driving directions, etc.), aircraft navigation, offshore survey and vessel navigation, fleet tracking, dredging, machine control, civil engineering, land surveying, GIS and mapping, deformation analysis, etc. The list is almost endless.

The day-to-day running of the GPS program and operation of the system rests with the US Department of Defense (DoD). Originally designed by the US Department of Defense (DoD), GPS comprises three main components: the *control segment*, the *space segment* and the *user segment*. Nowadays, GPS is used for topographic surveys and hydrographic surveys. The survey grade GPS receivers can achieve 2 mm accuracy for horizontal measurements and upto 5 mm accuracy for vertical measurements.

The control segments are stations situated on the ground on various parts of the world. These are the eyes and ears of the GPS and they keep tracking the satellites, which are rotating around the earth in various orbits in space at a height of 20,183 km above the earth. There are at present 24 such satellites, and they constitute the space segment.

The user segment comprises the receivers that have been designed to decode the signals transmitted from the satellites for the purposes of determining position, velocity or time. To compute the positional value from GPS satellite signals, a GPS receiver must perform the following tasks.

a. Selecting one or more satellites in view

b. Acquiring GPS signals

c. Measuring and tracking

d. Recovering navigational data

e. Calculation of the positional value

26.2.4 Automatic level

Automatic levels are extremely popular in present-day surveying operations and are very easy to be set up and can be used with any amount of precision.

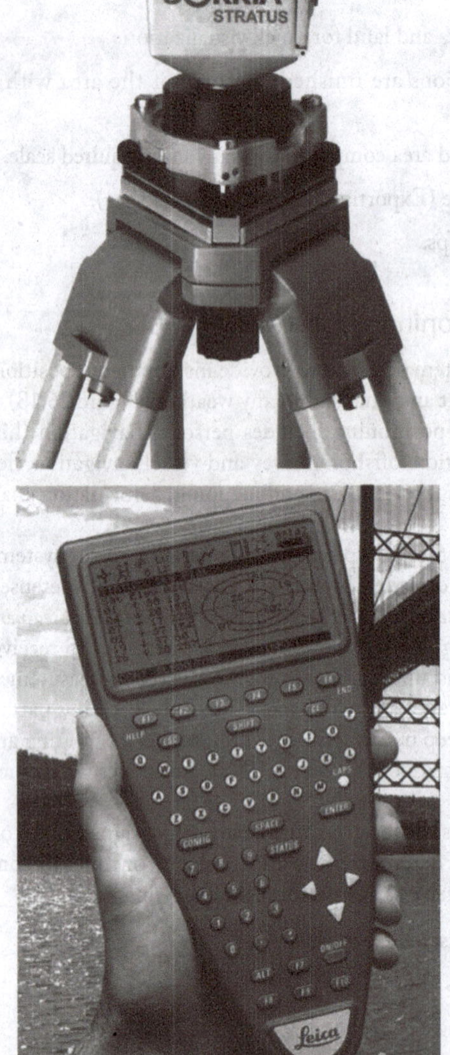

Figure 26.13 GPS and receiver (courtesy M/s. Sokkia and M/s. Leica)

They can be simply mounted on a tripod and can be set up and the circular bubble brought to the centre easily even on an angled and uneven surface. The instrument is equipped with an automatic compensator which is wire suspended and magnetically damped, which makes it suitable even under unstable conditions. Automatic focusing and weather proof telescope makes it useful even under unfavourable weather conditions (Figure 26.14).

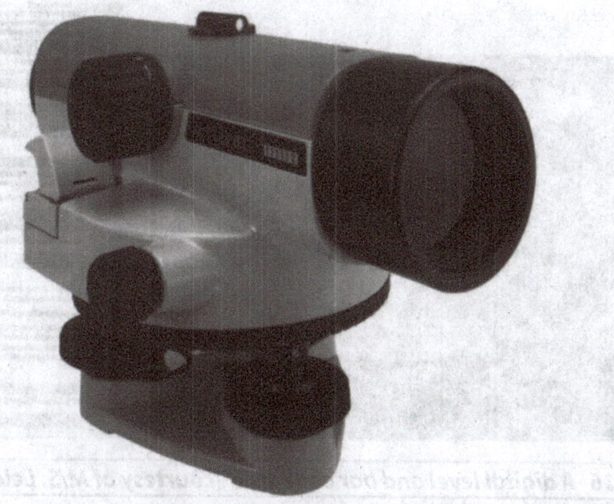

Figure 26.14 Automatic level (Pentax)

26.2.5 Digital level

Figures 26.15 and 26.16 show a digital level and a bar code levelling staff. This level features digital, electronic image processing for determining heights and distances, with the automatic recording of data for later transfer to the computer. The digital level is an automatic level (pendulum compensator) capable of normal optical levelling with rod graduated in feet or metres. When used in electronic mode, with the rod face graduated in barcode, this level will, with the press of a button, capture and process the image of the bar-code rod. This processed image of the rod reading is then compared with the image of the whole rod, which is stored permanently in the level's memory module, to determine height and distance values (Figures 26.15 and 26.16).

After the instrument has been levelled the operator must focus the image of the barcode properly. Next, the operator presses the measure button to begin the image processing, which takes about 4 seconds. Although the heights and distances are automatically determined and recorded, the horizontal angles must be read and recorded manually in this case. All the readings are digitally displayed and recorded for transfer to the computer.

Figure 26.15 A digital level (courtesy of Sokkia corp.)

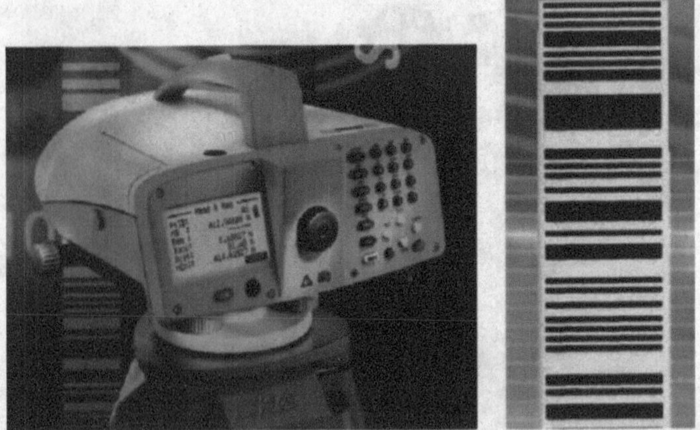

Figure 26.16 A digital level and bar code staff (courtesy of M/S. Leica geo-systems)

26.2.6 Microptic theodolite

It is a reliable, convenient, highly precise and easy to use instrument used for most surveying tasks including engineering and mining applications. It is equipped with a magnetically damped vertical circle compensator for fast and accurate zenith angle compensation. The optical plummet provided makes it easy to centre the instrument over the survey point with the help of a shifting tribrach. Coincident image projection is used to read the horizontal and vertical scales and thus possibility of any type of error in reading is eliminated. The readings can be directly read looking through the micrometer eye piece, which is illuminated using reflectors.

This instrument can be used in combination with an EDM (Figure 26.17).

26.2.7 Electronic digital theodolite

This functions in the same manner as that of an ordinary transit theodolite, except that it is more accurate, error free and easy to operate. There is an electronic compensator provided, which simplifies the levelling process and improves the accuracy by compensating the faulty adjustments. A laser plummet is provided for accurate centring. Electronic display of all the readings is automatically available. It works on rechargeable batteries.

26.2.8 Digital planimeter

Planimeter is an instrument used to measure the area of irregular shapes. The original version consists of tracing the boundaries with the arm of the planimeter and then converting the obtained reading into area of the figure, using the specified formula. Here, the possibility of error is very high and the procedure is not so easy (Figure 26.18).

Now, digital planimeters which work on rechargeable batteries and directly display the area required are available. These give a more accurate and quick measure of the area and in addition they allow calculations of line lengths, circumference, coordinates, angles, arc and circle radii and many other functions. LCD display, attached printers and computer compatibility are added advantages.

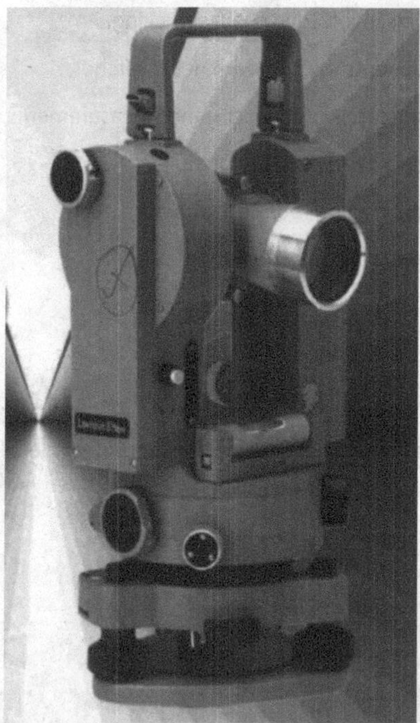

Figure 26.17 Microptic theodolite (Lawrence &Mayo)

Figure 26.18 Digital planimeter (PLACOM KP – 90 N)

REVIEW QUESTIONS

1. What are the basic principles of chain surveying?
2. What are the points to be considered while chain surveying?
3. Define tie stations, tie lines, base lines and offsets in chain surveying.
4. What are the different types of chains used in chain surveying?

5. Explain briefly the field works involved in chain surveying.
6. Explain the different components of chain with neat sketches.
7. Explain the different types of tapes used for distance measurements.
8. What are the advantages and disadvantages of tapes?
9. With a neat diagram explain ranging rods and arrows.
10. What is cross staff and give the classifications of cross staff?
11. What is the difference between direct and indirect ranging?
12. What are the advantages of total station?
13. What do you mean by Global Positioning System and how can it be used in surveying?
14. What is automatic level and what is its advantage over ordinary level?
15. Explain briefly
    a. Digital level
    b. Microptic theodolites
    c. Electronic theodolite
    d. Digital planimeter

chapter 27

Levelling

Levelling is the art of determining the relative elevations of different objects or points on the earth's surface. This is done by taking measurements in the vertical plane. Hence, this branch of surveying deals with measurements in vertical planes.

For the execution of civil engineering works such as major buildings, highways, dams, canals, water supply and sanitary schemes, it is necessary to determine elevations of different points along the alignments of a proposed project for the design and execution of the project. Success of the projects depends upon accurate determination of elevations of the ground control points as well as control points of the structures. Levelling is employed to provide an accurate network of heights, covering the entire area of the project. Levelling is of prime importance to the engineers, both in acquiring necessary data for the design of the project and also during execution.

27.1 PURPOSE OF LEVELLING

1. To find the elevations of given points with respect to a given or assumed datum (datum is an arbitrarily assumed level surface to which elevations are referred).
2. To establish points at a given elevation with respect to a given or assumed datum.

The instruments used in levelling are as follows:

1. A level
2. A levelling staff
3. A measuring tape

27.2 MAJOR PARTS OF LEVELLING INSTRUMENT

The level furnishes a horizontal line of sight. The level consists of the following main parts.

1. A telescope to provide the line of sight
2. A level tube to make the line of sight horizontal
3. A levelling head to bring the bubble in its centre of run
4. A tripod to support the instrument.

27.2.1 Telescope

Telescope is an optical instrument for magnifying and viewing the images of distant objects. The telescope of a levelling instrument is a metallic tube having an eyepiece at one end and an object glass at the other end. The telescope, which is fitted in levels, is generally of two types.

a. External focusing telescope

b. Internal focusing telescope

The external focusing telescopes were used in old model levels and the internal focusing telescopes are being used in modern survey instruments.

27.2.2 External focusing telescope

The telescope in which focusing is achieved by the external movement of either the objective glass or eyepiece is known as an external focusing telescope. In an external focusing telescope, the body is formed by two tubes at the ends of which the objective and eyepiece are fitted. One of the tubes is made to slide axially within the other by means of a rack and pinion arrangement attached to the focusing screw of the telescope (Figure 27.1).

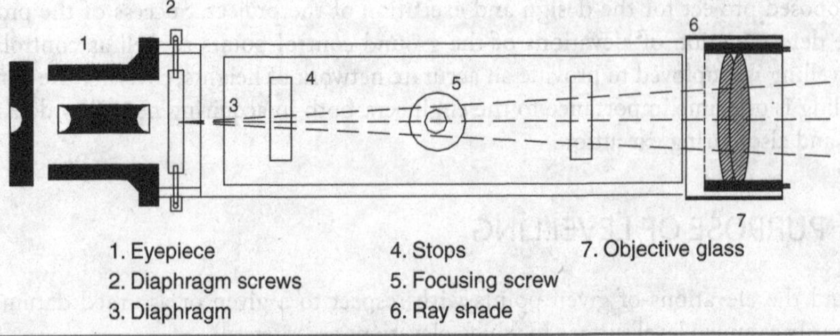

Figure 27.1 An external focusing telescope

27.2.3 Internal focusing telescope

The telescope in which focusing is achieved internally with a concave lens is known as internal focusing telescope. In an internal focusing telescope, the objective and eyepiece are kept at a fixed distance and focusing is achieved by a double concave lens mounted in a short tube capable of sliding axially between the eyepiece and the objective with a rack and pinion arrangement attached to the focusing screw (Figure 27.2).

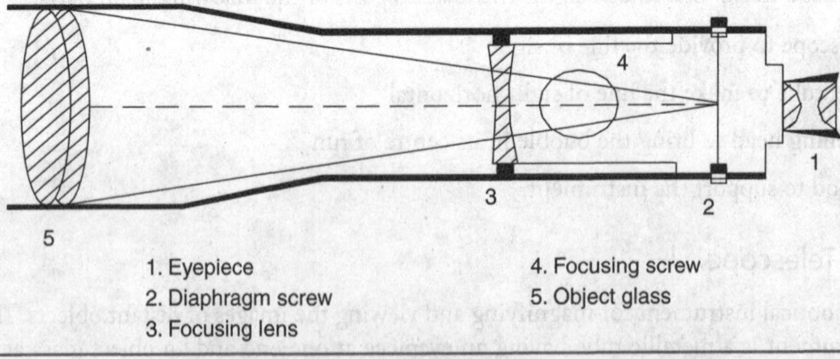

Figure 27.2 An internal focusing telescope

The objective is a compound lens consisting of a front double convex lens made of crown glass and the back concave-convex lens made of flint glass (Figure 27.3). The two lenses when cemented together with balsam at their common surface are generally known as achromatic lens. In such lenses, the spherical and chromatic aberrations known as optical serious defects are practically eliminated.

The eyepiece is composed of two plane convex lenses with a distance of two-third the local distance. The convex surfaces are turned towards one another. The Ramsden eyepiece is used in most surveying telescopes.

To provide a definite line of sight, horizontal and vertical cross hairs held in a flat metal ring called reticule are fitted into the diaphragm. The diaphragm is a flanged metal ring held in the telescope barrel by four capstan headed screws. With the help of the capstan headed screws, the position of the cross hairs inside the tube can be adjusted slightly, horizontally, vertically and rotationally. The hairs or lines are arranged in different ways as shown in Figure 27.4.

The telescope is used to read the levelling staff and the cross hairs enable the surveyor to take the staff reading.

27.2.4 Types of levelling instrument

The chief types of levels are

a. Dumpy level
b. Wye (or Y) level
c. Reversible level
d. Tilting level

27.2.5 Dumpy level

A dumpy level consists of a telescope tube firmly secured in two collars fixed by adjusting screws to the stage carried by a vertical spindle. In the modern form, the telescope tube and the vertical spindle are cast in one

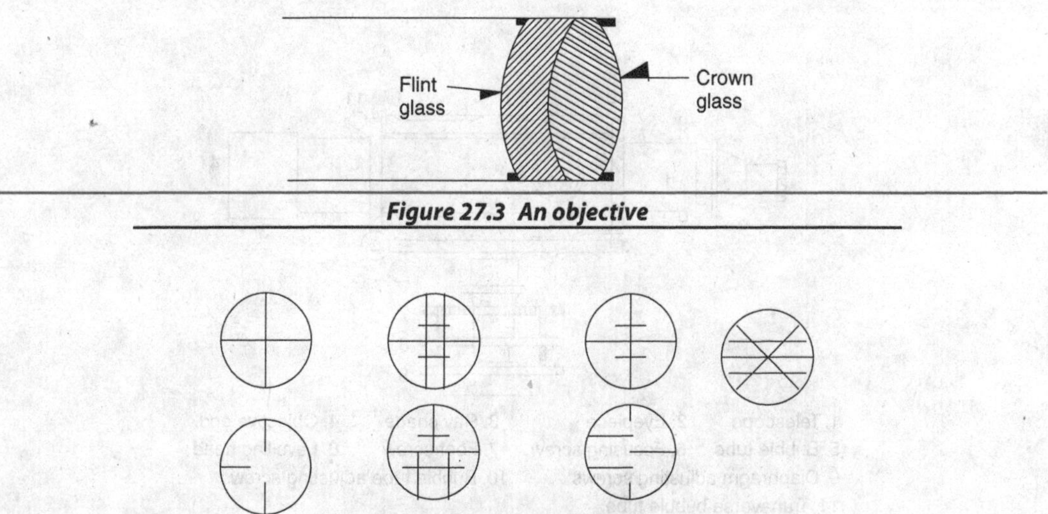

Figure 27.3 An objective

Figure 27.4 Different types of cross hairs

piece and a long bubble tube is attached to the top of the telescope. It can neither be rotated about its longitudinal axis nor can be removed from its supports. Figure 27.5 shows the elevation of a dumpy level showing different parts.

In some instruments, a clamp screw is provided to control the movement of the spindle about the vertical axis. For a small or precise movement, a slow motion screw or tangent screw is also provided. The levelling head generally consists of two parallel plates with three foot screws. The upper plate is known as tribrach and the lower plate is known as trivet, which can be screwed on to a tripod (Figure 27.6).

27.2.5.1 Advantages of a dumpy level

It is simple in construction with a few movable parts. It requires fewer permanent adjustments. Adjustments once carried out remain for a longer period.

Figure 27.5 A dumpy level

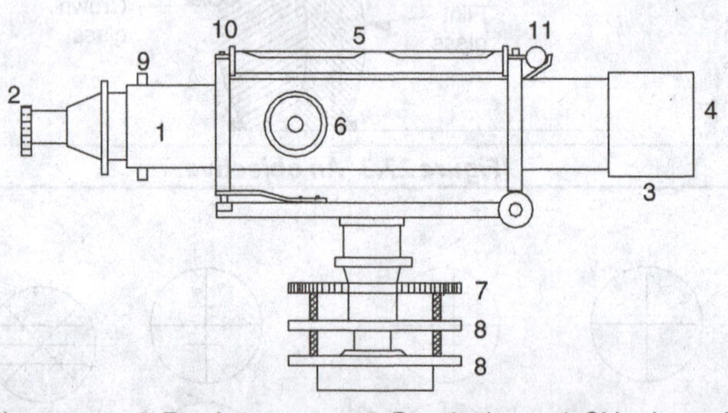

Figure 27.6 A dumpy level

27.3 LEVELLING STAFF

A straight rectangular wooden rod graduated in metres/feet and further smaller divisions is called a levelling staff. The bottom of the levelling staff represents the zero reading. The reading given by the line of sight on a levelling staff held vertically is the height of the line of collimation above the point on which the staff is held (Figure 27.7). The levelling staff may be divided into the following classes.

1. Self-reading staff
2. Target staff

27.3.1 Self-reading staff

A staff on which readings are directly read by the observer through the telescope is known as self-reading staff. Self-reading staffs are of three types as discussed below.

27.3.1.1 Solid staff

These are usually 3 m long. Due to the absence of a hinge or socket on these staffs, greater accuracy in reading is achievable, but on the other hand it is inconvenient to carry them in the field. Use of a solid staff is generally restricted to precise levelling work.

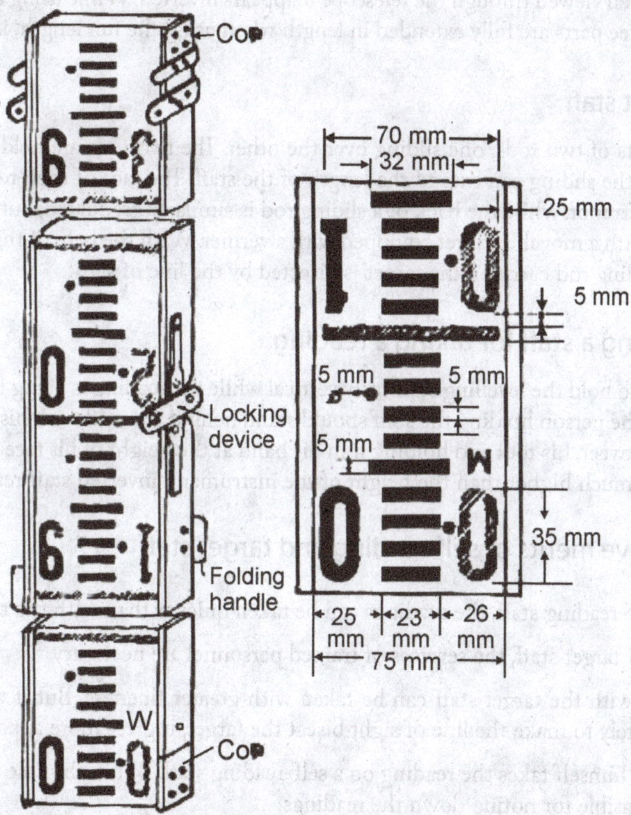

Figure 27.7 Levelling staff

27.3.1.2 *Folding or hinged staff*

A folding staff is made of well-seasoned timber. It is 4 m long and consists of two portions, each being 2 m long and hinged together. The width and thickness of the staff is kept at 75 mm and 18 mm, respectively. The foot of the staff is provided with a brass cap to avoid wear and tear due to usage. Sometimes, a plummet is also provided to test the verticality of the staff. Each metre length is subdivided into decimetres and each decimetre is further divided into 20 equal divisions of 5 mm width. Decimetre numerals 1–9 for each metre length are marked in black and metre numerals in red. The graduations are marked inverted so that they appear erect when viewed through the telescope. In modern levelling staff, the graduations are marked erect. The staff may be folded together so that one 2 m piece is capable of folding on the other when not in use and two pieces are detachable from one another so that one half may be used while working in plain areas.

27.3.1.3 *Telescopic or sopwith type staff*

It is made up of three pieces. The top piece is solid 1.25 m long, whereas the central piece of 1.25 m length and the lower piece of 1.5 m length are hollow. The top portion slides into the central portion telescopically. When fully extended, the total length of the staff is 4 m. The upper two pieces are held by brass spring catches. The smallest division of this type of levelling staff is also 5 mm. The metre numerals, which are shown on the left, are marked in red. The decimetre numerals 1–9 are shown on the right and marked in black. The decimetre number 10 of each metre length is omitted and letter M is marked to indicate the end of the metre length. Graduation is marked erect and when viewed through the telescope it appears inverted. While using a telescopic staff it may be ensured that the three parts are fully extended in length when using the full length, i.e., 4 m

27.3.2 Target staff

The target staff consists of two rods, one sliding over the other. The two rods are held together by means of brass clamps. Raising the sliding can extend the length of the staff. The face of each rod is graduated in feet, tens and hundredth of a foot, while the back of a sliding rod is similarly graduated, but from top downwards. The staff is provided with a movable target equipped with a vernier. With the help of this vernier, one can read up to 0.001 ft. The sliding rod carrying the target is bisected by the line of sight.

27.3.3 Holding a staff for taking a reading

Care should be taken to hold the levelling staff truly vertical while the reading is being taken. To hold the staff in a vertical position, the person holding the staff should stand behind the staff with his heels together, having the heel of the staff between his toes and holding it in his hand at the height of his face. When the level of the required point is very much higher than the height of the instrument, inverted staff reading is to be taken.

27.3.4 Relative merits of self-reading and target staff

a. With the self-reading staff, the readings can be taken quicker than with the target staff.

b. In the case of target staff, the services of trained personnel are necessary.

c. The reading with the target staff can be taken with greater fineness. But if the staff man does not direct accurately to make the line of sight bisect the target, it gives more apparent readings.

d. The surveyor himself takes the reading on a self-reading staff. But in the case of target staff, the staff man is responsible for noting down the readings.

e. It is tedious to adjust the target such that the line of sight bisects it accurately.

27.3.5 Fundamental axes of levelling instrument

a. Line of collimation or principal line of sight: It is the imaginary straight line which joins the optical centre of the object glass with the point of intersection of cross hairs of the diaphragm.

b. Axis of bubble tube: It is an imaginary line tangential to the longitudinal curve of the tube at its middle point. It is also known as bubble line. It is horizontal when the bubble is centred.

c. Axis of telescope: It is a line joining the optical centre of the object glass to the centre of the eyepiece.

d. The vertical axis: It is the imaginary line passing through the centre line of the axis of rotation.

e. The height of the instrument (HI): The height of the instrument is the elevation of the plane of collimation or plane of sight when the instrument is correctly levelled. When a level in adjustment is accurately levelled, the line of collimation will revolve in a horizontal plane known as the plane of collimation or plane of sight.

27.4 TECHNICAL TERMS USED IN LEVELLING

1. *Level surface:* A level surface is any surface parallel to the mean spherical surface of the earth. It is a curved surface, which at each point is perpendicular to the direction of gravity at that point. Every point on a level surface is equidistant from the centre of the earth (Figure 27.8).
2. *A level line:* A level line is a line lying in a level surface. It is normal to the plumb line at all points (Figure 27.9).

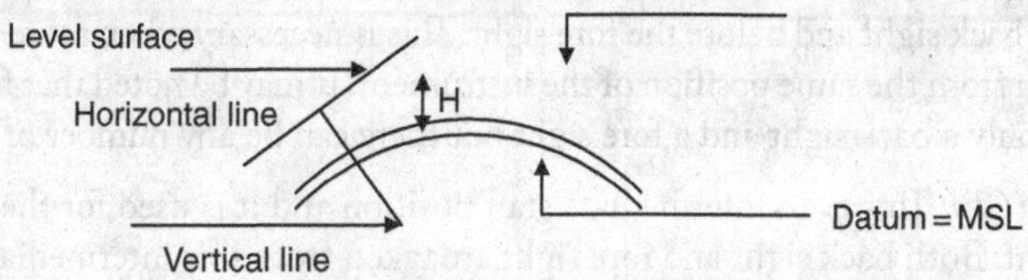

Figure 27.8 Level surface

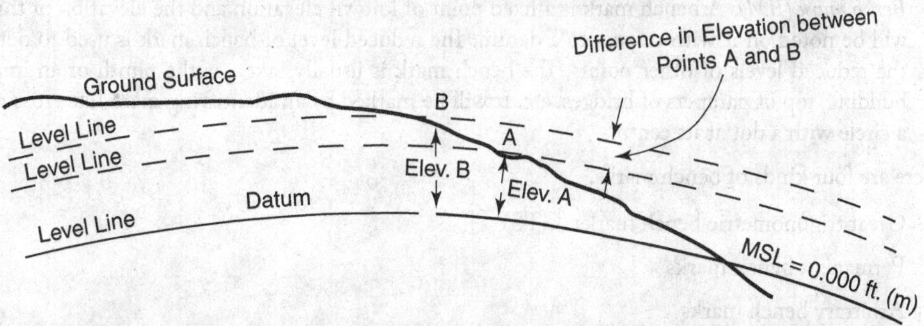

Figure 27.9 Figure showing level line, datum, mean sea level and ground surface

3. *A horizontal plane:* A horizontal plane through a point is a plane tangential to the level surface at that point.
4. *A horizontal line:* A horizontal line is any line lying in the horizontal plane.
5. *A vertical line:* A vertical line at any point is a line normal to the level surface through that point.
6. *Elevation:* The elevation of a particular point is the vertical distance above or below a reference surface. Usually the mean sea level is used as reference.
7. *Mean sea level (MSL):* Mean sea level is obtained from the average height of the sea's surface for all the stages of the tide, over a period of 18.6 years.
8. *Datum:* Datum is any arbitrarily assumed level surface to which elevations are referred.
9. *Reduced level (RL):* The reduced level of a place is its elevation or vertical distance above or below the datum or any fixed point.
10. *Line of sight:* It is the imaginary line joining the intersection of the cross hairs of the diaphragm to the optical centre of the object glass and its continuation.
11. *Back sight (BS):* A back sight is the first staff reading taken after setting up the instrument at any position. This will always be a reading on a point of known elevation. It ascertains the amount by which the line of sight is above or below the elevation of the point. Back sight enables the surveyor to obtain the height of the instrument.
12. *Fore sight (FS):* A fore sight is the last staff reading taken before shifting the instrument. This will always be a reading on a point whose elevation is to be determined. This reading indicates the shifting of the instrument. It is also generally known as minus sight as the fore sight reading is always subtracted from the height of the instrument (except when the staff is held inverted) to obtain the elevation.
13. *Intermediate sight (IS):* An intermediate sight is any staff reading, taken on a point of unknown elevation, after the back sight and before the fore sight. This is necessary when more than two staff readings are to be taken from the same position of the instrument. It may be noted that for one setting of a level there will be only a back sight and a fore sight but there can be any number of intermediate sights.
14. *Change point (CP):* This is an intermediate staff position and it is used for the purpose of shifting of the instrument. Both back sight and fore sight are taken from this intermediate staff position. Great care is necessary in taking readings at the change point since an error in reading affects every succeeding point of observation (elevation). Any firm point, which can be easily found, may serve as a change point.
15. *Bench mark (BM):* A bench mark is a fixed point of known elevation and the elevation of this point will be noted on it with respect to a datum. The reduced level of bench mark is used to determine the reduced levels of other points. The bench mark is usually taken as the plinth of an important building, top of parapets of bridges, etc. It will be marked by an identifying mark like cross mark or a circle with a dot at its centre.

There are four kinds of bench marks.

a. Great trigonometric bench marks (GTS)
b. Permanent bench marks
c. Arbitrary bench marks
d. Temporary bench marks

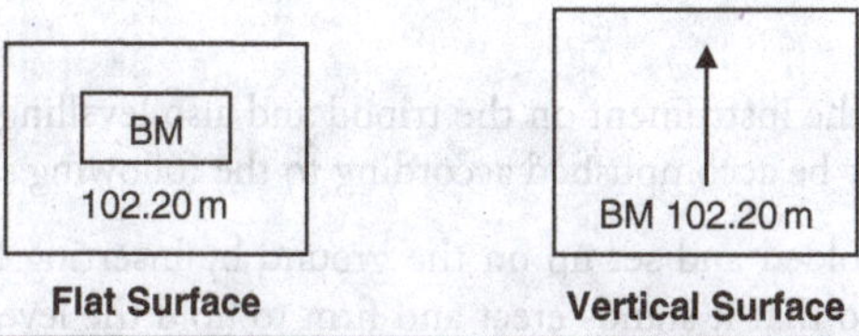

Figure 27.10 Figure showing a permanent bench mark

Great trigonometric bench marks (GTS) are established by the Survey of India Department. They will be marked with very high accuracy at intervals, all over the country by this department. Their positions and elevations above the standard datum are given in the catalogue published by that department.

Permanent bench marks are established when the distance between the GTS bench marks is very large. Hence, fixed reference points are established at closer intervals. They are established by the Government agencies such as the local surveying departments, on clearly defined and permanent points such as the top of a parapet wall of a bridge or culvert, a corner of the plinth of a building, a gate pillar and a kilometre stone. On a vertical surface there will be a broad arrow and a horizontal groove, the centre of which is the point of reference (Figure 27.10).

In small levelling works, the surveyor can assume a well-defined point for reference. The reduced level of that point is arbitrarily assumed (say as 100.00, 50.00 etc.). These are called arbitrary bench marks.

Temporary bench marks are the reference points established at the end of a day's work or when there is a break in the work. This becomes necessary when the whole levelling work cannot be completed at a stretch. The work when started again is continued with reference to these bench marks. They should be carefully established on definite and comparatively permanent objects, which can be easily described and found, such as the top of a stone gate, posts, spikes in the roots of the tree and highest point of solid rock.

27.5 LEVELLING – FIELD WORK

The survey work is mainly divided into two: (1) field work and (2) office work.

In the field, necessary adjustments are done and the results are recorded in a systematic manner as explained below.

27.5.1 Adjustments of a level

A level needs two type of adjustments, i.e.,

a. Temporary adjustments
b. Permanent adjustments

27.5.1.1 Temporary adjustments

The adjustments which are made for every setting of a levelling instrument are called temporary adjustments. These include the following:

i. Setting up the level
ii. Levelling up
iii. Elimination of parallax

Setting up the level

This operation includes fixing the instrument on the tripod and also levelling the instrument approximately by leg adjustments. Setting may be accomplished according to the following steps.

Step 1 The tripod is unfolded and set up on the ground by inserting the metal pointed leg bottoms into the ground so that it stands erect and firm to hold the level at a height convenient to the observer looking through the telescope.

Step 2 The box containing the level is opened and the position of the level as it suits in the box is noted. If required the edges of the box are marked so that the level may be replaced in the box correctly.

Step 3 The level is set on the stand by screwing it on or clamping it in the slot of the tribrach.

Step 4 One hand is placed lightly on the telescope and the other hand on one of the legs, pressed against the thigh, to avoid jerks and snaps, while the leg is given lateral to and fro motion to bring the bubble in the tube on the tribrach in the centre of its run, or to bring the bubble to the centre of a circular level on the tribrach, if such a level is provided on it.

Step 5 The movement of the legs is now locked with the help of any suitable mechanism provided with the tripod.

Levelling up

The vertical axis of the instrument is rendered truly vertical through the levelling up operation which may be accomplished in accordance with the following steps. In Figure 27.11 (a) and (b) three-foot screw arrangements of a dumpy level and in figures (c) and (d) four-foot screw arrangements of a dumpy level are shown.

Step 1 The telescope axis is placed parallel to the line joining one pair of screws as in Figure 27.11(a) and in Figure 27.11(c).

Step 2 Both the screws 1 and 2 are rotated simultaneously by moving them equally inwards or equally outwards till the bubble of the spirit level on the telescope is brought to the centre of its run.

Step 3 The telescope axis is now placed perpendicular to its previous position, i.e., perpendicular to 1-2 line.

Step 4 The third screw 3 in Figure 27.11(b) (in the case of a three-screw system) is rotated to bring the bubble of the spirit level on the telescope to the centre of its run.

In the case of four-screw arrangements, the screws 3 and 4 are to be moved equally inward or equally outward till the bubble of the telescope spirit level is brought to the centre of its run (see Figure 27.11(d)).

Step 1 through 4 have to be repeated several times before the bubble tube of the spirit level on the telescope remains central for all positions of the telescope when the vertical axis of the instrument becomes truly vertical. In reality, however, if the bubble remains central over the angle of rotation, of the telescope, required for reading the levelling staff, the purpose is served.

Elimination of parallax

An apparent change in the position of the object caused by the change in position of the observer's eye is known as parallax. In a telescope, parallax is caused when the image formed by the objective is not situated in the plane of the cross hairs. Unless the parallax is removed, accurate bisection and sighting of objects

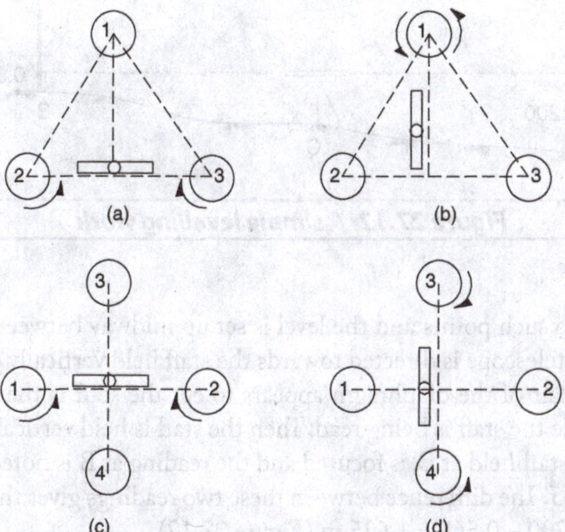

Figure 27.11 Levelling the instrument

become difficult. Elimination of parallax is done by focusing the eyepiece for distinct vision of cross hairs and focusing the objective to bring the image of the object in the plane of the cross hairs as explained below.

- *Focusing the eyepiece:* To focus the eyepiece for a distinct vision of cross hairs, either hold a white paper in front of the objective or sight the telescope towards the sky. Move the eyepiece in or out till the cross hairs are seen sharp and distinct.
- *Focusing the objective:* After the cross hairs have been properly focused, direct the telescope on a well-defined distinct object and intersect it with a vertical wire. Focus the objective till a sharp image is seen. Moving the eye slowly to one side may check removal of the parallax. If the object still appears intersected, there is no parallax. If on moving the eye laterally, the image of the object appears to move in the same direction as the eye and the observer's eye and the image of the object are on the opposite sides of the vertical wire, the image of the object and the eye are brought nearer to eliminate the parallax. This parallax is called far-parallax. If, on the other hand, the image appears to move in a reverse direction to the movement of the eye and the observer's eye and the image of the object are on the same side of the vertical wire, then the parallax is called near-parallax. It may be removed by increasing the distance between the image and the eye.

27.5.2 Classification of levelling

Levelling may be classified into two categories:

a. Simple levelling
b. Differential levelling

27.5.2.1 Simple levelling

When two points whose level difference is to be found out are situated in such a way that both of them are visible from a single position of level, this method is adopted.

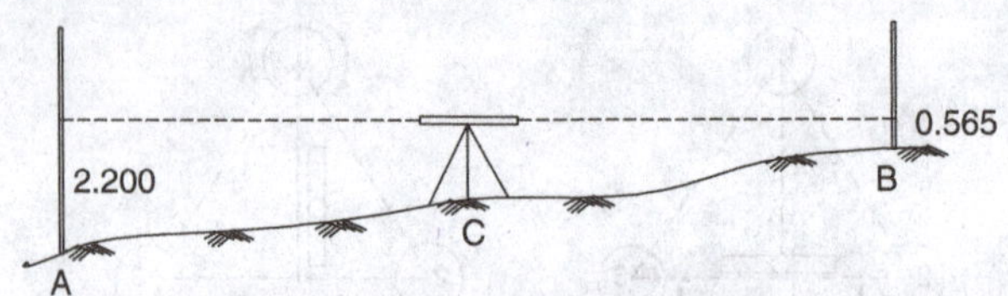

Figure 27.12 A simple levelling work

Suppose A and B are two such points and the level is set up midway between A and B. After the instrument is correctly levelled, the telescope is directed towards the staff held vertically on A and focused. The reading at which the horizontal hair of the diaphragm appears to cut the staff is then taken carefully. Make sure that the bubble is central while the staff is being read. Then the staff is held vertically on B and the telescope of the level is directed on to the staff held at B, is focused and the reading at B is noted. Let the respective reading on A and B be 2.200 and 0.565. The difference between these two readings gives the difference in level between A and B, which is equal to 2.200 – 0.565 = 1.635 m (Figure 27.12).

Assume the reduced level of A as 100.000 m.

Height of instrument at station C = Reduced level atof A + staff reading at A
= 100.00 + 2.200 = 102.200 m

Reduced Level of B = 102.200 – 0.565 = 101.635 m

27.5.2.2 Differential levelling

The method of levelling for determining the difference in elevation of two points either too far apart or obstructed by an intervening ground is known as differential levelling. In this method, the level is set at a number of points and the difference in elevation of successive points is determined as in the case of simple levelling. This levelling process is also known as fly levelling, compound levelling or continuous levelling.

Hence, this method is adopted when the points are too far apart, if the difference in elevation between them is too great or if there are obstacles between them.

In this case, the level is set up at different positions (points) for the execution of the levelling operations as in Figure 27.13 below. Consider two points A and E as in Figure 27.13; it is required to find the level difference between these two points. Set up the level at a convenient point, let it be at 'a'. The work is started by taking back sight to a bench mark or a known point and then fore sight is taken to a point to fix its level at A. Take a staff reading at A, let it be 'a1'. Select a firm point b, so that the distance from C to 'b' is approximately equal to the distance from A to 'a'. Hold the staff at B and take the staff reading. Let it be 'b1'. This forms first stage in the levelling series. Keeping the staff at B, shift the instrument to 'b'. Take a back sight at B, let it be 'b2'. With the levelling instrument at 'b' shift the levelling staff to a third position at 'C' and the work is continued till the point E is read.

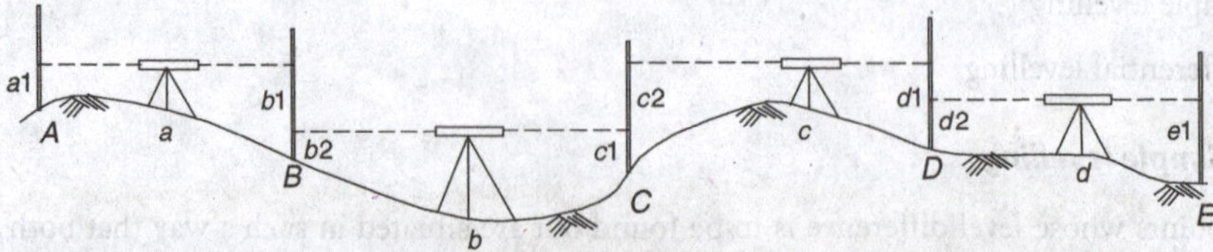

Figure 27.13 Execution of a direct levelling operation

The staff reading 'a1' taken at A from the instrument station 'a' is the back sight and the staff reading 'b1' taken at B is the fore sight. The staff reading 'b2' taken at B from the instrument station 'b' is the back sight and the staff reading 'c1' taken at C is the fore sight and so on.

Hence, the level difference between A and B = a1 – b1.

The level difference between B and C = b2 – c1 and so on.

The difference of level between A and E is equal to the algebraic sum of these differences or equals to the difference between the sum of the back sights and the sum of the fore sights.

27.6 FLY LEVELLING

Fly levelling is a process of finding the level difference between two points and the levelling consists of taking back sights and fore sights only and not intermediate sights. Differential levelling is the determination of level difference between two points; check levelling is finding the level difference between points in a way of checking the accuracy of levelling works already done; and fly levelling is the procedure of accomplishing the objective of finding out the level difference between two points, which are far apart. It can also be used for transferring a bench mark.

27.6.1 Fly levelling procedure

a. Set up the level on a firm ground and do the temporary adjustments. The instrument should be set up approximately midway between the change points.

b. Direct the telescope towards the staff, which is held vertically on the point.

c. Focus the telescope.

d. Bring the staff between the two vertical hairs.

e. Check the bubble. If it is not in the exact centre, use the screw in line with the telescope and bring it to the centre.

f. Read the staff when the horizontal hair of the diaphragm appears to cut it and record the correct reading.

g. Take the first reading on the bench mark and enter the reading in the back sight column of the field book.

h. Take fore sight reading on the change point, if the second bench mark is far away, and enter the reading in the fore sight column of the next horizontal line.

i. Shift the instrument and do all the temporary adjustments.

j. Take back sight on the same change point and enter the reading in the back sight column of the same horizontal line.

k. Take fore sight reading on another change point, if the next bench mark is far away, and enter the reading in the fore sight column of the next horizontal line.

l. Shift the instrument near to the next bench mark and again take readings on the change point and enter it in the back sight column of the same horizontal line.

m. Repeat the above process for a number of bench marks.

n. If possible close the fly level at the starting point.

o. Record all the reading and data in the field book.

27.7 CONTOUR

A map represents the relative positions of points in a plan. The value of the map is enhanced if the variation in the elevation of the earth's surface is also included along with their relative position in a plan. There are two methods by which the conformation of the ground may be presented on a map. One way is by delineating the surface slopes by hachure shading, etc., intended to give an impression of relative relief. The relative elevations of the points are not indicated in such a case. To a layman this method of portraying relief is very simple and legible and is very commonly used for geographical mapping. The other way, which is usually employed in plans for engineering purposes, is plotting the contour lines on maps. These lines are so arranged that the form of the earth's surface can be portrayed with greater accuracy and thoroughness and can readily be interpreted (Figure 27.14).

A contour may be defined as an imaginary line passing through the points of equal elevation. All the points on any one contour line have the same elevation above the datum surface and the contour may thus be defined as the line of same level. When the contours are drawn underwater, they are termed as submarine contours, fathoms or bathymetric curves. This is the best method of representation of features such as hills, depressions and undulations on a two-dimensional paper.

Contours are used in a variety of ways. Some of the engineering uses of contours are as follows:

1. With the help of a contour map, proper and precise location of engineering works such as roads and canals can be decided.
2. In the location of water supply, water distribution and to solve the problems of stream pollution, etc.
3. Planning and designing of dams, reservoirs, transmission lines, etc.
4. To select suitable sites for new industrial plants.
5. To ascertain the inter-visibility of stations.
6. To ascertain the profile of the land along any direction.
7. To estimate the quantity of cutting, filling and capacity of reservoirs.

27.7.1 Contour interval

The vertical distance between consecutive contours is termed as contour interval.

It is desirable to have a constant contour interval throughout the map.

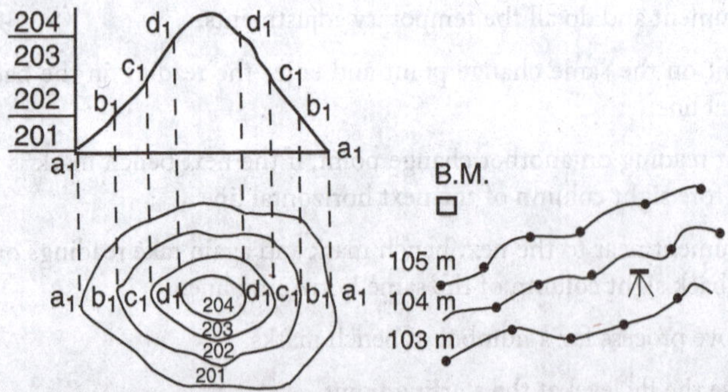

Figure 27.14 Contours

27.8 COMPUTING REDUCED LEVELS USING HEIGHT OF COLLIMATION METHOD (HI METHOD)

In this method, the reduced levels of points are computed by calculating the reduced levels of the plane of collimation for each set up of the instrument.

The height of collimation is obtained by adding the staff reading, which must be back sight to the known reduced level of the point on which the staff stands. Reduced levels of all the other points are obtained by subtracting the staff reading from the height of collimation. When the instrument is changed to a new station, a new height of collimation is obtained by adding the new back sight with the reduced level of the last point obtained from the previous set up of the instrument. The steps involved in booking and reading the level in the height of collimation method are illustrated with the help of an extract from a field book given below.

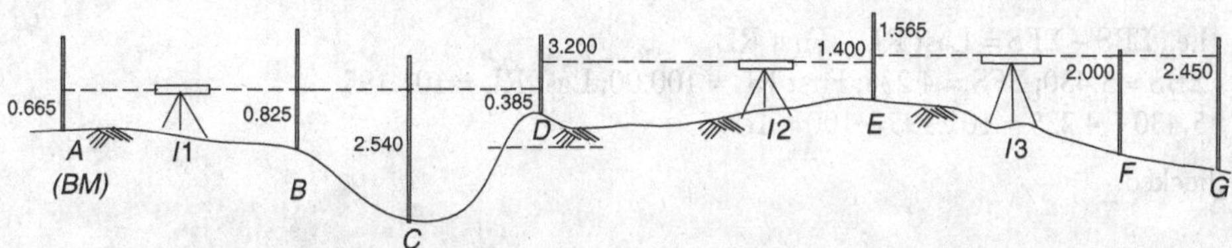

Station	BS	IS	FS	HI	RL	Remarks
A	0.665			100.665	100.00	BM
B		0.825			99.840	
C		2.540			98.125	
D	3.200		0.385	103.480	100.280	CP
E	1.565		1.400	103.645	102.080	CP
F		2.000			101.645	
G			2.450		101.195	

In this table, the back sight, intermediate sight and fore sight are the readings taken in the field and in the remarks column bench marks and change points are specified. Then, with the help of this available data the height of the instrument and reduced levels are calculated and the usual checks are carried out.

1. To begin with, the elevation of the plane of collimation of the first instrument station (I_1) is to be calculated. This is done by adding the back sight to the reduced level of the bench mark. The reduced level of the bench mark, i.e., point A, is taken as 100.00 and back sight is taken by holding a staff at point A (i.e., 0.665).

$$\text{Elevation of the plane of collimation} = 100 + 0.665$$
$$= 100.665$$

This is entered as the height of the instrument (HI) as seen in the table.

2. Then, from the same instrument station, readings are taken of the staff held at B and C and entered as the intermediate sight. Then the reduced level of B and C are calculated.

$$\text{For example, RL of B} = \text{HI} - \text{IS} = 100.665 - 0.825$$
$$= 99.840$$

3. Now the instrument is shifted to the next point I_2 and the back sight and fore sight are taken. Then, the HI and reduced level at this point are calculated.

$$
\begin{aligned}
RL &= \text{HI at first point} - FS \\
&= 100.665 - 0.385 \\
&= 100.280
\end{aligned}
$$

$$
\begin{aligned}
HI &= RL + BS \\
&= 100.280 + 3.200 \\
&= 103.480
\end{aligned}
$$

4. This procedure is followed till the last point is reached.
5. Now the checks are carried out to ascertain the correctness of the readings.

   i.e., $\Sigma BS - \Sigma FS$ = Last RL – First RL
   $\Sigma BS = 5.430$; $\Sigma FS = 4.235$; First RL = 100.00; Last RL = 101.195
   $5.430 - 4.235 = 101.195 - 100 = 1.195$

Hence checked.

Problem-1

Complete the levelling table given below.

Station	BS	IS	FS	HI	RL	Remarks
BM	3.10				193.62	BM
1		2.56				
2		1.07				
3	1.92		3.96			CP
4	1.20		0.67			CP
5		4.24				
6	0.22		1.87			CP
7		3.03				
8			1.41			

Solution: The levels are computed as below.

Station	BS	IS	FS	HI	RL	Remarks
BM	3.10			196.72	193.62	BM
1		2.56			194.16	
2		1.07			195.65	
3	1.92		3.96	194.68	192.76	CP
4	1.20		0.67	195.21	194.01	CP
5		4.24			190.97	
6	0.22		1.87	193.56	193.34	CP
7		3.03			190.53	
8			1.41		192.15	
	Σ 6.44		Σ 7.91			

Check: ΣBS – ΣFS = Last RL – First RL
ΣBS =6.44; ΣFS = 7.91; Last RL = 192.15; First RL = 193.62
i.e., 6.44 – 7.91 = –1.47 m
192.15 – 193.62 = –1.47 m

Hence checked.

Problem-2

Compute the levels from the levelling field book using HI method.

BS	IS	FS	HI	RL	Remarks
3.39				23.10	BM
	2.81				
	2.51				
	2.22				
2.61		1.88			CP
	2.32				
	1.92				
		1.54			

Solution. The levels are computed as below.

BS	IS	FS	HI	RL	Remarks
3.39			26.49	23.10	BM
	2.81			23.68	
	2.51			23.98	
	2.22			24.27	
2.61		1.88	27.22	24.61	CP
	2.32			24.90	
	1.92			25.30	
		1.54		25.68	
Σ 6.00		Σ 3.42			

Check: ΣBS – ΣFS = Last RL – First RL
ΣBS = 6.00; ΣFS = 3.42; Last RL = 25.68; First RL = 23.10
i.e., 6.00 – 3.42 = 2.58 m
25.68 – 23.10 = 2.58 m

Hence checked.

Problem-3

The following are the staff readings taken while making levels of a field. The back sights are underlined. Tabulate the levels in a field book and compute the level difference using HI method. <u>0.813</u>, 2.170, 2.908, 2.630, 3.133, <u>3.752</u>, 3.277, 1.899 and 2.390,

Here, in this problem, the readings have to be entered in a book form, entries checked and the reduced levels found.

BS	IS	FS	HI	RL	Remarks
0.813			40.376	39.563	BM
	2.170			38.206	
	2.908			37.468	
	2.630			37.746	
3.752		3.133	40.995	37.243	CP
	3.277			37.718	
	1.899			39.096	
		2.390		38.605	
Σ4.565		Σ5.523			

Check: ΣBS – ΣFS = Last RL – First RL
ΣBS = 4.565; ΣFS = 5.523; Last RL = 38.605; First RL = 39.563
4.565 – 5.523 = –0.958
38.605 – 39.563 = –0.958

Hence checked.

27.9 COMPUTING REDUCED LEVEL USING RISE AND FALL METHOD

This method consists in determining the difference of level between consecutive points by comparing each point, after the first, with that immediately preceding it. The difference between their staff readings indicates rise or fall depending on whether the staff reading at the point is smaller or greater than that at the preceding point. The reduced level of each point is then found by adding the rise to, or subtracting the fall from, the reduced level of the preceding point.

It is to be noted that the terms 'rise' and 'fall' always refer to rise or fall from the first point to the second point, second point to the third point and not conversely.

The steps involved are as follows:

1. It consists in determining the difference of levels between the consecutive points by comparing their staff readings.
2. Obtain the rise or fall by calculating the difference between the consecutive staff readings. Rise is indicated if the back sight is more than the fore sight, and a fall if the back sight is less than the fore sight.
3. Find out the reduced levels of each point by adding the rise to, or by subtracting the fall from, the reduced level of the preceding point.

Check: ΣBS – ΣFS = ΣRise – ΣFall = Last RL – First RL

Example-2

The following readings were extracted from a level field book.

Station	BS	IS	FS	Rise	Fall	RL	Remarks
A	0.665					100.000	BM
B		0.825			0.160	99.840	
C		2.540			1.715	98.125	
D	3.200		0.385	2.155		100.280	CP

(continued)

E	1.565		1.400	1.800		102.080	CP
F		2.000			0.435	101.645	
G			2.450		0.450	101.195	

$\Sigma BS = 5.430$; $\Sigma FS = 4.235$; $\Sigma Rise = 3.955$; $\Sigma Fall = 2.760$

Here also the reduced level of the bench mark is assumed as 100.00 and the back sight to the bench mark is taken. Then the staff is held at B and reading recorded as intermediate sight.

RL of B = BM – Fall

(As the intermediate sight at B is greater than the back sight at A)

Then RL of C = RL of B – Fall of C
(Fall of C = IS at C – IS at B)

Now there is a change point and so the back sight and fore sight are to be taken, and as the intermediate sight (i.e., 2.540) is greater than fore sight (i.e., 0.385) there is a rise.

Rise = 2.540 – 0.385 = 2.155
RL of D = RL of C + Rise = 98.125 + 2.155 = 100.280

Now there is another change point and here also, as the back sight (3.200) is greater than the F.S (1.400), there is a rise.

Rise = 3.200 – 1.400 = 1.800
RL of E = RL of D + Rise = 100.280 + 1.800 = 102.080

Intermediate sight is taken with the staff at F and here as the intermediate sight (i.e., 2.000) is greater than back sight (i.e., 1.565) there is a fall.

Fall = 2.000 – 1.565 = 0.435
RL of F = RL of E – Fall = 102.080 – 0.435 = 101.645

This procedure is repeated till the last point is reached and the usual checks are applied to assess the correctness of the calculations.

Check: $\Sigma BS - \Sigma FS = \Sigma Rise - \Sigma Fall$ = Last RL – First RL = 1.195

Problem-1

Given are the levels recorded in a levelling book. Complete the levels using rise and fall method.

Station	BS	IS	FS	Rise	Fall	RL	Remarks
BM	3.10					193.62	BM
1		2.56					
2		1.07					
3	1.92		3.96				CP
4	1.20		0.67				CP
5		4.24					
6	0.22		1.87				CP
7		3.03					
8			1.41				

Solution: The levels are computed as below using rise and fall method.

Station	BS	IS	FS	Rise	Fall	RL	Remarks
BM	3.10					193.62	BM
1		2.56		0.54		194.16	
2		1.07		1.49		195.65	
3	1.92		3.96		2.89	192.76	CP
4	1.20		0.67	1.25		194.01	CP
5		4.24			3.04	190.97	
6	0.22		1.87	2.37		193.34	CP
7		3.03			2.81	190.53	
8			1.41	1.62		192.15	
	Σ 6.44		Σ 7.91	Σ 7.27	Σ 8.74		

Check: ΣBS – ΣFS = Last RL – First RL = ΣRise – ΣFall
6.44 – 7.91 = –1.47, 192.15 – 193.62 = –1.47, 7.27 – 8.74 = –1.47

Hence checked.

Problem-2

Given are the levels recorded in a field book. Compute the levels using rise and fall method.

BS	IS	FS	Rise	Fall	RL	Remarks
3.39					23.10	BM
	2.81					
	2.51					
	2.22					
2.61		1.88				CP
	2.32					
	1.92					
		1.54				

Solution: The levels are computed as below using the rise and fall method.

BS	IS	FS	Rise	Fall	RL	Remarks
3.39					23.10	BM
	2.81		0.58		23.68	
	2.51		0.30		23.98	
	2.22		0.29		24.27	
2.61		1.88	0.34		24.61	CP
	2.32		0.29		24.90	
	1.92		0.40		25.30	
		1.54	0.38		25.68	
Σ6.00		Σ3.42	Σ2.58	Σ0		

Check: ΣBS – ΣFS = Last RL – First RL = ΣRise – ΣFall
Σ6.00 – Σ3.42 = 2.58, 25.68 – 23.10 = 2.58, 2.58 – 0.00 = 2.58

Hence checked.

27.10 THEODOLITE

The theodolite is the most accurate instrument used mainly for measuring horizontal and vertical angles. It can also be used for locating points on a line, prolonging survey lines, finding the differences in elevations, setting out grades, ranging curves, etc.

Depending upon the facilities provided for the reading of observations, the theodolites may be classified as simple vernier theodolite, micrometer theodolite, optical (glass arc) theodolite and electronic theodolite.

A modern theodolite is compact, light in weight, simple in design and can be used rough. All the movable parts and scales are fully enclosed and virtually dust and moisture proof. Its lower graduated circle defines the size of a theodolite. For example, a 20 cm theodolite means the diameter of the graduated circle of the lower plate is 20 cm. The size of the theodolites varies from 8 to 25 cm.

Theodolites may be classified into transit and non-transit theodolites. A theodolite is said to be a transit one when its telescope can be revolved through 180° in vertical plane about its horizontal axis, thus directing the telescope in exactly opposite directions.

A theodolite is said to be a non-transit one when its telescope cannot be revolved through 180° in a vertical plane about its horizontal axis. Such theodolites are obsolete nowadays. Examples are the Y-theodolite and everest theodolite (Figure 27.15).

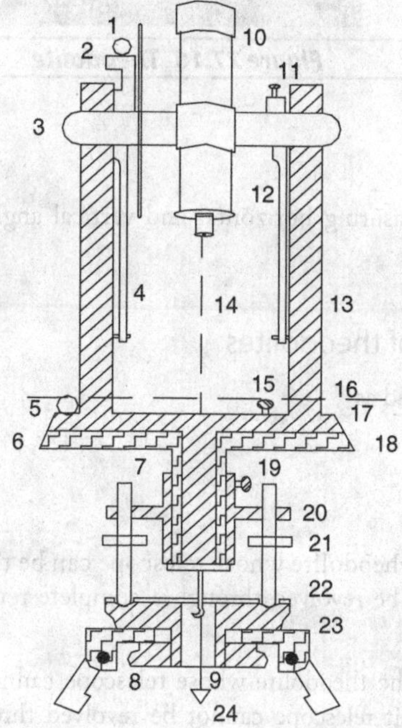

1. Vertical circle
2. Altitude bubble
3. Horizontal axis
4. Vernier arm
5. Plate bubble
6. Graduated arc
7. Spindle
8. Clamping nut
9. Vertical axis
10. Telescope
11. Vertical circle clamping screw
12. Arm of the vertical circle clamp
13. Standard
14. Line of collimation
15. Upper plate clamping screw
16. Axis of plate
17. Upper plate
18. Lower plate bubble
19. Lower plate
20. Tribrach
21. Foot screw clamping screw
22. Trivet
23. Tripod top
24. Plumb bob

Figure 27.15 Theodolite

Figure 27.16 Theodolite

An instrument used for measuring horizontal and vertical angles accurately is known as theodolite (Figure 27.16).

27.10.1 Classification of theodolites

Theodolites are primarily classified as

a. Transit theodolite

b. Non-transit theodolite

a. *Transit theodolite:* The theodolite whose telescope can be transited is called a transit theodolite. A transit telescope can be revolved through a complete revolution about its horizontal axis in a vertical plane.

b. *Non-transit theodolite:* The theodolite whose telescope cannot be transited is called a non-transit theodolite. A non-transit telescope cannot be revolved through a complete revolution about its horizontal axis in vertical plane.

Non-transit theodolites are inferior as compared to transit theodolites. These have become almost outdated nowadays.

The theodolites are classified as follows

a. Vernier theodolite: In this type of theodolites, verniers are provided for reading horizontal and vertical graduated circles.

b. Glass arc theodolite: In this type of theodolites, micrometers are provided for reading horizontal and vertical graduated circles.

27.10.2 Technical terms

a. *Vertical axis:* The axis about which the theodolite may be rotated in a horizontal plane is called vertical axis. Both upper and lower plates may be rotated about the vertical axis.

b. *Horizontal axis:* The axis about which the telescope along with the vertical circle of a theodolite may be rotated in a vertical plane is called horizontal axis. It is also called as transverse axis.

c. *Line of collimation:* The line that passes through the intersection of the cross hairs of the eyepiece and optical centre of the objective and its continuation is called line of collimation. The angle between the line of collimation and the line perpendicular to the horizontal axis is called error of collimation. The line passing through the eyepiece and any point on the objective is called line of sight.

d. *Axis of telescope:* The axis about which the telescope may be rotated is called axis of telescope.

e. *Axis of the level tube:* The straight line that is tangential to the longitudinal curve of the level tube at its centre is called axis of the level tube. When the bubble of the level tube is central, the axis of the level tube becomes horizontal.

f. *Centreing:* The process of setting up a theodolite exactly over the ground station mark is known as centreing. It is achieved when the vertical axis of the theodolite is made to pass through the ground station mark.

g. *Transiting:* The process of turning the telescope in a vertical plane through 180° about its horizontal axis is known as transiting. The process is also sometimes known as reversing or plunging.

h. *Swing:* A continuous motion of the telescope about the vertical axis in the horizontal plane is called swing. The swing may be in either direction, i.e., left or right. When the telescope is rotated in clockwise (right) direction, it is known as right swing. If it is rotated in the anticlockwise (left) direction, it is known as left swing.

i. *Face left observations:* When the vertical circle is on the left of the telescope at the time of observations, the observations of the angles are known as face left observations.

j. *Face right observations:* When the vertical circle is on the right of the telescope at the time of observations, the observations of the angles are known as face right observations.

k. *Changing face:* It is the operation of changing the face of the telescope from the right to left and vice versa.

l. *A measure:* It is the determination of the number of degrees, minutes and seconds or grades contained in an angle.

m. *A set:* A set of horizontal observation of any angle consists of two horizontal measures, one on the face left and the other on the face right.

n. *Telescope normal:* A telescope is said to be normal when its vertical circle is to its left and the bubble of the telescope is up.

o. *Telescope inverted:* A telescope is said to be inverted or reversed when its vertical circle is to its right and the bubble of the telescope is down.

27.10.3 Fundamental lines of a transit

The fundamental lines of a transit are as follows:

a. The vertical axis

b. The axis of plate bubble

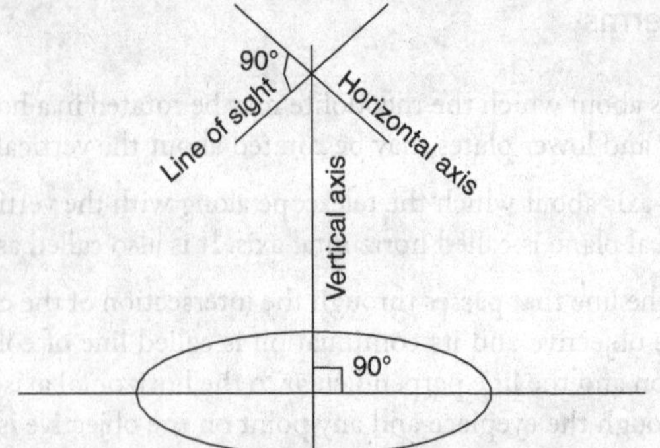

Figure 27.17 The fundamental lines of a transit

c. The line of collimation which is also sometimes called the line of sight
d. The horizontal axis
e. The bubble line of telescope bubble or altitude bubble (Figure 27.17).

27.10.4 Adjustments of a theodolite

The adjustments of a theodolite are of two types

a. Temporary adjustments
b. Permanent adjustments

27.10.4.1 Temporary adjustments

The adjustments which are required to be made at every instrument station before making observations are known as temporary adjustments. The temporary adjustments include mainly the following:

i. Setting up the theodolite over the station
ii. Levelling of the theodolite
iii. Elimination of parallax

Setting up

The operation of setting up of a theodolite includes the centreing of the theodolite over the ground mark and also the approximate levelling with the help of tripod legs.

Centreing The operation by which the vertical axis of the theodolite, represented by a plumb line, is made to pass through the ground station mark is called centreing.

The operation of centreing is carried out in the following steps:

- Suspend the plumb bob with a string attached to the hook fitted to the bottom of the instrument to define the vertical axis.

- Place the theodolite over the station mark by spreading the legs well apart so that the telescope is at a convenient height.
- The centreing may be done by moving the legs radially and circumferentially till the plumb bob hangs within 1 cm of the station mark.

It is necessary to ensure that the tripod is approximately levelled before centreing is done. The approximate levelling may be done by eye judgement.

Levelling of the theodolite

The process of making the vertical axis of the theodolite truly vertical is known as levelling. After having levelled approximately and centred accurately, accurate levelling is done with the help of plate levels (Figure 27.18).

The following steps are involved in levelling with a three-screw head:

- Turn the horizontal plate until the longitudinal axis of the plate level is approximately parallel to a line joining any two levelling screws.
- Bring the bubble to the centre of its run by turning both foot screws simultaneously in opposite directions.
- Turn the instrument through 180° in azimuth.
- Note the position of the bubble. If it occupies a different position, move it by means of the same foot screws to the approximate mean of the two positions.
- Turn the theodolite through 90° in azimuth so that the plate level becomes perpendicular to the previous position.
- With the help of the third foot screw, move the bubble to the approximate mean position already indicated.
- Repeat the process until the bubble retains the same position for every setting of the instrument in azimuth.

Elimination of parallax

An apparent change in the position of the object caused by the change in position of the observer's eye is known as parallax. In a telescope, parallax is caused when the image formed by the objective is not situated in the plane of the cross hairs. Unless the parallax is removed, accurate bisection and sighting of objects become difficult.

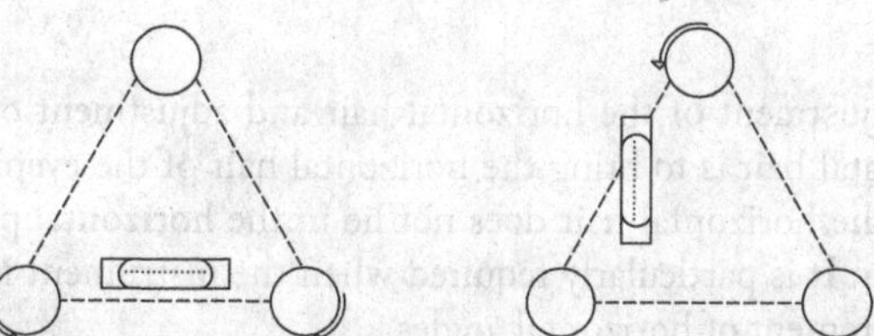

Figure 27.18 Levelling of a theodolite with three-screw head

Elimination of parallax may be done in two ways as discussed below:

- Focusing the eyepiece: To focus the eyepiece for distinct vision of cross hairs, either hold a white paper in front of the objective or sight the telescope towards the sky. Move the eyepiece in or out until the cross hairs are seen sharp and distinct.
- Focusing the objective: After cross hairs have been properly focused, direct the telescope towards a well-defined distant object and intersect it with a vertical wire. Focus the objective till a sharp image is seen. Removal of parallax may be checked by moving the eye slowly to one side. If the object still appears intersected there is no parallax.

If on moving the eye laterally, the image of the object appears to move in the same direction as the eye and the observer's eye and the image of the object are on the opposite sides of the vertical wire, then the image of the object and the eye are brought nearer to eliminate the parallax. This parallax is called far-parallax.

If, on the other hand, the image appears to move in a reverse direction to the movement of the eye and the observer's eye and the image of the object are on the same side of the vertical wire, this parallax is known as near-parallax. It is removed by increasing the distance between the image and the eye.

27.10.4.2 Permanent adjustments

The permanent adjustments include:

i. Adjustment of the horizontal plate level
ii. Adjustment of the horizontal axis
iii. Adjustment of the telescope
iv. Adjustment of the telescope level

Adjustment of the horizontal plate level

With this adjustment, the axis of the plate levels is made perpendicular to the vertical axis of the theodolite. This is necessary since the vertical axis should remain truly vertical for accurate measurement of vertical and horizontal angles. To test this, the theodolite must be set on a firm ground. Clamp the lower plate and turn the upper plate until the plate level becomes parallel to any pair of foot screws. Bring the bubble to the centre of its run by turning the foot screws. Now rotate the instrument about the vertical axis through 180°. If the bubble remains central, the vertical axis of the theodolite is perpendicular to the axis of the plate level.

Adjustment of the horizontal axis

With this, the horizontal axis is made perpendicular to the vertical axis. The object of this adjustment is to ensure that the line of sight revolves in a vertical plane perpendicular to the horizontal axis. This adjustment is very necessary for prolonging straight lines, by making observations on one face only.

Adjustment of the telescope

This adjustment includes adjustment of the horizontal hair and adjustment of the vertical hair. The object of adjustment of the horizontal hair is to bring the horizontal hair of the eyepiece into the horizontal plane through the optical axis. If the horizontal hair does not lie in the horizontal plane through the optical axis, vertical angles will have error. It is particularly required when the instrument is used in levelling operations. It has no effect in the measurement of horizontal angles.

The object of adjustment of the vertical hair is to make the line of collimation perpendicular to the horizontal axis. This adjustment is necessary for measuring horizontal angles between points at different elevations and also for prolonging the lines by making observations on one face only.

Adjustment of the telescope level

The object of this adjustment is that the line of collimation should remain horizontal when the bubble of the level tube, fitted on the telescope, is brought at the centre of its run. This adjustment is essential when the theodolite is used as a level and also when vertical angles are observed.

27.10.5 Measurement of horizontal angles

27.10.5.1 Direct method of measuring the angle

To measure the horizontal angle between BA and BC the following procedure is adopted (Figure 27.19):

i. Set up, centre and level the theodolite over the ground point B.

ii. Loosen the upper plate, set the vernier to read zero and clamp the upper plate.

iii. Loosen the lower plate and swing the telescope until the left point A is sighted. Tighten the lower clamp. Accurate bisection of the arrow held on station A is done by using the lower tangent screw. Read both the verniers and take the mean of the readings.

iv. Unclamp the upper plate and swing the telescope in clockwise direction until the point C is brought in the field of view. Tighten the upper clamp and bisect the arrow on station C accurately using the upper tangent screw.

v. Read both the verniers and take the mean of the readings. The difference of the means of the readings to stations C and A is the required angle ABC.

vi. Change the face of the instrument and repeat the whole procedure. The measure of the angle is again obtained by taking the difference of the means of the readings to C and A on face right.

vii. The means of the two measures of the angle ABC on the two faces is the required angle ABC.

27.10.5.2 Measurement of angle by method of repetition

Let ABC be the required angle between sides BA and BC to be measured. For accurate and precise work, the method of repetition is generally used. In this method, the value of the angle is added several times mechanically and the accurate value of the angular measure is obtained by dividing the accumulated reading by the number of repetitions (Figure 27.20).

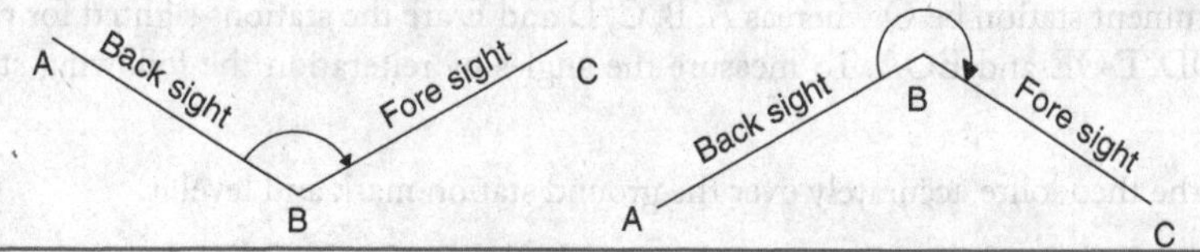

Figure 27.19 Measurement of horizontal angles

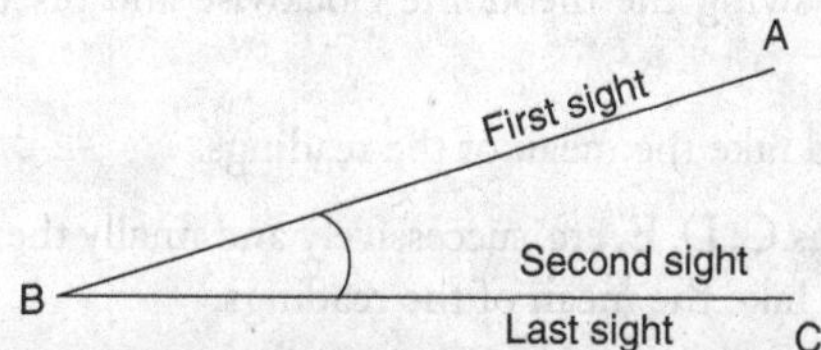

Figure 27.20 Repetition method

To measure a small horizontal angle ABC, the following procedure is adopted:

i. Keeping the face of the instrument left, centre and level it accurately over the ground point B.

ii. Set the vernier to read zero. Loosen the lower plate and swing the telescope in azimuth to sight the left-hand point A. Using the lower tangent screw, bisect the point A accurately.

iii. Read both the verniers and take the mean of the two readings.

iv. Loosen the upper plate and swing the telescope in clockwise direction until point C is brought in the field of view. Using the upper tangent screw bisect the point C accurately.

v. Read both the verniers and take the mean of the readings. The difference of the mean readings for points C and A gives the approximate value of the angle.

vi. Unclamp the lower plate and turn the telescope in clockwise direction until point A is again sighted. Clamp it and bisect it accurately with the lower tangent screw.

vii. Loosen the upper plate and swing the telescope in clockwise direction and again bisect point C accurately using the upper tangent screw. The verniers will now read double the value of the angle ABC.

viii. Repeat the process until the angle ABC is repeated the required number of times, say 5 times.

ix. Read both the verniers. The accumulated reading is obtained by taking the difference of the two mean readings to stations C and A.

x. Divide the accumulated angle by the number of repetitions to get the correct value of the angle ABC on face right.

xi. The mean of the two values on the angle obtained on face left and face right gives the required value of the angle ABC.

27.10.5.3 Measurement of the angle by reiteration method

This method is generally adopted when several angles having a common vertex are to be measured. In this method, angles are measured successively, starting from a reference station and closing on the same station. Making observations on the starting station twice provides a check on the sum of all angles around a station. The sum should invariably be equal to 360°. This method is sometimes known as direction method of observation of the horizontal angles.

Let the instrument station be O whereas A, B, C, D and E are the stations sighted for measuring angles AOB, BOC, COD, DOE and EOA. To measure the angles by reiteration the following steps are involved (Figure 27.21):

i. Centre the theodolite accurately over the ground station mark and level it.

ii. Bisect a well-defined distant station A using the lower clamp and make the vernier to read zero degrees.

iii. Unclamp the upper plate, swing the theodolite clockwise and bisect B accurately using the upper tangent screw.

iv. Read both the verniers and take the mean of the readings.

v. Similarly, bisect the stations C, D, E, etc. successively and finally the starting station A. In each case, read both the verniers and take the mean of the readings.

vi. Calculate the included angles by taking the differences between two consecutive readings.

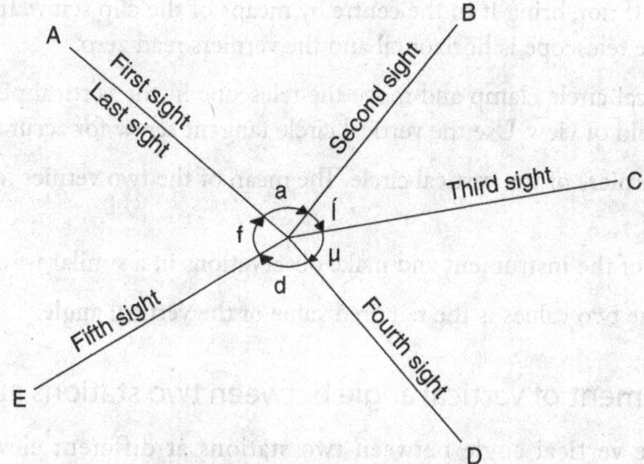

Figure 27.21 Reiteration method

vii. Transit the telescope, swing the instrument in an anticlockwise direction and make observations on the face right to get the measure of each angle.

viii. The mean of the two measures of each angle is the correct value of the angle.

27.10.6 Measurement of vertical angles

A vertical angle is defined as the angle subtended by the inclined line of sight and the horizontal line of sight in the vertical plane. If the point sighted is above the horizontal axis of the theodolite, the vertical angle is known as an angle of elevation, and if it is below it is known as an angle of depression (Figure 27.22).

To measure a vertical angle subtended by the station B at the instrument station A, the following steps are involved:

a. Set up the theodolite over the ground station mark A and level it.

b. Set the zero of the vertical vernier exactly in coincidence with the zero of the vertical scale using the vertical clamp and vertical tangent screw. Check whether the bubble of the altitude level is in the

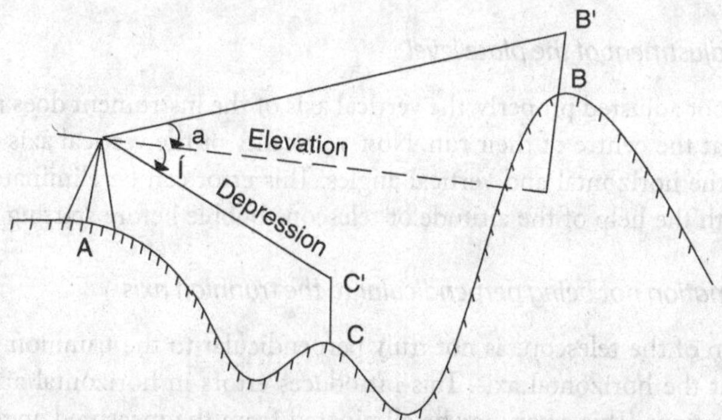

Figure 27.22 Measurement of vertical angles

centre of its run. If not, bring it to the centre by means of the clip screw. In this position the line of collimation of the telescope is horizontal and the verniers read zero.

c. Loosen the vertical circle clamp and move the telescope in the vertical plane until the station B is brought in the field of view. Use the vertical circle tangent screw for accurate bisection.

d. Read both the verniers of the vertical circle. The mean of the two vernier readings gives the value of the vertical angle.

e. Change the face of the instrument and make observations in a similar way.

f. The average of the two values is the required value of the vertical angle.

27.10.7 Measurement of vertical angle between two stations at different elevations

The measurement of the vertical angle between two stations at different elevations may be made as follows:

a. Measure the vertical angle of the higher station as explained earlier. Let it be α.

b. Measure the vertical angle of the lower station. Let it be β.

c. The required vertical angle between the stations may be calculated by finding the algebraic difference between the two readings, assuming the angles of elevation as positive and the angles of depression as negative.

27.10.8 Sources of error in theodolite work

The sources of error in the theodolite work may be broadly divided into three categories:

a. Instrumental errors

b. Personal errors

c. Natural errors

27.10.8.1 Instrumental errors

The theodolites are very delicate and sophisticated surveying instruments.

In spite of the best efforts during manufacturing, perfect adjustment of the fundamental axes of the theodolite may not be possible. Instrumental errors may be further subdivided as discussed below:

Error due to imperfect adjustment of the plate level

If the plate bubbles are not adjusted properly, the vertical axis of the instrument does not remain vertical even if the plate bubbles are at the centre of their run. Non-verticality of the vertical axis introduces errors in the measurements of both the horizontal and vertical angles. This error can be eliminated only by levelling the instrument carefully, with the help of the altitude or telescope bubble before starting the observations.

Error due to line of collimation not being perpendicular to the trunnion axis

If the line of collimation of the telescope is not truly perpendicular to the trunnion axis, it generates a cone when it is rotated about the horizontal axis. This introduces errors in horizontal angles measured between stations at different elevations. This error may be eliminated from the measured angle by taking the average of the two values of the horizontal angles measured on both the faces.

Error due to horizontal axis not being perpendicular to the vertical axis

If the horizontal axis is not perpendicular to the vertical axis, the line of collimation does not revolve in the vertical plane, when the telescope is raised or lowered. Due to this imperfect adjustment, errors are introduced in both the horizontal and vertical angles. The magnitude of the error depends on:

- The angle between the horizontal axis and the vertical axis.
- The vertical angle of the station sighted.
- Elevations of the stations sighted. It is considerable if the stations sighted are at different elevations.

For elimination of the error, observations must be made on both the faces. This is because the average of the two values of the horizontal angle observed on both the faces is equal to the correct value of the angle.

Error due to non-parallelism of the axis of the telescope level and line of collimation

If the axis of the telescope level is not parallel to the line of collimation, an error is introduced in the vertical angle, because the zero line of the vertical verniers does not represent the true line of reference. The error can be eliminated by taking the mean of the two observed values of the angle, one with the telescope normal and the other with the telescope inverted.

Error due to eccentricity of inner and outer vertical axes

If the centre of the graduated circle plate does not coincide with the centre of the vernier plate, the angle recorded by either of the verniers is incorrect. To eliminate the error due to this source, observe both the verniers and take the mean value.

Error due to eccentricity of verniers

If the line joining the zeros of the horizontal plate verniers does not pass through the centre of the vernier plate, an error in the measured horizontal angles is introduced. The error may be eliminated by taking the mean of the two values by reading both the verniers.

27.10.8.2 Personal errors

This includes the following two categories of errors:

i. Errors of manipulation
ii. Errors of sighting and reading

Errors of manipulation

This includes errors as explained below:

- ***Inaccurate centreing:*** If the centre of the theodolite does not coincide with the ground station mark, the horizontal angles measured will be in error, known as centreing error. The magnitude of the error depends upon the distance between the theodolite centre and the ground station mark, the direction and distance of the station sighted, etc.

  It may be noted that the error due to centreing cannot be eliminated unless accurate centreing is done. Also, the error due to defective centreing varies inversely as the length of sights.

- ***Error due to inadequate levelling:*** Inaccurate levelling introduces a serious error in the horizontal angles when the stations sighted are at considerable height differences. This error is similar to the error due to non-adjustment of the plate levels. If the stations sighted are at the same level, the error is small. For elimination of the error, accurate levelling should be done with the help of altitude bubble or telescope bubble which is generally more sensitive.

- ***Error due to manipulation of the wrong tangent screw:*** An inexperienced surveyor generally commits mistakes of using wrong tangent screws. It must be noted that manipulation of the upper tangent screw changes the graduated circle reading whereas manipulation of the lower tangent screw swings the theodolite without changing the readings.

Errors due to sighting and reading

These errors may arise due to the following reasons:

- ***Inaccurate bisection of signals:*** If the signal erected at the station sighted is not clearly visible, due to vegetative cover or intervening ground, the observer may bisect the signal wrongly. This introduces an error whose magnitude varies inversely with the length of sights. It may be eliminated by sighting the signal clearly and always at its lowest portion.
- ***Non-verticality of signals:*** If the signal is not truly vertical, an error is introduced. This error is inversely proportional to the length of sight. This error may be eliminated by erecting the signal truly vertical and also bisecting its lowest portion.
- ***Error due to parallax:*** If the objective and eyepiece are not properly focused before bisecting the station mark, this error is introduced. The error may be eliminated by properly focusing the eyepiece and objective before bisecting the station mark.

27.10.8.3 Natural errors

The errors included in this category are the errors occurring due to higher temperature, strong wind, blazing hot sun and unequal settlement of the tripod.

REVIEW QUESTIONS

1. What do you mean by levelling and what are its purposes?
2. What are the instruments used in levelling?
3. Explain briefly the major parts of a levelling instrument.
4. What is the difference between external focusing and internal focusing telescope?
5. What are the different types of levels?
6. What do you mean by a levelling staff?
7. How are staffs held while taking a reading?
8. What are the fundamental parts of the levelling instrument?
9. Define back sight, fore sight and intermediate sight.
10. What is the difference between bench mark and change point?
11. What are the different types of bench marks used? Explain.
12. Briefly discuss the adjustment of a level

13. Explain the field procedure of setting up of a level.
14. How is parallax eliminated?
15. Differentiate between simple levelling and differential levelling.
16. What is fly levelling?
17. What is contour and what are its uses?
18. What are theodolites and how are they classified?
19. What are temporary adjustments and permanent adjustments of theodolites?
20. What are the personal errors that may arise in measurements when using theodolites.

Problem-1

Compute the level difference from the following level book using rise and fall method

Station	BS	IS	FS	Rise	Fall	RL	Remarks
A	0.557					100.000	BM
B		0.225					
C		1.456					
D	3.000		0.485				CP
E	1.565		1.400				CP
F		2.5000					
G			2.650				

Problem-2

The following are the readings taken with a level and 3 m levelling staff on continuously sloping ground at a common interval of 15 m. 0.605, 1.235, 1.860, 2.575, 0.240, 0.915, 1.935, 2.875, 1.825 and 2.725. The reduced level of the first point is 100.00. Rule a page of a level field book and enter the above readings and calculate the reduced levels of the points.

Part-IV

Other Major Topics in Civil Engineering

chapter 28

Environmental Engineering

28.1 THE CONCEPT OF ENVIRONMENT

The environment comprises everything that creates natural conditions for the existence of organisms including man, and it is a precondition for their further development. Its components are mainly air, water, minerals, soil and living organisms.

The environment is born or created by nature and its growth or expansion is also set by nature itself. Man himself is a factor of the environment and has undergone the same processes of birth and evolution as undergone by the other co-factors, although he is a potent reactor to introduce large-scale changes in the environment. Environment not only includes the surface features and the flora and fauna on it, but also the zone of atmosphere and hydrosphere above and the mantle of the geological structure below. Hence, it has a volume or a zone in a cubical dimension. Environment has a regional and also a national aspect of analysis. The former is linked with natural factors, whereas the latter is associated with social or political factors. Environment includes the earth's surface with all its physical features and natural resources, the distribution of land and water, mountains and plains, minerals, plants and animals, the climates and all cosmic forces that play upon the earth and affect the life of man. Thus environment can be defined as the sum total of all conditions and influences that affect the development and life of organisms.

28.2 ENVIRONMENTAL ENGINEERING

Environmental engineering is the application of science and engineering principles to improve the environment to provide healthy water, air and land for human habitation and for other organisms, and to remediate polluted sites. Environmental engineering involves water and air pollution control, recycling, waste disposal and public health issues as well as knowledge of environmental engineering law. It also includes studies on the environmental impact of proposed construction projects.

Environmental engineering is the planning, design, construction, operation and maintenance of constructed facilities for the protection of human health and safety and the preservation of wildlife and the environment. It includes water supply and resources, modelling of environmental systems, environmental chemistry, wastewater management, solid waste management, hazardous-waste management and remediation, atmospheric systems and air pollution control and environmental and occupational health.

Environmental engineers conduct hazardous-waste management studies to evaluate the significance of any hazards, advise on treatment and containment and develop regulations to prevent mishaps. Environmental engineers also design municipal water supply and industrial wastewater treatment systems and are also concerned with local and worldwide environmental issues such as the effects of acid rain, ozone depletion and water and air pollution from automobile exhausts and industrial sources.

At many universities, environmental engineering programmes are offered by either the Department of Civil Engineering or the Department of Chemical Engineering. Environmental civil engineers focus on hydrology, water resources management and water treatment plant design. Environmental chemical engineers, on the other hand, focus on environmental chemistry, advanced air and water treatment technologies and separation processes.

28.3 DEVELOPMENT OF ENVIRONMENTAL ENGINEERING

Ever since people first recognized that their health and well-being were related to the quality of their environment, they have applied thoughtful principles to attempt to improve the quality of their environment. The ancient Harappan civilization utilized early sewers in some cities. The Romans constructed aqueducts to prevent drought and to create a clean, healthful water supply for the metropolis of Rome.

Modern environmental engineering began in the mid-nineteenth century. The introduction of drinking water treatment and sewage treatment in industrialized countries reduced water-borne diseases from leading causes of death to rarities. As societies grew, actions that were intended to achieve benefits for those societies had long-term impacts that reduced other environmental qualities. One example is the widespread application of dichlorodiphenyltrichloroethane (DDT) to control agricultural pests in the 1940s. While the agricultural benefits were outstanding and crop yields increased dramatically, thus reducing world hunger substantially, and malaria was controlled better than it ever had been, numerous species were brought to the verge of extinction due to the impact of DDT on their reproductive cycles.

28.4 SCOPE OF ENVIRONMENTAL ENGINEERING

Pollutants may be chemical, biological, thermal, radioactive or even mechanical. Environmental engineering emphasizes several areas: process engineering, environmental chemistry, water and sewage treatment (sanitary engineering), waste reduction/management and pollution prevention/clean-up. Environmental engineering is a synthesis of various disciplines, incorporating elements from the following:

1. Agricultural engineering
2. Biology
3. Chemical engineering
4. Chemistry
5. Civil engineering
6. Ecology
7. Geology
8. Mechanical engineering
9. Public health

Environmental engineering is the application of science and engineering principles to the environment. Some consider environmental engineering to include the development of sustainable processes. There are several divisions of the field of environmental engineering.

28.5 ENVIRONMENTAL IMPACT ASSESSMENT AND MITIGATION

In this division, engineers and scientists assess the impacts of a proposed project on environmental conditions. They apply scientific and engineering principles to evaluate if there are likely to be any adverse impacts on water quality, air quality, habitat quality, flora and fauna, and agricultural capacity and to study traffic impacts, social impacts, ecological impacts, noise impacts, visual impacts, etc. If impacts are expected, then they develop mitigation measures to limit or prevent such impacts. An example of a mitigation measure is the creation of wetlands in a nearby location to mitigate the filling in of wetlands necessary for road development if it is not possible to reroute the road.

28.5.1 Water supply and treatment

Engineers and scientists work to secure water supplies for potable and agricultural use. They evaluate the water balance within a watershed and determine the available water supply, the water needed for various needs in that watershed and the seasonal cycles of water movement through the watershed, and develop systems to store, treat and convey water for various uses. Water is treated to achieve water quality objectives for the end uses. In the case of potable water supply, water is treated to minimize the risk of infectious disease transmittal and the risk of non-infectious illness, and to create a palatable water flavour. Water distribution systems are designed and built to provide adequate water pressure and flow rates to meet various end-user needs such as domestic use, fire suppression and irrigation.

28.5.2 Wastewater conveyance and treatment

Most urban and many rural areas no longer discharge human waste directly to the land through outhouses, septic systems and honey bucket systems, but rather deposit such waste into water and convey it from households via sewer systems. Engineers and scientists develop collection and treatment systems to carry this waste material away from where people live and produce the waste and discharge it into the environment. In developed countries, substantial resources are applied to the treatment and detoxification of this waste before it is discharged into a river, lake or ocean system. Developing nations are striving to obtain the resources to develop such systems so that they can improve the water quality of their surface waters and reduce the risk of water-borne infectious diseases.

There are numerous wastewater treatment technologies. A wastewater treatment train can consist of a primary clarifier system to remove solid and floating materials, a secondary treatment system consisting of an aeration basin followed by flocculation and sedimentation basins or an activated sludge system and a secondary clarifier, and a tertiary biological nitrogen removal system for a final disinfection process.The aeration basin/ activated sludge system removes organic material by growing bacteria. The secondary clarifier removes the activated sludge from the water. The tertiary system, although not always included due to costs, is becoming more prevalent to remove nitrogen and phosphorus and to disinfect the water before discharge to a surface water stream or ocean outfall.

28.5.3 Air quality management

Engineers apply scientific and engineering principles to the design of manufacturing and combustion processes to reduce air pollutant emissions to acceptable levels. Scrubbers, electrostatic precipitators, catalytic converters and various other processes are utilized to remove particulate matter, nitrogen oxides, sulphur oxides, volatile organic compounds (VOC), reactive organic gases (ROG) and other air pollutants from flue gases and other sources prior to allowing their emission into the atmosphere. Scientists have developed air pollution dispersion models to evaluate the concentration of a pollutant at a receptor or the impact on overall air quality from vehicle exhausts and industrial flue gas stack emissions. To some extent, this field overlaps the desire to decrease carbon dioxide and other greenhouse gas emissions from combustion processes.

28.6 ENVIRONMENTAL SEGMENTS

The environment consists of four segments – atmosphere, hydrosphere, lithosphere and biosphere.

1. *Atmosphere:* The three layers of air that envelop the earth are known as the atmosphere. It is the protective blanket of gases surrounding the earth, which sustains life on earth and saves it from the hostile environment of outer space. It absorbs harmful radiation from the outer space and a major portion of the electromagnetic radiation from the sun. The atmosphere is essential for all living

organisms. The major components of the atmosphere are nitrogen (78.09 per cent) and oxygen (20.94 per cent), while the minor components are argon, carbon dioxide and some trace gases. Based on temperature, the earth's atmosphere can be divided into four major zones, namely, troposphere, stratosphere, mesosphere and thermosphere.

2. *Hydrosphere:* The aqueous envelope of the earth, including the oceans, all lakes, streams, rivers, subsurface waters, water vapour in the atmosphere and water in the form of ice, constitutes the hydrosphere. About 97 per cent of the earth's water supply is in the oceans, where the high salt content does not permit its use for human consumption. About 2 per cent of the water resources is locked in the polar ice caps and glaciers, while 1 per cent is available as freshwater (surface water such as rivers, lakes and streams, and groundwater) for human consumption and other uses.
3. *Lithosphere:* The soil component of the earth is called lithosphere. This is the outer mantle of the solid earth, consisting of minerals occurring in the earth's crust and the soil.
4. *Biosphere:* This denotes the realm of living organisms and their interactions with the environment, namely the atmosphere, hydrosphere and lithosphere. Both the biosphere and environment are influenced considerably by each other.

28.7 THE NATURAL CYCLES OF THE ENVIRONMENT

The biosphere in its widest sense consists of the earth's crust, the atmosphere and the various species of life that exist in the zone 600 m above and 10,000 m below sea level. The biosphere is very large and complex and, hence, is divided into smaller units called ecosystems. An ecosystem consists of plants, animals and micro-organisms that live in a definite zone along with the physical factors such as soil, water and air.

28.8 THE WATER OR HYDROLOGICAL CYCLE

The water or hydrological cycle is a continuous natural process that helps in the exchange of water between the atmosphere, the land, the sea, living plants and animals. The water cycle, also known as the hydrological cycle, describes the continuous movement of water on, above and below the surface of the earth. Since the water cycle is truly a 'cycle', there is no beginning or end. Water can change states among liquid, vapour and ice at various places in the water cycle.

The sun, which drives the water cycle, heats the water in the oceans. Water evaporates as vapour into the air. Ice and snow can sublimate directly into water vapour. Rising air currents take the vapour up into the atmosphere where cooler temperatures cause it to condense into clouds. Air currents move clouds around the earth; cloud particles collide, grow and fall out of the sky as precipitation. Some precipitation falls as snow and can accumulate as ice caps and glaciers, which can store frozen water for thousands of years. Snow-packs can thaw and melt, and the melted water flows over the land as snowmelt. Most precipitation falls back into the oceans or onto the land, where the precipitation flows over the ground as surface runoff. A portion of runoff enters the rivers in valleys in the landscape, with streamflow moving water towards the oceans. The runoff and groundwater are stored as freshwater in lakes. Not all runoff flows into rivers. Much of it soaks into the ground as infiltration. Some water infiltrates deep into the ground and replenishes aquifers, which store huge amounts of freshwater for long periods of time. Some infiltration stays close to the land surface and can seep back into surface water bodies (and the ocean) as groundwater discharge. Some groundwater finds openings in the land surface and comes out as freshwater springs. Over time, the water returns to the ocean, where the water cycle started.

28.9 DIFFERENT PROCESSES IN WATER CYCLE

Precipitation: Condensed water vapour that falls onto the earth's surface. Most precipitation occurs as rain, but also includes snow, hail, fog, drip and sleet. Approximately 505,000 km^3 of water falls as precipitation each year, 398,000 km^3 of it over the oceans.

Canopy interception: The precipitation that is intercepted by plant foliage and eventually evaporates back to the atmosphere rather than falling to the ground.

Snowmelt: The runoff produced by the melting snow.

Runoff: The variety of ways by which water moves across the land. This includes both surface runoff and channel runoff. As water flows, it may infiltrate into the ground, evaporate into the air, become stored in lakes or reservoirs, or be extracted for agricultural or other human uses.

Infiltration: The flow of water from the ground surface into the ground. Once infiltrated, the water becomes soil moisture or groundwater.

Subsurface flow: The flow of water underground, in the vadose zone and aquifers. Subsurface water may return to the surface (e.g. as a spring or by being pumped) or eventually seep into the oceans. Water returns to the land surface at a lower elevation than where it infiltrated, under the force of gravity or gravity-induced pressures. Groundwater tends to move slowly, and is replenished slowly, so it can remain in aquifers for thousands of years.

Evaporation: The transformation of water from liquid to gas phases as it moves from the ground or bodies of water into the overlying atmosphere. The source of energy for evaporation is primarily solar radiation. Evaporation often implicitly includes transpiration from plants, though together they are specifically referred to as evapotranspiration. Total annual evapotranspiration amounts to approximately 505,000 km^3 of water, 434,000 km^3 of which evaporates from the oceans.

Sublimation: The state change directly from solid water (snow or ice) to water vapour.

Advection: The movement of water – in solid, liquid or vapour states – through the atmosphere. Without advection, water that evaporated over the oceans cannot precipitate over land.

Condensation: The transformation of water vapour to liquid water droplets in the air, producing clouds and fog.

Transpiration: The release of water vapour from plants into the air. Water vapour is a gas that cannot be seen.

28.10 THE OXYGEN CYCLE

The oxygen cycle is the biogeochemical cycle that describes the movement of oxygen within and between its three main reservoirs: the atmosphere (air), the biosphere (living things) and the lithosphere (earth's crust). The main driving factor of the oxygen cycle is photosynthesis, which is responsible for the modern earth's atmosphere and life. The largest reservoir of the earth's oxygen is within the silicate and oxide minerals of the crust and mantle (99.5 per cent). Only a small portion has been released as free oxygen to the biosphere (0.01 per cent) and atmosphere (0.36 per cent). The main source of atmospheric oxygen is photosynthesis, which produces sugar and oxygen from carbon dioxide and water

$$6CO_2 + 6H_2O + \text{energy} \rightarrow C_6H_{12}O_6 + 6O_2$$

Photosynthesizing organisms include the plant life of the land areas as well as the phytoplankton of the oceans. The tiny marine cyanobacterium *Prochlorococcus* was discovered in 1986 and accounts for more than half of the photosynthesis of the open ocean. An additional source of atmospheric oxygen is photolysis, whereby high-energy ultraviolet radiation breaks down atmospheric water and nitrite into component atoms. The free H and N atoms escape into space leaving O_2 in the atmosphere

$$2H_2O + \text{energy} \rightarrow 4H + O_2$$
$$2N_2O + \text{energy} \rightarrow 4N + O_2$$

The main way oxygen is lost from the atmosphere is via respiration and decay, mechanisms in which animal life and bacteria consume oxygen and release carbon dioxide. Lithospheric minerals are oxidized in the presence of oxygen and chemical weathering of exposed rocks also consumes oxygen. An example of surface weathering chemistry is the formation of iron oxides (rust)

$$4FeO + O_2 \rightarrow 2Fe_2O_3$$

Oxygen is also cycled between the biosphere and lithosphere. Marine organisms in the biosphere create calcium carbonate shell material ($CaCO_3$) that is rich in oxygen. When the organism dies, its shell is deposited on the shallow seafloor and buried over time to create the limestone rock of the lithosphere. Weathering processes initiated by organisms can also free oxygen from the lithosphere. Plants and animals extract nutrient minerals from rocks and release oxygen in the process.

28.11 THE NITROGEN CYCLE

The nitrogen cycle is the biogeochemical cycle that describes the transformations of nitrogen and nitrogen-containing compounds in nature. This cycle includes gaseous components. The earth's atmosphere contains approximately 78.08 per cent nitrogen, making it the largest pool of nitrogen. Nitrogen is essential for many biological processes; it is crucial for any life on earth. Nitrogen is in all amino acids, is incorporated into proteins and is present in the bases that make up nucleic acids, such as DNA and RNA. In plants, much of the nitrogen is used in chlorophyll molecules, which are essential for photosynthesis and further growth.

Processing, or fixation, is necessary to convert gaseous nitrogen into forms usable by living organisms. Some fixation occurs in lightning strikes, but most of the fixation is done by free-living or symbiotic bacteria. These bacteria have the nitrogenous enzyme that combines gaseous nitrogen with hydrogen to produce ammonia, which is then further converted by the bacteria to make their own organic compounds. Some nitrogen-fixing bacteria, such as *Rhizobium*, live in the root nodules of legumes (such as peas or beans). Here they form a mutual relationship with the plant, producing ammonia in exchange for carbohydrates. Nutrient-poor soils can be planted with legumes to enrich them with nitrogen. A few other plants can form such symbioses. Nowadays, a very considerable portion of nitrogen is fixated in ammonia chemical plants. Other plants get nitrogen from the soil in the form of either nitrate ions or ammonium ions by absorption through their roots. All nitrogen obtained by animals can be traced back to the eating of plants at some stage of the food chain.

Due to their very high solubility, nitrates can enter groundwater. Elevated levels of nitrates in groundwater are a concern for drinking water use because nitrates can interfere with blood-oxygen levels in infants and cause blue-baby syndrome. Where groundwater recharges streamflow, nitrate-enriched groundwater can contribute to a process leading to high algal populations, especially blue-green algal populations, and the death of aquatic life due to excessive demand for oxygen. Although nitrates are not directly toxic to fish life like ammonia, they can have indirect effects on fish if they contribute to eutrophication. Nitrogen has contributed to severe eutrophication problems in some water bodies and hence its application as fertilizer is being controlled. This is occurring along the same lines as control of phosphorus fertilizer, the restriction of which is normally considered essential for the recovery of eutrophied water bodies.

Ammonia is highly toxic to fish, and the water discharge level of ammonia from wastewater treatment plants must often be closely monitored. To prevent loss of fish, nitrification prior to discharge is often desirable. Land application can be an attractive alternative to the mechanical aeration needed for nitrification.

During anaerobic or low-oxygen conditions, denitrification by bacteria occurs. This results in nitrates being converted to nitrogen gases (NO, N_2O, N_2) and returned to the atmosphere. Nitrate can also be reduced to nitrite and subsequently combine with ammonium in the anammox process, which also results in the production of dinitrogen gas.

28.12 THE PHOSPHATE CYCLE

Phosphorus enters the environment from rocks or deposits laid down on the earth many years ago. Phosphate rock in the commercially available form is called apatite. Other deposits may be from fossilized bone or bird droppings called guano. Weathering and erosion of rocks gradually releases phosphorus as phosphate ions, which are soluble in water. Land plants need phosphate as a fertilizer or nutrient.

Phosphate is incorporated into many molecules essential for life, such as ATP (adenosine triphosphate), which is important in the storage and use of energy. Phosphate is also in the backbone of DNA and RNA, molecules involved in genetic coding.

When plant materials and waste products decay through bacterial action, the phosphate is released and returned to the environment for reuse.

Much of the phosphate eventually is washed into the water from erosion and leaching. Again water plants and algae utilize the phosphate as a nutrient. Studies have shown that phosphate is the limiting agent in the growth of plants and algae. If phosphate is not present in sufficient quantity, the plants exhibit slow growth or will be stunted. If too much phosphate is present, excess growth may occur, particularly in algae.

A large percentage of the phosphate in water is precipitated from the water as iron phosphate, which is insoluble. If the phosphate is in shallow sediments, it may be readily recycled back into the water for further reuse. In deeper sediments in water, it is available for use only as part of a general uplifting of rock formations for the cycle to repeat itself.

28.13 ECOSYSTEM

An ecosystem is a natural unit consisting of all plants, animals and micro-organisms (biotic factors) in an area functioning together with all of the non-living physical (abiotic) factors of the environment. An ecosystem is a completely independent unit of interdependent organisms that share the same habitat. Ecosystems usually form a number of food webs, which show the interdependence of the organisms within the ecosystem. Examples of ecosystems are rain forest, savanna, desert, coral reef and urban ecosystem.

Similar to an ecosystem is a biome, which is a climatically and geographically defined area of ecologically similar climatic conditions such as communities of plants, animals and soil organisms, often referred to as ecosystems. Biomes are defined based on factors such as plant structures (such as trees, shrubs and grasses), leaf types (such as broadleaf and needle-leaf), plant spacing (forest and woodland) and climate. Unlike ecozones, biomes are not defined by genetic, taxonomic or historical similarities. Biomes are often identified with particular patterns of ecological succession and climax vegetation.

28.14 FUNCTION AND BIODIVERSITY

From an anthropological point of view, many people see ecosystems as production units similar to those that produce goods and services. Among some of the most common goods produced by ecosystems is wood by the forest ecosystems and grass for cattle by the natural grasslands. Meat from wild animals, often referred to as bush meat in Africa, has proven to be extremely successful under well-controlled management schemes in South Africa and Kenya. Much less successful has been the discovery and commercialization of substances of wild organisms for pharmaceutical purposes. Services derived from ecosystems are referred to as ecosystem services. They may include (1) facilitating the enjoyment of nature, which may generate many forms of income and employment in the tourism sector, often referred to as ecotourism, (2) water retention, thus facilitating a more evenly distributed release of water and (3) soil protection, open-air laboratory for scientific research, etc.

A greater degree of species or biological diversity – popularly referred to as biodiversity – of an ecosystem may contribute to greater resilience of an ecosystem, because there are more species present at a location to respond to change and thus 'absorb' or reduce its effects. This reduces the effect before the ecosystem's structure is fundamentally changed to a different state. This is not universally the case and there is no proven relationship between the species diversity of an ecosystem and its ability to provide goods and services on a sustainable level. Humid tropical forests produce very few goods and direct services and are extremely vulnerable to change, while many temperate forests readily grow back to their previous state of development within a lifetime after felling or a forest fire. Some grassland has been sustainably exploited for thousands of years.

28.15 THE STUDY OF ECOSYSTEMS

Introduction of new elements, whether biotic or abiotic, into an ecosystem tends to have a disruptive effect. In some cases, this can lead to ecological collapse or trophic cascading and the death of many species within the ecosystem. Under this deterministic vision, the abstract notion of ecological health attempts to measure the robustness and recovery capacity for an ecosystem; i.e. how far the ecosystem is away from its steady state. Ecosystems have the ability to rebound from a disruptive agent. The difference between collapse and a gentle rebound is determined by two factors, namely the toxicity of the introduced element and the resiliency of the original ecosystem.

Ecosystems are primarily governed by stochastic (chance) events, the reactions these events provoke on non-living materials and the responses by organisms to the conditions surrounding them. Thus, an ecosystem results from the sum of individual responses of organisms to stimuli from elements in the environment. The presence or absence of populations merely depends on reproductive and dispersal success, and population levels fluctuate in response to stochastic events. As the number of species in an ecosystem is higher, the number of stimuli is also higher. Since the beginning of life, organisms have survived continuous change through natural selection of successful feeding, reproductive and dispersal behaviour. Through natural selection, the planet's species have continuously adapted to change through variation in their biological composition and distribution. Mathematically it can be demonstrated that greater numbers of different interacting factors tend to dampen fluctuations in each of the individual factors.

28.16 ECOSYSTEM ECOLOGY

Ecosystem ecology is the integrated study of biotic and abiotic components of ecosystems and their interactions within an ecosystem framework. This science examines how ecosystems work and relates this to their components such as chemicals, bedrock, soil, plants and animals. Ecosystem ecology examines physical and biological structure and examines how these ecosystem characteristics interact.

28.17 SYSTEMS ECOLOGY

Systems ecology is an interdisciplinary field of ecology, taking a holistic approach to the study of ecological systems, especially ecosystems. Systems ecology can be seen as an application of general systems theory to ecology. Central to the systems ecology approach is the idea that an ecosystem is a complex system exhibiting emergent properties. Systems ecology focuses on interactions and transactions within and between biological and ecological systems, and is especially concerned with the way the functioning of ecosystems can be influenced by human interventions. It uses and extends concepts from thermodynamics and develops other macroscopic descriptions of complex systems.

REVIEW QUESTIONS

1. Define the concept of environment.
2. What do you mean by environmental engineering? Describe the role of an environmental engineer in our society.
3. Explain the scope of environmental engineering.
4. Briefly discuss the scope of environmental engineering.
5. Briefly discuss the several divisions of environmental engineering.
6. What are the four segments of environment? Explain briefly.
7. What is water cycle and what are the different processes in it?
8. Write short notes on:

   a. Oxygen cycle
   b. Nitrogen cycle
   c. Phosphate cycle
   d. Ecosystem
   e. Biodiversity
   f. Ecosystem ecology

chapter 29

Geotechnical Engineering

29.1 INTRODUCTION TO GEOTECHNICAL ENGINEERING

Soil engineering, soil mechanics or geotechnical engineering is one of the youngest disciplines of civil engineering involving the study of soil, its behaviour and application as an engineering material. Geotechnical engineering is the application of laws of mechanics and hydraulics to engineering problems dealing with sediments and other unconsolidated accumulations of solid particles produced by the mechanical or chemical disintegration of rocks regardless of whether they contain an admixture of organic constituents or not.

Geotechnical engineering can also be defined as a branch of civil engineering concerned with the engineering behaviour of the earth's materials. It includes investigating existing subsurface conditions and materials; determining their physical/mechanical and chemical properties that are relevant to the project considered and assessing the risks posed by site conditions; designing earthworks and structure foundations and monitoring site conditions, earthwork and foundation construction.

A typical geotechnical engineering project begins with a review of project needs to define the required material properties. This is followed by a site investigation of soil, rock, fault distribution and bedrock properties on and below an area of interest to determine their engineering properties, including how they will interact with, on or in a proposed construction. Site investigations are needed to gain an understanding of the area in or on which the engineering will take place. Investigations can include assessment of the risk to humans, property and the environment from natural hazards, such as earthquakes, landslides, sinkholes, soil liquefaction, debris flows and rock falls.

A geotechnical engineer then determines and designs the type of foundations, earthworks and/or pavement subgrades required for the intended man-made structures to be built. Foundations are designed and constructed for structures of various sizes such as high-rise buildings, bridges, medium to large commercial buildings and smaller structures where the soil conditions do not allow code-based design. Foundations built for above-ground structures include shallow and deep foundations. Retaining structures include earth-filled dams and retaining walls. Earthworks include embankments, tunnels, dikes, channels, reservoirs, deposition of hazardous waste and sanitary landfills. Geotechnical engineering is also related to coastal and ocean engineering. Coastal engineering can involve the design and construction of wharves, marinas and jetties. Ocean engineering can involve foundation and anchor systems for offshore structures such as oil platforms. The fields of geotechnical engineering and engineering geology are closely related and have large areas of overlap. However, the field of geotechnical engineering is a specialty of engineering, whereas the field of engineering geology is a specialty of geology.

29.2 THE HISTORY OF GEOTECHNICAL ENGINEERING

Knowledge of the use of soil extends into prehistoric times, when man started constructing dwellings for living and roads for transportation. In the more primitive civilizations, soil was used by man as a construction material for foundations of structures and for the structures themselves. The knowledge of soils for the foundations, bunds and roads was gained by trial and error experiences. Through ancient times and even within the last few generations, practically all improvement was the result of a continuous broadening by empirical knowledge. Humans have historically used soil as a material for flood control, irrigation purposes, burial sites, building foundations and as construction material for buildings. First activities were linked to irrigation and flood control, as demonstrated by traces of dykes, dams and canals dating back to at least 2000 BC that were found in ancient Egypt, ancient Mesopotamia and the Fertile Crescent, as well as around the early settlements of Mohenjo Daro and Harappa in the Indus valley. As the cities expanded, structures were erected, supported by formalized foundations; ancient Greeks notably constructed pad footings and strip-and-raft foundations. Until the eighteenth century, however, no theoretical basis for soil design had been developed and the discipline was more of an art than a science, relying on past experience.

Many structures were built in the medieval period (about AD 400–1400). One of the main problems they had was about the compression of soil and the consequent settlement of buildings. During the past centuries, the compressible soil upon which heavy structures such as cathedrals were built had enough time to consolidate, causing large settlements. The Leaning Tower of Pisa is an example. In India, the Taj Mahal was constructed between 1632 and 1650. It had unique foundation problems because of its proximity to the river Jamuna. Several foundation-related engineering problems, such as for the Leaning Tower of Pisa, prompted scientists to begin taking a more scientific-based approach to examining the subsurface.

Classical geotechnical mechanics began in 1773 with Charles Coulomb's introduction of mechanics to soil problems. Using the laws of friction and cohesion to determine the true sliding surface behind a retaining wall, Coulomb inadvertently defined failure criteria for soil. By combining Coulomb's theory with Christian Otto Mohr's theory of a 2-D stress state, the Mohr–Coulomb theory was developed – a very useful graphical construction still used today. A rudimentary soil classification system was also developed based on a material's unit weight, which is no longer considered a good indication of soil type.

29.3 GEOTECHNICAL ENGINEERING APPLICATIONS

The civil engineer has many diverse important encounters with soil. Apart from the testing and classification of various types of soils in order to determine its physical properties, the knowledge of soil mechanics is particularly helpful in the following problems in civil engineering.

1. *Foundation design:* Foundation is a very important element of all civil engineering structures. All civil engineering structures like buildings, dams, bridges, retaining walls, walls, canals, tunnels or pillars are founded in or on the surface of the earth. Hence, it is necessary to understand the bearing capacity of the soil, the pattern of stress distribution in the soil beneath the loaded area, the probable settlement of the soil, effect of groundwater and vibration, etc.
2. *Pavement design:* Pavement can be either flexible or rigid, and its performance depends upon the subsoil on which it rests. The thickness of a pavement and its component parts depends upon certain characteristics of the subsoil, which should be determined before the design is made. On busy pavements, where the intensity of traffic is very high, the effect of repetition of loading and the consequent fatigue failure has to be taken into account.
3. *Design of earth-retaining structures and underground structures:* The design and construction of underground and earth-retaining structures constitute an important phase of engineering. The examples of underground structures include tunnels, underground buildings, drainage structures and pipelines. A thorough knowledge of geotechnical engineering is essential to design gravity-retaining walls, tunnels, underground buildings, etc. subjected to soil loadings.
4. *Design of embankments and excavations:* When the surface of the soil structure is not horizontal, the component of gravity tends to move the soil downward, and may disturb the stability of the earth structure. The possibility of seeping groundwater reducing the soil strength while excavating must also be taken into account. Sometimes, it is required to drain the subsoil water to increase the soil

strength and to reduce the seepage forces. Deep excavation requires lateral braces and sheet walls to prevent caving in.

5. *Design of earth dams:* The construction of an earth dam requires a very thorough knowledge of geotechnical engineering. As soil is used as the only construction material in an earth dam, which may be either homogeneous or of composite section, its design involves the determination of the physical properties of soil such as the index properties, such as density, plasticity characteristics and specific gravity, particle size distribution and gradation of the soil, permeability, consolidated and compaction characteristics and shear strength parameters under various drainage conditions. The knowledge of theoretical geotechnical engineering, assuming the soil to be an ideal elastic isotropic and homogeneous material, helps in predicting the behaviour of the soil in the field.

Geotechnical engineers are typically graduates of a four-year civil engineering programme and often hold a masters degree.

29.4 SOIL MECHANICS

Soil mechanics is a discipline that applies the principles of engineering mechanics, e.g., kinematics, dynamics, fluid mechanics and mechanics of material, to predict the mechanical behaviour of soils. Together with rock mechanics, it is the basis for solving many engineering problems in civil engineering (geotechnical engineering), geophysical engineering and engineering geology. Some of the basic theories of soil mechanics are related to the basic description and classification of soil, effective stress, shear strength, consolidation, lateral earth pressure, bearing capacity, slope stability and permeability. Foundations, embankments, retaining walls, earthworks and underground openings are all designed in part with theories from soil mechanics.

29.5 THE BASIC CHARACTERISTICS OF SOIL

Soil is usually composed of three phases: solid, liquid and gas. The mechanical properties of soils depend directly on the interactions of these phases with each other and with applied potentials (e.g., stress, hydraulic head, electrical potential and temperature difference).

The solid phase of soils contains various amounts of crystalline clay and non-clay minerals, non-crystalline clay material, organic matter and precipitated salts. These minerals are commonly formed by atoms of elements such as oxygen, silicon, hydrogen and aluminium, organized in various crystalline forms. These elements along with calcium, sodium, potassium, magnesium and carbon comprise over 99 per cent of the solid mass of soils. Although the amount of non-clay material is greater than that of clay and organic material, clay and organic material have a greater influence on the behaviour of soils. Solid particles are classified by size as clay, silt, sand, gravel, cobbles or boulders.

The liquid phase in soils is commonly composed of water containing various types and amounts of dissolved electrolytes. Organic compounds, both soluble and immiscible, are present in soils from chemical spills, leaking wastes and contaminated groundwater. The gas phase, in partially saturated soils, is usually air, although organic gases may be present in zones of high biological activity or in chemically contaminated soils. Soil mineralogy controls the size, shape and physical and chemical properties of soil particles and, thus, its load-carrying ability and compressibility.

The structure of a soil is the combined effects of fabric (particle association, geometrical arrangement of particles, particle groups and pore spaces in a soil), composition and interparticle forces. The structure of soils is also used to account for differences between the properties of natural (structured) and remoulded soils (destructured). The structure of a soil reflects all facets of the soil composition, history, present state and environment. Initial conditions dominate the structure of young deposits at high porosity or freshly compacted soils, whereas older soils at lower porosity reflect the post-depositional changes more.

Soil, like any other engineering material, distorts when placed under a load. This distortion is of two kinds – shearing or sliding distortion and compression. In general, soils cannot withstand tension. In some situations, the particles can be cemented together and a small amount of tension may be withstood, but not for long periods.

Particles of sands and many gravels consist overwhelmingly of silica. They can be rounded due to abrasion while being transported by wind or water, or sharp-cornered, or anything in between, and are roughly equidimensional. Clay particles arise from weathering of rock crystals like feldspar and commonly consist of aluminosilicate minerals. They are generally flake shaped with a large surface area compared with their mass. As their mass is extremely small, their behaviour is governed by forces of electrostatic attraction and repulsion on their surfaces. These forces attract and adsorb water to their surfaces, with the thickness of the layer being affected by dissolved salts in the water.

29.5.1 Sieve analysis

Sieve analysis is the process of determining the size of soil particles by passing the soil sample through a number of different sieves having different openings (hole sizes).

Sieve Number	Length of one side of opening (mm)
4	4.75
10	2.0
20	0.850
40	0.425

29.5.2 Effective stress

The concept of effective stress is one of Karl Terzaghi's most important contributions to soil mechanics. It is a measure of the stress on the soil skeleton (the collection of particles in contact with each other), and determines the ability of soil to resist shear stress. It cannot be measured in itself, but must be calculated from the difference between two parameters that can be measured or estimated with reasonable accuracy. Effective stress (σ') on a plane within a soil mass is the difference between total stress (σ) and pore water pressure (u).

29.5.3 Total stress

The total stress σ is equal to the overburden pressure or stress, which is due to the weight of soil vertically above the plane, together with any forces acting on the soil surface (e.g., the weight of a structure). Total stress increases with increasing depth in proportion to the density of the overlying soil.

29.5.4 Pore water pressure

The pore water pressure u is the pressure of the water on that plane in the soil, and is most commonly calculated as the hydrostatic pressure. For stability calculations in conditions of dynamic flow (under sheet piling, beneath a dam toe or within a slope, for instance), u must be estimated from a flownet. In the situation of a horizontal water table, pore water pressure increases linearly with increasing depth below it.

29.5.5 Shear strength

Most problems in geotechnics, like bearing capacity of shallow and deep foundations, slope stability, retaining wall design, penetration resistance and soil liquefaction, are affected by the soil shear strength. Analytical and numerical analyses use values of shear strength for solving these engineering problems.

Shearing strength in soils is the result of resistance to movement at interparticle contacts, due to particle interlocking, physical bonds formed across the contact areas (resulting from surface atoms sharing electrons at interparticle contacts) and chemical bonds (i.e., cementation particles connected through a solid substance such as recrystallized calcium carbonate).

Different criteria can be used to define the point of failure in a stress–strain curve of a particular material. Failure and yield should not be confused. There is no unique way of defining failure. For some materials, failure can be assumed to be the yield point. For soils, 'failure' is usually considered to be occurring at 15–20 per cent strain. This deformation usually implies that the function of a particular structure, e.g., a building foundation, might be impaired but not have failed. Failure of the soil does not imply failure of the system. In this sense, the shear strength of soils can be defined as the maximum stress applied on any plane in a soil mass at some strain considered as failure.

Different failure criteria are applied to define failure. The Mohr–Coulomb failure criterion is the most common empirical failure criterion used in soil mechanics.

The stress–strain relationship of soils, and therefore the shearing strength, is affected by the following:

a. *Soil composition* (basic soil material): Mineralogy, grain size and grain size distribution, shape of particles, pore fluid type and content, ions on grain and in pore fluid.

b. *State* (initial): Defined by the initial void ratio, effective normal stress and shear stress (stress history). State can be described by terms such as loose, dense, over-consolidated, normally consolidated, stiff, soft, contractive and dilative.

c. *Structure:* Refers to the arrangement of particles within the soil mass, the manner in which the particles are packed or distributed. Features such as layers, joints, fissures, slickensides, voids, pockets and cementation are part of the structure. The structure of soils is described by terms such as undisturbed, disturbed, remoulded, compacted, cemented; flocculent, honey-combed, single-grained; flocculated, deflocculated; stratified, layered, laminated; isotropic and anisotropic.

d. *Loading conditions:* Effective stress path – drained or undrained, type of loading, magnitude, rate (static and dynamic) and time history (monotonic and cyclic).

In reality, a complete shear strength formulation would account for all these factors. Laboratory tests, e.g., direct shear test, triaxial shear test, simple shear test, using different drainage conditions (drained or undrained), rate of loading, range of confining pressures and stress history, are used for determining values of shear strength, unconfined compressive strength, drained shear strength, undrained shear strength, peak strength, critical state shear strength and residual strength.

29.5.6 Consolidation

Consolidation is a process by which soils decrease in volume. It occurs when stress is applied to a soil that causes the soil particles to pack together more tightly, therefore reducing volume. When this occurs in a soil that is saturated with water, water will be squeezed out of the soil. The magnitude of consolidation can be predicted by many different methods. In the classical method, developed by Karl Terzaghi, soils are tested with an oedometer test to determine their compression index. This can be used to predict the amount of consolidation.

When stress is removed from a consolidated soil, the soil will rebound, regaining some of the volume it had lost in the consolidation process. If the stress is reapplied, the soil will consolidate again along a recompression curve, defined by the recompression index. The soil that had its load removed is considered to be over-consolidated. This is the case for soils that previously had glaciers on them. The highest stress that a soil has been subjected to is termed the pre-consolidation stress. A soil that is currently experiencing its highest stress is said to be normally consolidated.

29.5.7 Lateral earth pressure

Lateral earth stress theory is used to estimate the amount of stress the soil can exert perpendicular to gravity. This is the stress exerted on retaining walls. A lateral earth stress coefficient, K, is defined as the ratio of lateral (horizontal) stress to vertical stress for cohesionless soils ($K = \sigma_h/\sigma_v$). There are three coefficients: at-rest, active and passive. At-rest stress is the lateral stress in the ground before any disturbance takes place. The active stress state is reached when a wall moves away from the soil under the influence of lateral stress, and results from shear failure due to reduction of lateral stress. The passive stress state is reached when a wall is pushed into the soil far enough to cause shear failure within the mass due to increased lateral stress. There are many theories for estimating lateral earth stress; some are empirically based and some are analytically derived.

29.5.8 Bearing capacity

The bearing capacity of soil is the average contact stress between a foundation and the soil that will cause shear failure in the soil. Allowable bearing stress is the bearing capacity divided by a factor of safety. Sometimes, on soft soil sites, large settlements may occur under loaded foundations without actual shear failure occurring; in such cases, the allowable bearing stress is determined with regard to the maximum allowable settlement.

Three modes of failure are possible in soil: general shear failure, local shear failure and punching shear failure.

29.5.9 Slope stability

The field of slope stability encompasses the analysis of static and dynamic stability of slopes of earth and rock-fill dams, slopes of other types of embankments, excavated slopes and natural slopes in soil and soft rock. As seen to the right, earthen slopes can develop a cut-spherical weakness zone. The probability of this happening can be calculated in advance using a simple 2-D circular analysis package. A primary difficulty with analysis is locating the most probable slip plane for any given situation. Many landslides have been analyzed only after this fact.

29.5.10 Permeability and seepage

Seepage is the flow of a fluid through soil pores. After measuring or estimating the intrinsic permeability (κ_i), one can calculate the hydraulic conductivity (K) of a soil, and the rate of seepage can be estimated. K has the unit m/s and is the average velocity of water passing through a porous medium under a unit hydraulic gradient. It is the proportionality constant between average velocity and hydraulic gradient in Darcy's law. In most natural and engineering situations, the hydraulic gradient is less than 1, so the value of K for a soil generally represents the maximum likely velocity of seepage. A typical value of hydraulic conductivity for natural sands is around 1×10^{-3} m/s, while K for clays is similar to that of concrete. The quantity of seepage under dams and sheet piling can be estimated using the graphical construction known as a flownet.

When the seepage velocity is great enough, erosion can occur because of the frictional drag exerted on the soil particles. Vertically upwards seepage is a source of danger on the downstream side of sheet piling and beneath the toe of a dam or levee. Erosion of the soil, known as 'piping', can lead to failure of the structure and to sinkhole formation. Seeping water removes soil, starting from the exit point of the seepage, and erosion advances upgradient. The term sand boil is used to describe the appearance of the discharging end of an active soil pipe.

Seepage in an upward direction reduces the effective stress within the soil. In cases where the hydraulic gradient is equal to or greater than the critical gradient (i.e., when the water pressure in the soil is equal to the total vertical stress at a point), effective stress is reduced to zero. When this occurs in a non-cohesive soil, a 'quick' condition is reached and the soil becomes a heavy fluid (i.e., liquefaction has occurred). Quicksand was so named because the soil particles move around and appear to be 'alive'. (Note that it is not possible to be 'sucked down' into quicksand.) In geotechnical engineering, soils are considered a three-phase material composed of rock or mineral particles, water and air. The voids of a soil, the spaces in between mineral particles, contain water and air.

The engineering properties of soils are affected by four main factors: the predominant size of the mineral particles, the type of mineral particles, the grain size distribution and the relative quantities of mineral, water and air present in the soil matrix. Fine particles (fines) are defined as particles less than 0.075 mm in diameter.

29.6 SOIL PROPERTIES

The following properties of soils are used by geotechnical engineers in the analysis of site conditions and design of earthworks, retaining structures and foundations.

Unit Weight

Total unit weight: Cumulative weight of the solid particles, water and air in the material per unit volume. Note that the air phase is often assumed to be weightless.

Dry unit weight: Weight of the solid particles of the soil per unit volume.

Saturated unit weight: Weight of the soil when all voids are filled with water such that no air is present per unit volume. Note that this is typically assumed to occur below the water table.

Porosity

Ratio of the volume of voids (containing air and/or water) in a soil to the total volume of the soil expressed as a percentage. A porosity of 0 per cent implies that there is neither air nor water in the soil.

Void ratio is the ratio of the volume of voids to the volume of solid particles in a soil. Void ratio is mathematically related to the porosity and is more commonly used in geotechnical formulae than porosity.

Permeability

A measure of the ability of water to flow through the soil, expressed in units of velocity.

Consolidation

As a noun, the state of the soil with regard to prior loading conditions; soils can be under-consolidated, normally consolidated or over-consolidated.

As a verb, the process by which water is forced out of a soil matrix due to loading, causing the soil to deform, or decrease in volume, with time.

Shear strength

Amount of shear stress a soil can resist without failing.

Atterberg limits

Liquid limit, plastic limit and shrinkage limit related to the plasticity of a soil. It is used in estimating other engineering properties of a soil and in soil classification.

29.7 GEOTECHNICAL INVESTIGATION

Geotechnical investigations are performed by geotechnical engineers or engineering geologists to obtain information on the physical properties of soil and rocks around a site to design earthworks and foundations for proposed structures and for repair of distress to earthworks and structures caused by subsurface conditions. A geotechnical investigation will include surface exploration and subsurface exploration of a site. Sometimes, geophysical methods are used to obtain data about sites. Subsurface exploration usually involves soil sampling and laboratory tests of the soil samples retrieved.

Surface exploration can include geologic mapping, geophysical methods and photogrammetry, or it can be as simple as a geotechnical professional walking around on the site to observe the physical conditions at the site. To obtain information about the soil conditions below the surface, some form of subsurface exploration is required. Methods of observing the soils below the surface, obtaining samples and determining physical properties of the soils and rocks include test pits, trenching (particularly for locating faults and slide planes), boring and in-situ tests.

29.7.1 Soil sampling

Borings come in two main varieties, large-diameter and small-diameter borings. Large-diameter borings are rarely used due to safety concerns and expense, but are sometimes used to allow a geologist or engineer to visually and manually examine the soil and rock stratigraphy in situ. Small-diameter borings are frequently used to allow a geologist or engineer examine soil or rock cuttings from the drilling operation, to retrieve soil samples at depth and to perform in-place soil tests.

Soil samples are obtained in either 'disturbed' or 'undisturbed' condition; however, 'undisturbed' samples are not truly undisturbed. A disturbed sample is one in which the structure of the soil has been changed sufficiently such that tests of structural properties of the soil will not be representative of in-situ conditions, and only properties of the soil grains can be accurately determined. An undisturbed sample is one where the condition of the soil in the sample is close enough to the conditions of the soil in situ to allow tests of structural properties of the soil to be used to approximate the properties of the soil in situ.

29.7.2 Soil samplers

Soil samples are taken using a variety of samplers; some provide only disturbed samples, while others can provide relatively undisturbed samples.

a. *Shovel:* Samples can be obtained by digging out soil from the site. Samples taken this way are disturbed samples.

b. *Hand/Machine-driven auger:* This sampler typically consists of a short cylinder with a cutting edge attached to a rod and handle. The sampler is advanced by a combination of rotation and downward force. Samples taken this way are disturbed samples.

c. *Continuous flight auger:* A method of sampling using an auger as a corkscrew. The auger is screwed into the ground and then lifted out. Soil is retained on the blades of the auger and kept for testing. Soil sampled this way is considered disturbed.

d. *Split-spoon/SPT sampler:* Utilized in the 'standard test method for standard penetration test (SPT) and split-barrel sampling of soils' (ASTM D 1586), this sampler is typically a 18"–30" long, 2.0" outside diameter (OD) hollow tube split in half lengthwise. A hardened metal drive shoe with a 1.375" opening is attached to the bottom end, and a one-way valve and drill rod adapter at the sampler head. It is driven into the ground with a 140-pound hammer falling 30". The blow counts (hammer strikes) required to advance the sampler a total of 18" are counted and reported. It is generally used for non-cohesive soils, and samples taken this way are considered disturbed.

e. *Modified California sampler:* Similar in concept to the SPT sampler, the sampler barrel has a larger diameter and is usually lined with metal tubes to contain samples. Samples from the modified California sampler can be considered undisturbed if the soil is not excessively soft, does not contain gravel or is not a very dense sand.

f. *Shelby tube sampler:* Utilized in the 'standard practice for thin-walled tube sampling of soils for geotechnical purposes' (ASTM D 1587), this sampler consists of a thin-walled tube with a cutting edge at the toe. A sampler head attaches the tube to the drill rod, and contains a check valve and pressure vents. Generally used in cohesive soils, this sampler is advanced into the soil layer, generally 6" less than the length of the tube. The vacuum created by the check valve and cohesion of the sample in the tube cause the sample to be retained when the tube is withdrawn. Standard ASTM dimensions are 2" OD, 36" long, 18 gauge thickness; 3" OD, 36" long, 16 gauge thickness and 5" OD, 54" long, 11 gauge thickness. It should be noted that ASTM allows other diameters as long as they are proportional to the standardized tube designs, and the tube length is to be suited for field conditions. Soil sampled in this manner is considered undisturbed.

g. *Piston samplers:* These samplers are thin-walled metal tubes that contain a piston at the tip. The samplers are pushed into the bottom of a borehole, with the piston remaining at the surface of the soil while the tube slides past it. These samplers will return undisturbed samples in soft soils, but are difficult to advance in sands and stiff clays, and can be damaged (compromising the sample) if gravel is encountered. The Livingstone corer, developed by D. A. Livingstone, is a commonly used piston sampler. A modification of the Livingstone corer with a serrated coring head allows it to be rotated to cut through subsurface vegetable matter such as small roots or buried twigs.

h. *Pitcher barrel sampler:* This sampler is similar to piston samplers, except that there is no piston. There are pressure-relief holes near the top of the sampler to prevent pressure build up of water or air above the soil sample.

29.7.3 In-situ tests

A standard penetration test (SPT) is an in-situ dynamic penetration test designed to provide information on the properties of soil, while also collecting a disturbed soil sample for grain size analysis and soil classification.

A cone penetration test (CPT) is performed using an instrumented probe with a conical tip, pushed into the soil hydraulically at a constant rate. A basic CPT instrument reports tip resistance and shear resistance along the cylindrical barrel. CPT data have been correlated to soil properties. Sometimes, instruments other than the basic CPT probe are used, including:

a. **CPTu – piezocone penetrometer:** This probe is advanced using the same equipment as a regular CPT probe, but the probe has an additional instrument that measures the groundwater pressure as the probe is advanced.

b. **SCPTu – seismic piezocone penetrometer:** This probe is advanced using the same equipment as a CPT or CPTu probe, but the probe is also equipped with either geophones or accelerometers to detect shear waves and/or pressure waves produced by a source at the surface.

c. **Full flow penetrometers – T-bar, ball and plate:** These probes are used in extremely soft clay soils (such as seafloor deposits) and are advanced in the same manner as the CPT. As their names imply, the T-bar is a cylindrical bar attached at right angles to the drill string forming what look likes a T, the ball is a large sphere and the plate is a flat circular plate. In soft clays, soil flows around the probe similar to a viscous fluid. The pressure due to overburdened stress and pore water pressure is equal on all sides of the probes (unlike with CPTs), so no correction is necessary, reducing the source of error and increasing accuracy. It is especially desired in soft soils due to the very low loads on the measuring sensors. Full flow probes can also be cycled up and down to measure remoulded soil resistance. Ultimately, the geotechnical professional can use the measured penetration resistance to estimate undrained and remoulded shear strengths.

Flat plate dilatometer test (DMT) is a flat plate probe often advanced using CPT rigs, but can also be advanced from conventional drill rigs. A diaphragm on the plate applies a lateral force to the soil materials and measures the strain induced for various levels of applied stress at the desired depth interval.

29.7.4 Laboratory tests

A wide variety of laboratory tests can be performed on soils to measure a wide variety of soil properties. Some soil properties are intrinsic to the composition of the soil matrix and are not affected by sample disturbance, while other properties depend on the structure of the soil as well as its composition, and can be effectively tested only on relatively undisturbed samples. Some soil tests measure direct properties of the soil, while others measure index properties that provide useful information about the soil without directly measuring the property desired.

Atterberg limits: The Atterberg limits define the boundaries of several states of consistency for plastic soils. The boundaries are defined by the amount of water a soil needs to be at one of those boundaries. The boundaries are called the plastic limit and the liquid limit, and the difference between them is called the plasticity index. The shrinkage limit is also a part of the Atterberg limits. The results of this test can be used to help predict other engineering properties.

California bearing ratio: This test is used to determine the aptitude of a soil or aggregate sample as a road subgrade. A plunger is pushed into a compacted sample and its resistance is measured. This test was developed by Caltrans, but it is no longer used in the Caltrans pavement design method. It is still used as a cheap method to estimate the resilient modulus.

Direct shear test: The direct shear test determines the consolidated, drained strength properties of a sample. A constant strain rate is applied to a single shear plane under a normal load and the load response is measured. If this test is performed with different normal loads, the common shear strength parameters can be determined.

Expansion index test: This test uses a remoulded soil sample to determine the expansion index (EI), an empirical value required by building design codes, at a water content of 50 per cent for expansive soils, like expansive clays.

Hydraulic conductivity tests: There are several tests available to determine a soil's hydraulic conductivity. They include the constant head, falling head and constant flow methods. The soil samples tested can be of any type, including remoulded, undisturbed and compacted samples.

Oedometer test: This test is used to determine consolidation and swelling parameters.

Particle size analysis: This is done to determine the soil gradation. Coarser particles are separated in the sieve analysis portion and the finer particles are analysed with a hydrometer. The distinction between coarse and fine particles is usually made at 75 μm. The sieve analysis shakes the sample through progressively smaller meshes to determine its gradation. The hydrometer analysis uses the rate of sedimentation to determine particle gradation.

R-value test (California test 301): This test measures the lateral response of a compacted sample of soil or aggregate to a vertically applied pressure under specific conditions. This test is used by Caltrans for pavement design, replacing the California bearing ratio test.

Soil compaction tests: (Standard Proctor test (ASTM D698), Modified Proctor test (ASTM D1557) and California Test 216.) These tests are used to determine the maximum unit weight and optimal water content a soil can achieve for a given compaction effort.

Triaxial shear tests: This is a type of test that is used to determine the shear strength properties of a soil. It can simulate the confining pressure a soil would see deep in the ground. It can also simulate drained and undrained conditions.

Unconfined compression test: This test compresses a soil sample to measure its strength. The modifier 'unconfined' contrasts this test to the triaxial shear test.

Water content: This test provides the water content of the soil, normally expressed as a percentage of the weight of water to the dry weight of the soil.

29.8 GEOPHYSICAL EXPLORATION

Geophysical methods are used in geotechnical investigations to evaluate a site's behaviour in a seismic event. By measuring a soil's shear wave velocity, the dynamic response of that soil can be estimated. A number of methods are used to determine a site's shear wave velocity:

1. Crosshole method
2. Downhole method (with a seismic CPT or a substitute device)
3. Surface wave reflection or refraction
4. Suspension logging (also known as P-S logging)
5. Spectral analysis of surface waves (SASW)
6. Reflection microtremor (ReMi)

Methods of observing the soils below the surface, obtaining samples and determining physical properties of the soils and rock include test pits, trenching (particularly for locating faults and slide planes), borings and CPT or SPT. CPT allows continuous recording of soil changes with depth, whereas SPT records only major changes at discrete steps of 150 mm (6"); however, SPT allows soil sampling for laboratory testing.

A CPT is typically performed using an instrumented probe with a conical tip, pushed into the soil hydraulically. A basic CPT instrument reports tip resistance and frictional resistance along the friction sleeve, which is located just above the tip. CPT data have been correlated to soil properties. Sometimes, instruments other than the basic CPT probe are used.

Geophysical exploration is also sometimes used; geophysical techniques used for subsurface exploration include measurement of seismic waves (pressure, shear and Rayleigh waves), using surface wave methods and/or downhole methods, and electromagnetic surveys (magnetometer, resistivity and ground-penetrating radar).

29.9 GEOSYNTHETICS

Geosynthetics is the term used to describe a range of synthetic products used to aid in solving some geotechnical problems. The term is generally regarded to encompass four main products: geotextiles, geogrids, geomembranes and geocomposites. The synthetic nature of the products make them suitable for use in the ground where high levels of durability are required–this is not to say that they are indestructible.

Geosynthetics are available in a wide range of forms and materials, each to suit a slightly different end use. These products have a wide range of applications and are currently used in many civil and geotechnical engineering applications including roads, airfields, railroads, embankments, retaining structures, reservoirs, canals, dams, landfills, bank protection and coastal engineering.

REVIEW QUESTIONS

1. What are the roles of geotechnical engineers in our society?
2. How is geotechnical engineering helpful in solving the problems of foundation design and pavement design?
3. What do you mean by soil mechanics?
4. What are the basic characteristics of soil?
5. Define the shear strength of soils?
6. What are the factors that affect the shearing strength of soil?
7. What is the bearing capacity of soil? Explain about consolidation of soil.
8. Define permeability and seepage of soil.
9. What are the properties of soil? Explain.
10. What is geotechnical investigation? Why is it performed?
11. Briefly discuss the laboratory tests of soil.
12. Briefly discuss soil samplers.
13. What is geophysical exploration?
14. Explain briefly about geosynthetics.

chapter 30

Transport, Traffic and Urban Engineering

30.1 INTRODUCTION

Transport engineering or transportation engineering is the science of safe and efficient movement of people and goods. It is a sub-discipline of civil engineering. Transportation contributes to the economic, industrial, social and cultural development of any country. Transportation is vital for the economic development of any region since every commodity produced, whether it is food, clothing, agricultural products, industrial products or medicine, needs transportation at all stages from production to distribution. In the production stage, transportation is required for carrying raw materials like seeds, manure, coal, steel, oil, etc. In the distribution stage, transportation is required from the production centres, namely, farms and factories to the marketing centres and later to the retailers and consumers for distribution.

30.2 THE PLANNING AND DESIGN ASPECTS OF TRANSPORT ENGINEERING

The planning aspects of transport engineering relate to urban planning and involve technical forecasting decisions and political factors. Technical forecasting of passenger travel usually involves an urban transportation planning model, requiring the estimation of trip generation (how many trips for what purpose), trip distribution (destination choice), mode choice (such as what mode is being taken) and route assignment (such as which streets or routes are being used). More sophisticated forecasting can include other aspects of traveller decisions, including auto ownership, trip chaining and the choice of residential or business location. Passenger trips are the focus of transport engineering because they often represent the peak of demand on any transportation system.

The design aspects of transport engineering include the sizing of transportation facilities (how many lanes or how much capacity the facility has), determining the materials and thickness used in pavement and designing the geometry such as vertical and horizontal alignment of the roadway or track.

Operations and management involve traffic engineering, so that vehicles move smoothly on the road or track. Older techniques include signs, signals, markings and tolling. Newer technologies involve intelligent transportation systems, including advanced traveller information systems (such as variable message signs), advanced traffic control systems and vehicle infrastructure integration. Human factors are an aspect of transport engineering, particularly concerning driver–vehicle interface and user interface of road signs, signals and markings. Transportation engineering is related to design and analysis of highways, railways, airports, urban and suburban road networks, parking lots and traffic control signal systems.

30.3 DIFFERENT MODES OF TRANSPORT

The basic modes of transport are by land, water and air. Land has given scope for development of road and rail transport. Water and air have developed waterways and airways, respectively. The roads or highways not only include the modern highway system but also the city streets feeder roads and village roads, catering to a wide range of road vehicles and pedestrians. Railways have been developed both for long distance transportation and for urban travel. Waterways include oceans, rivers, canals and lakes for the movement of ships and boats.

Aircrafts and helicopters use the airways. Apart from these major modes of transportation, other modes include pipelines, elevators, belt conveyors, cable cars, aerial ropeways and monorails.

The four major modes of transportation are:

1. Roadways or highways
2. Railways
3. Waterways
4. Airways

Transport by air is the fastest among the four modes. Air travel also provides more comfort apart from saving in transportation time for the passengers and goods between the airports. Transportation by water is the slowest among the four modes, but it is the most economical mode of transport. Water transport needs minimum energy to haul unit load through unit distance. Transportation by water is possible between the ports on the sea routes or along the rivers or canals where inland transportation facilities are available.

Transportation through the railways could be advantageous between stations both for the passengers and goods, particularly for long distances. Railway tracks serve as arteries for transportation by land and the roads could serve as feeder systems for transportation to the interior parts and to the intermediate localities between the railway stations. The energy requirement to haul unit load through unit distance by the railway is only a fraction (one fourth to one sixth) of that required by road. Transportation by road is the only mode that could give maximum service to one and all. This mode also has the maximum flexibility for travel with reference to route, direction, time, speed of travel, etc. through any mode of road vehicle.

30.4 HIGHWAY ENGINEERING

Highway engineering handles the planning, design, construction and operation of highways, roads and other vehicular facilities as well as their related bicycle and pedestrian realms. It estimates the transportation needs of the public and then secures the funding for the project and analyzes locations of high traffic volumes and high collisions for safety and capacity. Highway engineering uses civil engineering principles to improve the transportation system. A highway is defined as the main road intended for travel by the public between important cities and towns.

30.5 RAIL ENGINEERING

Railway engineers handle the design, construction and operation of railroads and mass transit systems that use a fixed guideway (such as light rail or even monorails). Typical tasks would include determining horizontal and vertical alignment design, station location and design and construction cost estimation. Railroad engineers can also move into the specialized field of train dispatching, which focusses on train movement control.

30.6 PORT AND HARBOUR ENGINEERING

Port and harbour engineers handle the design, construction and operation of ports, harbours, canals and other maritime facilities. This is not to be confused with marine engineering.

30.7 AIRPORT ENGINEERING

Airport engineers design and construct airports. Airport engineers must account for the impacts and demands of aircrafts in their design of airport facilities. One such example is the analysis of predominant wind direction to determine runway orientation.

30.8 TRAFFIC ENGINEERING

Traffic engineering is a branch of civil engineering that uses engineering techniques to achieve the safe and efficient movement of people and goods. It focusses mainly on research and construction of the immobile infrastructure necessary for this movement, such as roads, railway tracks, bridges, traffic signs and traffic lights.

Increasingly, however, instead of building additional infrastructure, dynamic elements are also introduced into road traffic management (they have long been used in rail transport). These use sensors to measure traffic flows and automatic, interconnected guidance systems (e.g. traffic signs that open a lane in different directions depending on the time of the day) to manage traffic, especially in peak hours.

The relationship between lane flow (Q) (vehicles per hour), maximum speed (V) (kilometres per hour) and density (K) (vehicles per kilometre) is $Q = KV$. Observation on limited access facilities suggests that up to a maximum flow, speed does not decline while density increases, but above a critical threshold, increased density reduces speed, and beyond a further threshold, increased density reduces flow as well. Therefore, managing traffic density by limiting the rate that vehicles enter the highway during peak periods can keep both speeds and lane flows at bottlenecks high. Ramp meters, signals on entrance ramps that control the rate at which vehicles are allowed to enter the mainline facility, provide this function (at the expense of increased delay for those waiting at the ramps).

Highway safety engineering is a branch of traffic engineering that deals with reducing the frequency and severity of crashes. It uses physics and vehicle dynamics, as well as road-user psychology and human factors engineering, to reduce the influence of factors that contribute to crashes.

Traffic engineering is closely associated with other disciplines such as:

1. Transport engineering
2. Highway engineering
3. Transportation planning
4. Urban planning
5. Human factors engineering

Highway engineering is the process of design and construction of efficient and safe highways and roads. It became prominent in the twentieth century and has its roots in the discipline of civil engineering. Standards of highway engineering are continuously being improved. Concepts such as grade, surface texture, sight distance and radii of horizontal bends and vertical slopes in relation to design speed and in addition to road junction design (intersections and interchanges) are all important elements of highway engineering. Most developed nations have extensive highway networks. Transportation planning is the field involved with the development of transportation facilities such as streets, highways, sidewalks, bike lanes and public transport lines. Transportation planning historically has followed the rational planning model of defining goals and objectives, identifying problems, generating alternatives, evaluating alternatives and developing the plan.

30.9 MUNICIPAL OR URBAN ENGINEERING

Municipal engineering is concerned with municipal infrastructure. This involves specifying, designing, constructing and maintaining streets, sidewalks, water supply networks, sewers, street lighting, municipal solid waste management and disposal, storage depots for various bulk materials used for maintenance and public works, such as salt and sand, public parks and bicycle paths. In the case of underground utility networks, it may also include the civil portion of the local distribution networks of electrical and telecommunication services. It can also include the optimizing of garbage collection and bus service networks. Some of these disciplines overlap with other civil engineering specialities; however, municipal engineering focusses on the coordination of these infrastructure networks and services, in as much as they are often built simultaneously and managed by the same municipal authority.

Municipal or urban engineering combines elements of environmental engineering, water resources engineering and transport engineering. Municipal engineering may be confused with urban design or urban planning. Whereas the urban planner may design the general layout of streets and public places, the municipal engineer is concerned with the detailed design. For example, in the case of the design of a new street, the urban planner may specify the general layout of the street, including landscaping, surface finishing and urban accessories, but the municipal engineer will prepare the detailed plans and specifications for the roads, sidewalks, municipal services and street lighting.

REVIEW QUESTIONS

1. What do you mean by transportation engineering?
2. What are the planning and design aspects of transportation engineering?
3. What are the different modes of transport? Explain briefly.
4. Briefly discuss traffic engineering.
5. What do you mean by urban engineering?

chapter 31

Irrigation and Water Supply Engineering

31.1 INTRODUCTION

Irrigation is defined as the process of artificial supply of water to soil for raising crops. It is a science of planning and designing an efficient, low-cost, economic irrigation system tailored to fit natural conditions. It is the engineering of controlling and harnessing the various natural sources of water, by constructing dams and reservoirs, canals and headworks, and finally distributing the water to the agricultural fields. Irrigation engineering includes the study and design of works in connection with river control, drainage of waterlogged areas and generation of hydroelectric power. India is basically an agricultural country and all its resources depend on the agricultural output.

The scope of irrigation is not limited to the application of water to the soil. It deals with all aspects and problems extending from watershed management to the agricultural field. It deals with the design and construction of all works, such as dams, weirs, head regulators, etc., in connection with the storage and diversion of water as well as the problems of subsoil drainage, soil reclamation and water–soil–crop relationship. The scope of irrigation can be divided into two heads.

1. Engineering aspect
2. Agricultural aspect

The engineering aspect deals with the following:

1. Storage, diversion or lifting of water
2. Conveyance of water to agricultural fields
3. Application of water to agricultural fields
4. Drainage and relieving waterlogging
5. Development of water power

The agricultural aspect deals with the study of:

1. Proper depths of water necessary in a single application of water for various crops
2. Distribution of water uniformly and periodically
3. Capacities of different soils for irrigation water and the flow of water in soils
4. Reclamation of waste and alkaline lands

31.2 BENEFITS OF IRRIGATION

Apart from increase in the production of food, there are many indirect benefits or advantages of irrigations such as:

1. Protection from famine
2. Cultivation of cash crops

3. Addition to the wealth of the country
4. Increase in prosperity of people
5. Generation of hydroelectric power
6. Domestic and industrial water supply
7. Inland navigation
8. Canal plantations
9. Improvement in the groundwater storage
10. General development of the country
11. Storage of water for various purposes

31.3 TYPES OF IRRIGATION

Irrigation has the following main classes:

Flow Irrigation: Flow irrigation is that type of irrigation in which the supply of irrigation water available is at such a level that it is conveyed onto the land by gravity flow. Flow irrigation may be further divided into perennial irrigation system and inundation or flood irrigation system. In perennial irrigation system, the water required for irrigation is supplied in accordance with the crop requirements throughout the crop period. Inundation irrigation is carried out by deep flooding and through saturation of the land to be cultivated, which is then drained off prior to the planting of the crop.

Lift Irrigation: Lift irrigation is practised when the water supply is at too low a level to run by gravity onto the land. In such systems, water is lifted up by mechanical means.

31.4 DAM AND WEIRS

A dam is a barrier that impounds water or underground streams. Dams generally serve the primary purpose of retaining water, while other structures such as floodgates, levees and dikes are used to manage or prevent water flow into specific land regions.

31.5 HISTORY OF DAMS

Early dam building took place in Mesopotamia and the Middle East. Dams were used to control the water level, for Mesopotamia's weather affected the Tigris and Euphrates rivers, and could be quite unpredictable. The earliest known dam is situated in Jawa, Jordan, 100 km northeast of the capital Amman. The Ancient Egyptian Sadd Al-Kafara at Wadi Al-Garawi, located about 25 km south of Cairo, was 102 m long at its base and 87 m wide. The structure was built around 2800 or 2600 BC as a diversion dam for flood control, but was destroyed by heavy rain during construction or shortly afterwards. The Romans were also great dam builders, with many examples such as the three dams at Subiaco on the river Anio in Italy. Many large dams also survive at Mérida in Spain.

The oldest surviving and standing dam in the world is believed to be the Quatinah barrage in modern-day Syria. The dam is assumed to date back to the reign of the Egyptian pharaoh Sethi (1319–1304 BC), and was enlarged in the Roman period and between 1934 and 1938. It still supplies the city of Homs with water.

The Kallanai is a massive dam of unhewn stone, over 300 m long, 4.5 m high and 20 m wide, across the main stream of the Kaveri river in India. The basic structure dates to the second century AD. The purpose of the dam was to divert the waters of the Kaveri across the fertile delta region for irrigation via canals.

In the Netherlands, a low-lying country, dams were often built to block rivers in order to regulate the water level and to prevent the sea from entering the marshlands. Such dams often marked the beginning of a town or city because it was easy to cross the river at such a place, and often gave rise to the respective place's names in Dutch. For instance, the Dutch capital Amsterdam started with a dam through the river Amstel in the late twelfth century, and Rotterdam started with a dam through the river Rotte, a minor tributary of the Nieuwe Maas. The central square of Amsterdam, believed to be the original place of the 800-year-old dam, still carries the name Dam Square or simply the Dam.

31.6 TYPES OF DAMS

Dams can be formed by human agency, natural causes or even by the intervention of wildlife such as beavers. Man-made dams are typically classified according to their size or height and intended purpose or structure.

31.6.1 By size

International standards define large dams as higher than 15–20 m and major dams as over 150–250 m in height. The tallest dam in the world is the 300 m high Nurek Dam in Tajikistan.

31.6.2 By purpose

The intended purposes include providing water for irrigation to a town or city water supply, improving navigation, creating a reservoir of water to supply for industrial uses, generating hydroelectric power, creating recreation areas or habitat for fish and wildlife, retaining wet season flow to minimize downstream flood risk and containing effluents from industrial sites such as mines or factories. Few dams serve all of these purposes but some multi-purpose dams serve more than one.

A saddle dam is an auxiliary dam constructed to confine the reservoir created by a primary dam either to permit a higher water elevation and storage or to limit the extent of a reservoir for increased efficiency. An auxiliary dam is constructed in a low spot or saddle through which the reservoir would otherwise escape. On occasion, a reservoir is contained by a similar structure called a dike to prevent inundation of nearby land. Dikes are commonly used for reclamation of arable land from a shallow lake. This is similar to a levee, which is a wall or embankment built along a river or stream to protect adjacent land from flooding.

An overflow dam is designed to be over-topped. A weir is a type of small overflow dam that is often used within a river channel to create an impoundment lake for water abstraction purposes and which can also be used for flow measurement.

A check dam is a small dam designed to reduce flow velocity and control soil erosion. Conversely, a wing dam is a structure that only partly restricts a waterway, creating a faster channel that resists the accumulation of sediment.

A dry dam is a dam designed to control flooding. It normally holds back no water and allows the channel to flow freely, except during periods of intense flow that would otherwise cause flooding downstream.

A diversionary dam is a structure designed to divert all or a portion of the flow of a river from its natural course.

31.6.3 By structure

Based on the structure and material used, dams are classified as timber dams, arch-gravity dams, embankment dams or masonry dams with several subtypes.

31.6.3.1 Masonry dams

Masonry dams can be classified into arch dams, gravity dams, embankment dams, rock-fill dams, earth-fill dams, concrete dams, etc.

31.6.3.2 Arch dams

In the arch dams, stability is obtained by a combination of arch and gravity action. If the upstream face is vertical the entire weight of the dam must be carried to the foundation by gravity, while the distribution of the normal hydrostatic pressure between vertical cantilever and arch action will depend upon the stiffness of the dam in a vertical and horizontal direction. When the upstream face is sloped the distribution is more complicated. The normal component of the weight of the arch ring may be taken by the arch action, while the normal hydrostatic pressure will be distributed as described above. For this type of dam, firm reliable supports at the abutments are more important. The most desirable place for an arch dam is a narrow canyon with steep side walls composed of sound rock. The safety of an arch dam is dependent on the strength of the side wall abutments, hence not only should the arch be well seated on the side walls but also the character of the rock should be carefully inspected.

Two types of single-arch dams are in use, namely the constant-angle and the constant-radius dam. The constant-radius type employs the same face radius at all elevations of the dam, which means that as the channel grows narrower towards the bottom of the dam the central angle subtended by the face of the dam becomes smaller. In a constant-angle dam, also known as a variable radius dam, this subtended angle is kept a constant and the variation in distance between the abutments at various levels is taken care of by varying the radii. Constant-radius dams are much less common than constant-angle dams.

A similar type is the double-curvature or thin-shell dam. This method of construction minimizes the amount of concrete necessary for construction but transmits large loads to the foundation and abutments. The appearance is similar to a single-arch dam but with a distinct vertical curvature to it as well lending it the vague appearance of a concave lens when viewed from downstream. The multiple-arch dam consists of a number of single-arch dams with concrete buttresses as the supporting abutments. The multiple-arch dam does not require as many buttresses as the hollow gravity type, but requires good rock foundation because the buttress loads are heavy.

31.6.3.3 Gravity dams

In a gravity dam, stability is secured by making it of such a size and shape that it will resist overturning, sliding and crushing at the toe. The dam will not overturn provided that the moment around the turning point caused by the water pressure is smaller than the moment caused by the weight of the dam. This is the case if the resultant force of water pressure and weight falls within the base of the dam. However, in order to prevent tensile stress at the upstream face and excessive compressive stress at the downstream face, the dam cross section is usually designed so that the resultant falls within the middle at all elevations of the cross section. For this type of dam, impervious foundations with high bearing strength are essential.

When situated on a suitable site, gravity dams can prove to be a better alternative to other types of dams. When built on a carefully studied foundation, the gravity dam probably represents the best-developed example of dam building. Since the fear of flood is a strong motivator in many regions, gravity dams are being built in some instances where an arch dam would have been more economical.

Gravity dams are classified as solid or hollow. The solid form is the more widely used of the two, although the hollow dam is frequently more economical to construct. Gravity dams can also be classified as overflow and non-overflow. A gravity dam can be combined with an arch dam, an arch-gravity dam, for areas with massive amounts of water flow but less material available for a purely gravity dam.

31.6.3.4 Embankment dams

Embankment dams are made from compacted earth and are of two main types, rock-fill and earth-fill dams. Embankment dams rely on their weight to hold back the force of water, like the gravity dams made from concrete.

31.6.3.5 Rock-fill dams

Rock-fill are embankments of compacted free-draining granular earth with an impervious zone. The earth utilized often contains a large percentage of large particles, hence the term rock fill. The impervious zone may be on the upstream face and made of masonry, concrete, plastic membrane, steel sheet piles, timber or other material. The impervious zone may also be within the embankment in which case it is referred to as a core. In the instances where clay is utilized as the impervious material, the dam is referred to as a composite dam. To prevent internal erosion of clay into the rock fill due to seepage forces, the core is separated using a filter. Filters are specifically graded soil designed to prevent the migration of fine grain soil particles. When suitable material is at hand, transportation is minimized leading to cost savings during construction. Rock-fill dams are resistant to damage from earthquakes. However, inadequate quality control during construction can lead to poor compaction and sand in the embankment, which can lead to liquefaction of the rock fill during an earthquake. Liquefaction potential can be reduced by keeping susceptible material from being saturated and by providing adequate compaction during construction.

31.6.3.6 Earth-fill dams

Earth-fill dams are constructed as a simple embankment of well-compacted earth. A homogeneous rolled-earth dam is entirely constructed of one type of material but may contain a drain layer to collect seep water. A zoned-earth dam has distinct parts or zones of dissimilar material, typically a locally plentiful shell with a watertight clay core. Most modern zoned-earth embankments employ filter and drain zones to collect and remove seep water and preserve the integrity of the downstream shell zone. Rolled-earth dams may also employ a watertight facing or core in the manner of a rock-fill dam. An interesting type of temporary earth dam occasionally used in high latitudes is the frozen-core dam, in which a coolant is circulated through pipes inside the dam to maintain a watertight region of permafrost within it. Earthen dams can be constructed from materials found on-site or nearby and, hence, they can be very cost effective.

31.6.3.7 Asphalt-concrete core

Another type of embankment dam is built with asphalt concrete core. Such dams are built with rock and/or gravel as the main fill material. Almost 100 dams of this design have now been built worldwide since the first such dam was completed in 1962. All asphalt-concrete core dams built so far have an excellent performance record. The type of asphalt used is a viscoelastic plastic material that can adjust to the movements and deformations imposed on the embankment as a whole, and to settlements in the foundation. The flexible properties of the asphalt make such dams especially suited in earthquake regions.

31.6.4 Cofferdams

Cofferdam is a temporary barrier constructed to exclude water from an area that is normally submerged. Made commonly of wood, concrete or steel sheet piling, cofferdams are used to allow construction on the foundation of permanent dams, bridges and similar structures. When the project is completed, the cofferdam may be demolished or removed. Common uses for cofferdams include construction and repair of offshore oil platforms. In such cases, the cofferdam is fabricated from sheet steel and welded into place under water. Air is pumped into the space, displacing the water, allowing a dry work environment below the surface. Upon completion, the cofferdam is usually deconstructed unless the area requires continuous maintenance.

31.6.5 Timber dams

Timber dams were widely used in the early part of the industrial revolution and in frontier areas due to ease and speed of construction. Rarely built in modern times by humans due to relatively short lifespan and limited height to which they can be built, timber dams must be kept constantly wet in order to maintain their water retention properties and limit deterioration by rot, similar to a barrel. The locations where timber dams are most economical to build are those where timber is plentiful, cement is costly or difficult to transport and either a low-head diversion dam is required or longevity is not an issue. Timber crib dams were erected of heavy timbers or dressed logs in the manner of a log house and the interior filled with earth or rubble. The heavy crib structure supported the dam's face and the weight of the water. Timber plank dams were more elegant structures that employed a variety of construction methods utilizing heavy timbers to support a water-retaining arrangement of planks. Very few timber dams are still in use.

31.6.6 Steel dams

A steel dam is a type of dam that uses steel plating and load-bearing beams as the structure. Intended as permanent structures, steel dams were an experiment to determine if a construction technique could be devised that was cheaper than masonry, concrete or earthworks, but sturdier than timber crib dams. The spillway can be gradually eroded by water flow, including cavitations or turbulence of the water flowing over the spillway, leading to its failure. Erosion rates are often monitored, and the risk is ordinarily minimized, by shaping the downstream face of the spillway into a curve that minimizes turbulent flow, such as an ogee curve.

31.7 PURPOSES FOR CONSTRUCTION OF DAMS

The common purposes for the construction of dams are as follows:

1. **Power generation:** Hydroelectric power is a major source of electricity in the world. Many countries have rivers with adequate water flow that can be dammed for power generation purposes.
2. **Water supply:** Many urban areas of the world are supplied with water abstracted from rivers pent up behind low dams or weirs. Other major sources include deep upland reservoirs contained by high dams across deep valleys.
3. **Stabilize water flow/irrigation:** Dams are often used to control and stabilize water flow, often for agricultural purposes and irrigation.
4. **Flood prevention:** Dams that are created for flood control.
5. **Land reclamation:** Dams are used to prevent ingress of water to an area that would otherwise be submerged, allowing its reclamation for human use.
6. **Water diversion:** Dams that are constructed for diverting water for various purposes.
7. **Recreation:** Dams built for any of the above purposes may find themselves displaced by the time of their original use. Nevertheless, the local community may have come to enjoy the reservoir for recreational and aesthetic reasons.

31.8 IDENTIFYING A LOCATION FOR THE CONSTRUCTION OF A DAM

One of the best places for building a dam is a narrow part of a deep river valley; the valley sides can then act as natural walls. The primary function of the dam's structure is to fill the gap in the natural reservoir line left by the stream channel. The sites are usually those where the gap becomes a minimum for the required storage capacity. The most economical arrangement is often a composite structure such as a masonry dam flanked by earth embankments. The current use of the land to be flooded should be dispensable. Significant other engineering and engineering geology considerations when building a dam include:

1. Permeability of the surrounding rock or soil
2. Earthquake faults
3. Landslides and slope stability
4. Water table
5. Peak flood flows
6. Reservoir silting
7. Environmental impacts on river fisheries, forests and wildlife (see Section 31.13, Functions of a weir)
8. Impacts on human habitations
9. Compensation for land being flooded as well as population resettlement
10. Removal of toxic materials and buildings from the proposed reservoir area

31.9 IMPACT ASSESSMENT

Impact is assessed in several ways:

1. The benefits to human society arising from dams, such as agriculture, water, floods and hydroelectric power.
2. Harm to nature and wildlife (especially rare species) and impact on the geology of an area.
3. Whether the change to water flow and levels will increase or decrease stability, and the disruption to human lives.

A large dam can cause the loss of entire ecospheres, including endangered and undiscovered species in the area, and the replacement of the original environment by a new inland lake.

31.10 HUMAN SOCIAL IMPACT

The impact on human society is also significant. For example, the Three Gorges Dam on the Yangtze River in China will create a reservoir 600 km long, to be used for hydro-power generation. Its construction required the loss of over a million people's homes and their mass relocation, the loss of many valuable archaeological and cultural sites, as well as significant ecological change. It is estimated that to date 40–80 million people worldwide have been physically displaced from their homes as a result of dam construction.

31.11 ECONOMICS

Construction of a hydroelectric plant requires a long lead time for site studies, hydrological studies and environmental impact assessment, and are large-scale projects by comparison to traditional power generation based upon fossil fuels. The number of sites that can be economically developed for hydroelectric production is limited; new sites tend to be far from population centres and usually require extensive power transmission lines. Hydroelectric generation can be vulnerable to major changes in the climate, including variation of rainfall, ground and surface water levels and glacial melt, causing additional expenditure for the extra capacity to ensure that sufficient power is available in low water years.

31.12 WEIR

A weir, also known as a low-head dam is a small overflow-type dam commonly used to raise the level of a river or stream. Weirs have traditionally been used to create mill ponds in such places. Water flows over the top of a weir, although some weirs have sluice gates, which release water at a level below the top of the weir. The crest of an overflow spillway on a large dam is often called a weir.

31.13 FUNCTIONS OF A WEIR

Weirs are used in conjunction with locks, to render a river navigable and to provide even flow for navigation. In this case, the weir is made significantly longer than the width of the river by forming it in a 'U' shape or running it diagonally, instead of the short perpendicular path. Since the weir is the portion where water overflows, a long weir allows a lot more water with a small increase in overflow depth. This is done in order to minimize fluctuation in the depth of the river upstream with changes in the flow rate of the river. Doing so avoids unnecessary complication in designing and using the lock or irrigation diversion devices.

A weir allows a simple method of measuring the rate of fluid flow in small- to medium-sized streams, or in industrial discharge locations. Since the geometry of the top of the weir is known, and all water flows over the weir, the depth of water behind the weir can be converted to a rate of flow. The calculation relies on the fact that fluid will pass through the critical depth of the flow regime in the vicinity of the crest of the weir. If water is not carried away from the weir, it can make flow measurement complicated or even impossible. A weir may be used to maintain the vertical profile of a stream or channel, and is then commonly referred to as a grade stabilizer.

A weir will typically increase the oxygen content of the water as it passes over the crest, and hence it can have a detrimental effect on the local ecology of a river system. A weir will artificially reduce the upstream water velocity, which can lead to an increase in siltation. The weir may pose a barrier to migrating fish. Fish ladders provide a way for fish to get between the water levels. Mill ponds provide a water mill with the power it requires, using the difference in water level above and below the weir to provide the necessary energy.

A walkway over the weir is likely to be useful for the removal of floating debris trapped by the weir, or for working staunches and sluices on it as the rate of flow changes. This is sometimes used as a convenient pedestrian crossing point for the river. Even though the water around weirs can often appear relatively calm, they are dangerous places to boat, swim or wade; the circulation patterns on the downstream side can submerge a person indefinitely.

31.14 TYPES OF WEIRS

There are different types of weirs. It may be a simple metal plate with a V-notch cut into it, or it may be a concrete and steel structure across the bed of a river. A weir that causes a large change of water level behind it, compared to the error inherent in the depth measurement method, will give an accurate indication of the flow rate.

1. Sharp crested weir
2. Broad crested weir (or broad-crested weir)
3. Crump weir (named after the designer)
4. Needle dam
5. Proportional weir
6. Combination weir
7. MF weir
8. V-notch weir
9. Rectangular weir
10. Cipolletti (trapezoidal) weir
11. Labyrinth weir

REVIEW QUESTIONS

1. Define irrigation and explain the scope of irrigation.
2. What are the benefits of irrigation?
3. Explain the different types of irrigation.
4. What is a dam and how are dams categorized?
5. Write short notes on:
   a. Arch dams
   b. Masonry dams
   c. Gravity dams
6. How are dams categorized by their purpose?
7. Briefly discuss cofferdams.
8. What are the purposes for constructing a dam?
9. What is a weir? What are the different types of weirs?
10. Briefly discuss identifying a location for the construction of a dam.

chapter 32

Computer-Aided Design (CAD)

32.1 INTRODUCTION

Computer-aided design or CAD is a technique of design of objects, which is real or virtual with the help of computer technology. The design of geometric models for object shapes, in particular, is often called computer-aided geometric design (CAGD). In the manual drafting of technical and engineering drawings, the output of CAD often must convey symbolic information such as materials, processes, dimensions and tolerances, according to application-specific conventions. CAD can be used to design curves and figures in two-dimensional (2D) space and as three-dimensional (3D) objects. CAD is an important industrial art extensively used in many applications including automotive, shipbuilding and aerospace industries, industrial and architectural design, and many more. CAD is also widely used to produce computer animation for special effects in movies, advertising and technical manuals. Due to its enormous economic importance, CAD has been a major driving force for research in computational geometry, computer graphics and discrete differential geometry.

Current CAD software packages range from 2D vector-based drafting systems to 3D solid and surface modelling. CAD packages can frequently allow rotations in three dimensions, allowing the view of a designed object from any desired angle, even from the inside looking out. Some CAD softwares are capable of dynamic mathematic modelling. CAD is used in the design of tools and machinery, and in the drafting and design of all types of buildings, e.g. from small residential types houses to the largest commercial and industrial structures.

CAD is mainly used for detailed engineering of 3D models and 2D drawings of physical components, and also used throughout the engineering process from conceptual design and layout of products through the strength and dynamic analysis of assemblies to the definition of manufacturing methods of components. CAD has become an especially important technology within the scope of computer-aided technologies, with benefits such as lower product development costs and a greatly shortened design cycle. CAD enables designers to design a layout and develops work on screen, print it out and save it for future editing, thereby saving time on their drawings. Originally, software for CAD systems was developed with computer languages such as FORTRAN, but with the advancement of object-oriented programming methods this has radically changed. Typical modern parametric feature based modeller and free form surface systems are built around a number of key C-programming language modules with their own Application Program Interfaces (APIs).

32.2 CAD COMPUTERS

Most CAD computers are Windows based PCs. Some CAD systems also run on one of the Unix operating systems and with Linux operating systems. Some CAD systems such as QCad, NX or CATIA V5 provide multiplatform support including Windows, Linux, UNIX and Mac OS X. Generally, no special hardware is required with the possible exception of a good graphics card, depending on the CAD software used. However, for complex product design, machines with high-speed CPUs and large amounts of RAM are recommended. CAD was an application that benefited from the installation of a numeric coprocessor especially in early personal computers. The human-machine interface is generally via a computer mouse but can also be via a pen and digitizing graphics tablet. Some systems also support stereoscopic glasses for viewing the 3D model.

32.3 CAPABILITY OF CAD

Since 1980, the development of readily affordable CAD programs that could be run on personal computers began a trend of massive downsizing in drafting departments in many small- to mid-size companies. As a general rule, one CAD operator could readily replace at least three to five drafters using traditional methods. Additionally, many engineers began to do their own drafting work, further eliminating the need for traditional drafting departments. This trend mirrored the elimination of many office jobs traditionally performed by a secretary. Now a days word processors, spreadsheets, databases, etc. became standard software packages that everyone was expected to learn. In the field of product development, there are often immense costs associated with the testing of a new product. Every new product must undergo at least a small measure of physical testing, not only to ensure that it meets minimum safety standards but also to ensure that it will successfully operate under the range of conditions to which it can expect to be exposed. For instance, wing of an aeroplane must undergo stress tests to ensure that it will retain its integrity even under the most gruelling weather and turbulence conditions before it is approved for use.

32.4 NUMERICALLY CONTROLLED MACHINES

Before the development of CAD, the manufacturing world adopted tools controlled by numbers and letters for manufacturing complex shapes in an accurate and repeatable manner. During 1950s, these numerically controlled (NC) machines used the existing technology of paper tapes with regularly spaced holes punched in them to feed numbers into controller machines that were wired to the motors positioning the work on machine tools. The electro-mechanical nature of the controllers allowed digital technologies to be easily incorporated as they were developed. In late 1960s, NC machining centres were commercially available, incorporating a variety of machining processes and automatic tool changing. Such tools were capable of doing work on multiple surfaces of a work piece, moving the work piece to positions programmed in advance and using a variety of tools -- all automatically. What is more, the same work could be done over and over again with extraordinary precision and a very little additional human input. NC tools immediately raised automation of manufacturing to a new level once feedback loops were incorporated. What finally made NC technology enormously successful was the development of the universal NC programming language called automatically programmed tools (APT).

32.5 AUTOCAD

AutoCAD is a CAD software application for 2D and 3D design and drafting, developed and sold by Autodesk, Inc. Initially released in late 1982, AutoCAD was one of the first CAD programs to run on personal computers, and notably the IBM PC. Most CAD software at the time ran on graphics terminals connected to the mainframe computers or mini-computers.

In earlier releases, AutoCAD used primitive entities, such as lines, poly-lines, circles, arcs, etc., as the foundation for more complex objects. Since the mid-1990s, AutoCAD has supported custom objects through its C++ Application Program Interfaces. Modern AutoCAD includes a full set of basic solid modelling and 3D tools. With the release of AutoCAD 2007, improved 3D modelling functionality came, which meant better navigation when working in 3D. Moreover, it became easier to edit 3D models. Through AutoCAD 2010, Autodesk introduced parametric functionality and mesh modelling. AutoCAD supports a number of APIs for customization and automation. These include AutoLISP, Visual LISP, VBA, .NET and ObjectARX. ObjectARX is a C++ class library, which was also the base for products extending AutoCAD functionality to specific fields, to create products such as AutoCAD Architecture, AutoCAD Electrical, AutoCAD Civil 3D or a third-party AutoCAD-based applications.

AutoCAD's native file format, Drawing (DWG), and, to a lesser extent, its interchange file format, Drawing Interchange Format, or Drawing Exchange Format (DXF), have become *de facto* standards for CAD data interoperability. AutoCAD in recent years has included support for DWF, a format developed and promoted by Autodesk for publishing CAD data. In 2006, Autodesk estimated a number of active DWG files to be in excess of one billion. In past, Autodesk estimated the total number of DWG files in existence to be more than three billion. AutoCAD currently runs exclusively on Microsoft Windows desktop operating systems. Versions for UNIX and Mac OS were released in the 1980s and 1990s, but these were later dropped. AutoCAD can run on an emulator or compatibility layer like VMware Workstation or Wine, albeit subject to various performance issues that can often arise when working with 3D objects or large drawings.

AutoCAD and AutoCAD LT are available for German, French, Italian, Spanish, Japanese, Korean, Chinese Simplified (No LT), Chinese Traditional, Russian, Czech, Polish, Hungarian (No LT), Brazilian Portuguese (No LT), Danish, Dutch, Swedish, Finnish, Norwegian and Vietnamese. The extent of localization varies from full translation of the product to documentation only.

32.6 AUTOCAD LT

AutoCAD LT is a version of AutoCAD with more limited capabilities. It is cheaper when compared to AutoCAD full version. Besides being sold directly by Autodesk, it is available for purchase at computer stores, unlike full AutoCAD which has to be purchased from an official Autodesk dealer. It was developed so that Autodesk could have an entry-level CAD package available to compete in that price class. AutoCAD LT is marketed as a CAD package for those only who need 2D functionality. Compared to the full edition of AutoCAD, AutoCAD LT lacks several features: most notably, it has no 3D modelling capabilities (although it has a full suite of 3D viewing functions for looking at 3D models created in other CAD packages) and blocks the use of any programming interfaces, such as support for most third party programs and does not allow AutoLISP programs. A full listing of differences is on the Autodesk website. AutoCAD LT originated by taking the code base of AutoCAD and commenting out substantial portions, which allowed AutoCAD and AutoCAD LT to be developed simultaneously.

32.7 AUTOCAD STUDENT VERSIONS

The student version of AutoCAD is functionally identical to the full commercial version, with one exception that the DWG files created or edited by a student version have an internal bit-flag set (the educational flag). The student version of AutoCAD is cheap when compared to the full version. When such a DWG file is printed by any version of AutoCAD (commercial or student version), the output will include a plot stamp/banner on all four sides of the print. Objects created in the student version cannot be used for commercial use. These student version objects will infect a commercial version DWG file if imported. The Autodesk student community provides registered students with free access to different Autodesk applications.

32.8 VERTICAL PROGRAMS

Autodesk has also developed a few vertical programs for discipline-specific enhancements. AutoCAD Architecture (formerly architectural desktop), for example, permits architectural designers to draw 3D objects such as walls, doors and windows, with more intelligent data associated with them, rather than simple objects such as lines and circles. The data can be programmed to represent specific architectural products sold in the construction industry, or extracted into a data file for pricing, materials estimation and other values related to the objects represented. Additional tools allow designers to generate standard 2D drawings, such as elevations and sections from a 3D architectural model. Similarly, Civil Design, Civil Design 3D and Civil Design Professional allow data-specific objects to be used, allowing standard civil engineering calculations to be made and represented easily. AutoCAD Electrical, AutoCAD Civil 3D, AutoCAD Map 3D, AutoCAD Mechanical, AutoCAD MEP, AutoCAD P&ID and AutoCAD Structural Detailing are other examples of industry-specific CAD applications built on the AutoCAD platform.

32.9 DIFFERENT VERSIONS OF AUTOCAD

Different versions released from 1982 are shown below.

Name	Version	Date of Release
AutoCAD Version 1.0	1.0	1982, December
AutoCAD Version 1.2	1.2	1983, April
AutoCAD Version 1.3	1.3	1983, August
AutoCAD Version 1.4	1.4	1983, October
AutoCAD Version 2.0	2.0	1984, October
AutoCAD Version 2.1	2.1	1985, May
AutoCAD Version 2.5	2.5	1986, June
AutoCAD Version 2.6	2.6	1987, April
AutoCAD Release 9	N/A	1987, September
AutoCAD Release 10	N/A	1988, October
AutoCAD Release 11	N/A	1990, October
AutoCAD Release 12	N/A	1992, June
AutoCAD Release 13	N/A	1994, November
AutoCAD Release 14	N/A	1997, February
AutoCAD 2000	15.0	1999, March
AutoCAD 2000i	15.1	2000, July
AutoCAD 2002	15.6	2001, June
AutoCAD 2004	16.0	2003, March
AutoCAD 2005	16.1	2004, March
AutoCAD 2006	16.2	2005, March
AutoCAD 2007	17.0	2006, March
AutoCAD 2008	17.1	2007, March
AutoCAD 2009	17.2	2008, March
AutoCAD 2010	18.0	2009, March 24

32.10 DESIGN WEB FORMAT

Design web format (DWF) is a secure file format developed by Autodesk for the efficient distribution and communication of rich design data to anyone who needs to view, review or print design files. Because DWF files are highly compressed, they are smaller and faster to transmit than design files, without the overhead associated with complex CAD drawings (or the management of external links and dependencies). With DWF functionality, publishers of design data can limit the specific design data and plot styles to only what they want recipients to see and can publish multisheet drawing sets from multiple AutoCAD drawings in a single DWF file. They can also publish 3D models from most Autodesk design applications.

DWF files are not a replacement for native CAD formats such as AutoCAD DWG. The sole purpose of DWF is to allow designers, engineers, project managers and their colleagues to communicate design information and design content to anyone needing to view, review or print design information – without these team members needing to know AutoCAD or the other design software. DWF is a file format developed by Autodesk for representing design data in a manner that is independent of the original application software, hardware and operating system used to create that design data. A DWF file can describe design data containing any combination of text, graphics and images in a device independent and resolution independent format. These files can be one sheet or multiple sheets, very simple or extremely complex with a rich use of fonts, graphics, colour and images. The format also includes intelligent metadata that captures the design intent of the data being represented.

32.11 ALTERNATIVES FOR DWF

PDF is an internationally recognized open, secure file format developed by Adobe Systems for the efficient distribution and communication of rich design data to anyone who needs to view, review or print design files.

Scalable vector graphics (SVG) is an open, XML-based file format. It is suitable for use both as a format for creating and editing drawings and as a format viewing and publication. For instance, Inkscape uses SVG as its native format, and both the Firefox and Opera browsers natively display SVG.

32.12 DWG (DRAWING) (FILE EXTENSION IS .DWG)

DWG is a format used for storing 2D and 3D design data and metadata. It is the native format for several CAD packages including AutoCAD, Intellicad (and its variants), Caddie and DWG are supported non-natively by many other CAD applications. DWG (denoted by the .dwg filename extension) was the native file format for the Interact CAD package, developed by Mike Riddle in the late 1970s, and subsequently licensed by Autodesk in 1982 as the basis for AutoCAD. From 1982 to 2007, Autodesk created versions of AutoCAD which wrote not less than 18 major variants of the DWG file format, none of which is publicly documented. The DWG format is probably the most widely used format for CAD drawings. Autodesk estimates that in 1998 there were in excess of two billion DWG files in existence.

There are several claims to control of the DWG format. It is Autodesk who designs, defines and iterates the DWG format as the native format for their CAD applications. Autodesk sells a read/write library, called RealDWG, under selective licensing terms for use in non-competitive applications. In the year 1998, Autodesk added a file verification to AutoCAD R14.01, through a function called DWGCHECK. This function was supported by an encrypted checksum and a product code (called a "watermark" by Autodesk), written into DWG files created by the program. In 2006, in response to Autodesk users experiencing bugs and incompatibilities in files written by reverse-engineered DWG read/write libraries, Autodesk modified AutoCAD 2007 to include "TrustedDWG technology", a function which would embed a text string within DWG files written by the program: Autodesk DWG. This file is a trusted DWG last saved by an Autodesk application or Autodesk licensed application. This helped Autodesk software users to ensure that the files they were opening were created by an Autodesk, or a RealDWG application, reducing risk of incompatibilities. AutoCAD would pop up a message, warning of potential stability problems, if a user opened a 2007 version DWG file which did not include this text string.

In 2008, the Free Software Foundation asserted the need for an open replacement for the DWG format by placing 'Replacement for OpenDWG libraries' in the 9th place on their high-priority free software projects list. In 2008, Autodesk and Bentley (the company which made the famous "Microstation" software) agreed on exchange of software libraries, including Autodesk RealDWG, to improve the ability to read and write the companies' respective DWG and DGN formats in mixed environments with greater fidelity.

In addition, the two companies will facilitate work process interoperability between their AEC applications through supporting the reciprocal use of available APIs.

32.13 AUTOCAD DXF (FILE EXTENSION IS .DXF)

AutoCAD DXF is a CAD data file format developed by Autodesk for enabling data interoperability between AutoCAD and other programs. DXF was originally introduced in December 1982 as part of AutoCAD 1.0, and intended to provide an exact representation of the data in the AutoCAD native file format, DWG, for which Autodesk for many years did not publish specifications. Because of this, correct imports of DXF files have been difficult. Versions of AutoCAD Version Release 10 and above support both ASCII and binary forms of DXF. Earlier versions support only ASCII. As AutoCAD has become more powerful, supporting more complex object types, DXF has become less useful. Many CAD applications use the DWG format which can be licensed from Autodesk or non-natively from the Open Design Alliance.

32.14 FILE STRUCTURE OF DXF

ASCII versions of DXF can be read with a text-editor. The basic organization of a DXF file is as follows.

1. **HEADER** section – This section contains the general information about the drawing. Each parameter has a variable name and an associated value.
2. **CLASSES** section – This section holds the information for application-defined classes whose instances appear in the BLOCKS, ENTITIES and OBJECTS sections of the database. Generally does not provide sufficient information to allow interoperability with other programs.
3. **TABLES** section – This section contains definitions of named items.
   a. Application ID (APPID) table
   b. Block Record (BLOCK_RECORD) table
   c. Dimension Style (DIMSTYPE) table
   d. Layer (LAYER) table
   e. Linetype (LTYPE) table
   f. Text style (STYLE) table
   g. User Coordinate System (UCS) table
   h. View (VIEW) table
   i. Viewport configuration (VPORT) table
4. **BLOCKS** section – This section contains Block Definition entities describing the entities comprising each block in the drawing.
5. **ENTITIES** section – This section contains the drawing entities including any block references.
6. **OBJECTS** section – This section contains the data that apply to non-graphical objects, used by AutoLISP and ObjectARX applications.
7. **THUMBNAILIMAGE** section – This section contains the preview image for the DXF file.
8. **END OF FILE.**

The data format of a DXF is called a "tagged data" format which means that each data element in the file is preceded by an integer number that is called a group code. A group code's value indicates what type of data element follows. This value also indicates the meaning of a data element for a given object (or record) type. Virtually all user-specified information in a drawing file can be represented in DXF format.

32.15 SOFTWARE WHICH SUPPORTS DXF

The following software supports .dxf format:

A9Cad, Adobe Illustrator, AGI32, AI4CAD 3D, Alibre Design, Altium, ArchiCAD, ArcMap, AutoCAD, Blender – Using an import script , BRL-CAD, Cadwork, Corel Draw, DWGeditor, Drawbase, Easy-PC, Epanet, Eye-Sys, EnRoute, FASTechnologies NC-CAM, Geosoft Oasis montaj, Google SketchUp, GraphCalc – Export only, Inkscape - export only as of version 0.46, IntelliCAD. Harness Expert, Hevacomp, Kabeja, Leica Geo Office, Lenel OnGuard, LD Assistant, Lightcalc, Manifold System, Maple 12, Mathematica, MetaCAM, Microlux, Microsoft Visio, Microstation, miniPLAN, Modo (software), OmniWin Cadnest, Paint Shop Pro, PETRA, Photopia, Pro/Engineer, pdf2cad, Qcad, RackTool, Rhinoceros 3D, Solid Edge, Solidworks, VariCAD, VectorWorks, ViaCAD and Visual

32.16 THE INITIAL GRAPHICS EXCHANGE SPECIFICATION (IGES)

The Initial Graphics Exchange Specification (IGES) (pronounced *eye-jess*) defines a neutral data format that allows the digital exchange of information among CAD systems. The official title of IGES is Digital Representation for Communication of Product Definition Data, first published in January, 1980 by the National Bureau of Standards as NBSIR 80-1978. Using IGES, a CAD user can exchange product data models in the form of circuit diagrams, wireframe, freeform surface or solid modelling representations. Applications supported by IGES include traditional engineering drawings, models for analysis and other manufacturing functions.

The IGES project was started in 1979 by a group of CAD users and vendors, including Boeing, General Electric, Xerox, Computervision and Applicon, with the support of the National Bureau of Standards (now known as NIST) and the US Department of Defense (DoD). The name was carefully chosen to avoid any suggestion of a database standard that would compete with the proprietary databases then used by the different CAD vendors.

An ANSI standard since 1980, IGES has generated warehouses full of magnetic tapes and CD-ROMs of digital PMI for the automotive, aerospace and shipbuilding industries, as well as for weapons systems from missile guidance systems to entire aircraft carriers. These part models may have to be used years after the vendor of the original design system has gone out of business. IGES files provide a way to access these data decades from now. Today, plugin viewers for Web browsers allow IGES files created 20 years ago to be viewed from anywhere in the world.

An IGES file is composed of 80-character ASCII records, a record length derived from the punch card era. Text strings are represented in "Hollerith" format, the number of characters in the string, followed by the letter "H", followed by the text string, *e.g.*, "4HSLOT" (this is the text string format used in early versions of the FORTRAN language). Early IGES translators had problems with IBM mainframe computers because the mainframes used EBCDIC encoding for text, and some EBCDIC-ASCII translators would either substitute the wrong character or improperly set the Parity bit, causing a misreading.

32.17 COMPUTER-AIDED ENGINEERING (CAE)

Computer-aided engineering (often referred to as CAE) refers to the use of information technology to support engineers in tasks such as analysis, simulation, design, manufacture, planning, diagnosis and repair. Software tools that have been developed to support these activities are considered CAE tools. CAE tools are being used, for example, to analyze the robustness and performance of components and assemblies. The term encompasses simulation, validation and optimization of products and manufacturing tools. In future, CAE systems will be major providers of information to help support design teams in decision making. In regard to information networks, CAE systems are individually considered as a single node on a total information network and each node may interact with other nodes on the network. CAE systems can provide support to businesses. This is achieved by the use of reference architectures and their ability to place information views on the business process. Reference architecture is the basis from which information models such as product models and manufacturing models are developed.

CAE tools are very widely used in the automotive industry. In fact, their use has enabled the automakers to reduce the product development cost and time while improving the safety, comfort and durability of the vehicles they produce. The predictive capability of CAE tools has progressed to the point where much of the design verification is now done using computer simulations rather than physical prototype testing. Even though there have been many advances in CAE and it is widely used in the engineering field. Physical testing is still used as a final confirmation for subsystems due to the fact that CAE cannot predict all variables in complex assemblies (i.e. metal stretch and thinning).

Softwares such as Altair RADIOSS, LSTC's LS-DYNA, Cranes CAE suite and ESI's PAM-CRASH are used for automotive crashworthiness and occupant safety. Tools like eta/VPG, NISA, Altair HyperWorks, BETA CAE Systems', MSC's Patran, MSC's ADAMS, LMS's Virtual.Lab, SIMPACK, NEi Nastran and UGS's Scenario and Nastran packages are used in a variety of structural and dynamic analysis tasks. Other tools like LMS's AMESim are used to analyze functional performance of multi-disciplinary systems.

32.18 COMPUTER-AIDED MANUFACTURING (CAM)

Computer-aided manufacturing (CAM) is the use of computer-based software tools that assist engineers and machinists in manufacturing or prototyping product components. CAM is a programming tool that makes it possible to manufacture physical models using computer-aided design (CAD) programs. CAM creates real life versions of components designed within a software package. CAM was first used in 1971 for car body design and tooling. Traditionally, CAM was considered to be a numerical control (NC) programming tool, wherein three-dimensional (3D) models of components generated in CAD software are used to generate CNC code to drive NC machine tools. Although this remains the most common CAM function, CAM functions have expanded to integrate CAM more fully with CAD/CAM/CAE PLM solutions.

As with other 'Computer-Aided' technologies, CAM does not eliminate the need for skilled professionals such as manufacturing engineers and NC programmers. CAM, in fact, both leverages the value of the most skilled manufacturing professionals through advanced productivity tools, while building the skills of new professionals through visualization, simulation and optimization tools.

The first commercial applications of CAD were in large companies in the automotive and aerospace industries, for example, UNISURF in 1971 at Renault for car body design and tooling. Integration of CAD with other components of CAD/CAM/CAE PLM environment requires an effective CAD data exchange. Usually, it had been necessary to force the CAD operator to export the data in one of the common data formats, such as IGES or STL, that are supported by a wide variety of software. The output from the CAM software is usually a simple text file of G-code, sometimes many thousands of commands long, that is then transferred to a machine tool using a direct numerical control (DNC) program.

CAM packages could not, and still cannot, reason as a machinist can. They could not optimize toolpaths to the extent required for mass production. Users would select the type of tool, machining process and paths to be used. While an engineer may have a working knowledge of G-code programming, small optimization and wear issues compound over time. Mass-produced items that require machining are often initially created through casting or some other non-machine method. This enables hand-written, short and highly optimized

G-code that could not be produced in a CAM package. Over time, the historical shortcomings of CAM are being attenuated, both by providers of niche solutions and by providers of high-end solutions. This is occurring primarily in three arenas:

1. Ease of use
2. Manufacturing complexity
3. Integration with PLM and the extended enterprise

For the user who is just getting started as a CAM user, out-of-the-box capabilities, providing process wizards, templates, libraries, machine tool kits, automated feature based machining and job function specific tailorable user interfaces, build user confidence and speed up the learning curve. User confidence is further built on 3D visualization through a closer integration with the 3D CAD environment, including error-avoiding simulations and optimizations.

The manufacturing environment is increasingly complex. The need for CAM and PLM tools of the manufacturing engineer, NC programmer or machinist is similar to the need for computer assistance of the pilot of modern aircraft systems. The modern machinery cannot be properly used without this assistance. Today's CAM systems support the full range of machine tools, including turning, five axis machining and wire EDM. Today's CAM user can easily generate streamlined tool paths, optimized tool axis tilt for higher feed rates and optimized Z axis depth cuts as well as driving non-cutting operations such as the specification of probing motions.

The largest CAM software companies (by revenue by the year 2005) are UGS Corp (now owned by Siemens and called Siemens PLM Software, Inc) and Dassault Systèmes, both with over 10 per cent of the market; CAMWorks, PTC, Hitachi Zosen and Delcam have over 5 per cent each; while Planit-Edgecam, Tebis, TopSolid, CATIA, CNC (Mastercam), SolidCAM, DP Technology's ESPRIT, OneCNC, and Sescoi between 2.5 per cent and 5 per cent each. The remaining 35 per cent is accounted for by other niche suppliers like T-Flex, Dolphin CAD/CAM, MecSoft Corporation, SurfCAM, BobCAD, Metamation, GibbsCAM and SUM3D.

32.19 ELECTRONIC DESIGN AUTOMATION

Electronic design automation (EDA) is the category of tools for designing and producing electronic systems ranging from printed circuit boards (PCBs) to integrated circuits. This is sometimes referred to as ECAD (electronic computer-aided design) or just CAD.

The term EDA is also used as an umbrella term for computer-aided engineering, computer-aided design and computer-aided manufacturing of electronics in the discipline of electrical engineering. The segments of the industry that must use EDA are chip designers at semiconductor companies. Large chips are too complex to be designed by hand. EDA for electronics has rapidly increased in importance with the continuous scaling of semiconductor technology. EDA tools are also used for programming design functionality into FPGAs.

Before EDA, integrated circuits were designed by hand and manually laid out. Some advanced shops used geometric software to generate the tapes for the Gerber photoplotter, but even those copied digital recordings of mechanically-drawn components. The process was fundamentally graphic, with the translation from electronics to graphics done manually. By the mid-70s, developers had started to automate the design and not just the drafting. The first placement and routing (place and route) tools were developed. The proceedings of the Design Automation Conference cover much of this era.

The next era began more or less with the publication of 'Introduction to VLSI Systems' by Carver Mead and Lynn Conway in 1980. This groundbreaking text advocated chip design with programming languages that compiled to silicon. The immediate result was a hundredfold increase in the complexity of the chips that could be designed, with improved access to design verification tools that used logic simulation. Often the chips were not just easier to lay out but more correct as well, because their designs could be simulated more thoroughly before construction. The earliest EDA tools were produced academically, and were in the public domain. One of the most famous was the 'Berkeley VLSI Tools Tarball', a set of UNIX utilities used to design early VLSI systems. Another crucial development was the formation of MOSIS, a consortium of universities and fabricators that developed an inexpensive way to train student chip designers by producing real integrated circuits. The basic idea was to use reliable, low-cost, relatively low-technology IC processes and pack a large number of projects per wafer, with just a few copies of each projects' chips. Cooperating fabricators either donated the processed wafers, or sold them at cost, finding the program helpful for their own long-term growth.

1981 marked the beginning of EDA as an industry. For many years, the larger electronic companies, such as Hewlett Packard, Tektronix and Intel, had pursued EDA internally. In 1981, managers and developers spun out of these companies to concentrate on EDA as a business. Daisy Systems, Mentor Graphics and Valid Logic Systems were all founded around this time, and collectively referred to as DMV. Within a few years there were many companies specializing in EDA, each with a slightly different emphasis.

In 1986, Verilog, a popular high-level design language, was first introduced as a hardware description language by Gateway. In 1987, the US Department of Defense funded the creation of VHDL as a specification language. Simulators quickly followed these introductions permitting direct simulation of chip designs—executable specifications. In a few more years, back-ends were developed to perform logic synthesis. Many of the EDA companies acquire small companies with software or other technology that can be adapted to their core business. Most of the market leaders are rather incestuous amalgamations of many smaller companies. This trend is helped by the tendency of software companies to design tools as accessories that fit naturally into a larger vendor's suite of programs. (On digital circuitry, many new tools incorporate analog design and mixed systems. This is happening because there is now a trend to place entire electronic systems on a single chip.)

Current digital flows are extremely modular. The front ends produce standardized design descriptions that compile into invocations of 'cells', without regard to the cell technology. Cells implement logic or other electronic functions, using a particular integrated circuit technology. Fabricators generally provide libraries of components for their production processes, with simulation models that fit standard simulation tools. Analog EDA tools are much less modular, since many more functions are required, they interact more strongly, and the components are less ideal.

32.20 MULTIDISCIPLINARY DESIGN OPTIMIZATION

Multidisciplinary design optimization (MDO) is a field of engineering that uses optimization methods to solve design problems, incorporating a number of disciplines. It is also known as multidisciplinary optimization and multidisciplinary system design optimization (MSDO). MDO allows designers to incorporate all relevant disciplines simultaneously. These techniques have been used in a number of fields, including automobile design, naval architecture, electronics, computers and electricity distribution. However, the largest number of applications has been in the field of aerospace engineering, such as aircraft and spacecraft design. For example, the Boeing blended wing body (BWB) aircraft concept has used MDO extensively in the conceptual and preliminary design stages. The disciplines considered in the BWB design are aerodynamics, structural analysis, propulsion, control theory and economics.

32.21 3D COMPUTER GRAPHICS SOFTWARE

3D computer graphics software refers to programs used to create 3D computer-generated imagery. There are typically many stages in the 'pipeline' that studios use to create 3D objects for film and games, and this chapter only covers some of the software used. Note that most of the 3D packages have a very plugin-oriented architecture and high-end plugins costing tens or hundreds of thousands of dollars are often used by studios. Larger studios usually create enormous amounts of proprietary software to run alongside these programs. Many 3D modellers are general purpose based and can be used to produce models of various real-world entities, from plants to automobiles to people. Some are specially designed to model certain objects, such as chemical compounds or internal organs. 3D modellers allow users to create and alter models via their 3D mesh. Users can add, subtract, stretch and otherwise change the mesh to their desire. Models can be viewed from a variety of angles, usually simultaneously. Models can be rotated and the view can be zoomed in and out. 3D modellers can export their models to files, which can then be imported into other applications as long as the metadata is compatible. Many modellers allow importers and exporters to be plugged-in, so they can read and write data in the native formats of other applications.

Most 3D modellers contain a number of related features, such as ray tracers and other rendering alternatives and texture mapping facilities. Some also contain features that support or allow animation of models. Some may be able to generate full-motion video of a series of rendered scenes (i.e. animation). 3D modelers are used in a wide variety of industries. The medical industry uses them to create detailed models of organs. The movie industry uses them to create and manipulate characters and objects for animated and real-life motion pictures. The video game industry uses them to create assets for video games. The science sector uses them to create highly detailed models of chemical compounds. The architecture industry uses them to create models of proposed buildings and landscapes. The engineering community uses them to design new devices, vehicles and structures as well as a host of other uses.

32.22 POPULAR PACKAGES USED IN 3D MODELLING

Popular packages used in 3D-Modelling are:

1. 3DS Max (Autodesk),
2. AC3D (Inivis)
3. Aladdin4D (DiscreetFX),
4. Blender (Blender Foundation)
5. Cinema 4D (Maxon)
6. Electric Image Animation System (EI Technology Group)
7. form•Z (AutoDesSys, Inc.)
8. Houdini (Side Effects Software)
9. Hypershot
10. Inventor (Autodesk)
11. LightWave 3D (NewTek),
12. MASSIVE
13. Maya (Autodesk)
14. Modo (Luxology)
15. Silo (Nevercenter)
16. SketchUp Pro (Google)

17. Softimage (Autodesk)
18. solidThinking (solidThinking Ltd).
19. SolidWorks (SolidWorks Corporation)
20. trueSpace (Caligari Corporation)
21. Vue 7 (E-on Software)
22. ZBrush (Pixologic)
23. RealFlow
24. Realsoft3D Real3D
25. Rhinoceros 3D
26. Seamless3d NURBS
27. Terragen and Terragen 2 (Both are freeware scenery generators)
28. Wings 3D

32.23 BASIC COMMANDS IN AUTOCAD

You can start the AutoCAD by using the following procedure:

1. In Windows: on task bar click start and then choose programs then choose AutoCAD from the menu
2. Otherwise, it is possible to open AutoCAD by double clicking the short cut placed in the desktop

32.24 THE AUTOCAD INTERFACE

When you first start AutoCAD, the initial screen contains the menu bar at the top, the status bar at the bottom, the drawing window, the command window and several tool bars. Tool bars contain icons that represent commands. The menu bar contains the menu. The status bar displays the cursor coordinates and the status the modes such as Grip, Snap, OSnap, OTrack, etc. Mode names are always visible in the status bar as selectable buttons. Double click to turn it on.

32.24.1 Utility commands

32.24.1.1 New

New lets you start a new drawing from scratch or use an existing drawing as a template for a new drawing.

To create a new drawing—Command line: New, Ctrl+N

Menu: File → New

Options

- Open a Drawing: Use Open a Drawing to locate and open the first AutoCAD drawing in your session.

- Start from Scratch: Use the Start from Scratch button for a drawing setup that allows you to start a drawing quickly with either English or Metric settings.
- Use a Template: Starts a drawing based on a template.
- Use a Wizard: Click Use a Wizard and choose between quick setup or advance setup. The quick setup wizard contains two pages: units and area. The advanced setup box offers three additional pages: angle, angle measures and angle direction. Each page displays a preview window showing the results of your selection.

32.24.1.2 Save

Save stores your currently open file to disk.

Command line: Save

Menu: File → Save

32.24.1.3 Save As

Save As saves an unnamed drawing with a filename or allows you to rename your drawing, making it to another drawing.

Command line: Save As

Menu: File → Save As

32.24.1.4 Open

Open lets you open an existing drawing.

Command line: Open, Ctrl+O

Menu: File → Open

Options

- Read only: Check this box if you wish to open a drawing in read-only mode. Drawings that are opened in read-only mode can be edited and saved to a new name. You cannot save changes to the drawing in the original name.
- Partial open: Opens the partial open dialog, allowing you to partially load layers and a saved view from a drawing saved in the current release of AutoCAD.

32.24.1.5 Line

Line draws simple lines – either a single line or a series of line segments end to end.

To draw a line—Command line: Line L

Menu: Draw Line

i. Specify first point: Select a point to begin the line.
ii. Specify next point: Select the line end point, or press U to remove the last line segment drawn.

iii. Specify next point or (Close/Undo): Continue to select points to draw consecutive lines, press to exit or enter an option.

C: Closes a series of lines, connecting the last start point and the last end point with a line.

32.24.1.6 X-Line

X-Line allows you to create construction lines anywhere in 3D space. By default X-Line creates an infinite line based on two input points.

To draw an X-Line—Command line: X-Line, xl

Menu: construction line

i. Specify a point or (Hor/Ver/Ang/Bisect/Offset): Pick a first point or select an option like mid-point.

ii. Specify through point: Pick a second point or enter coordinates to orient the X-line. Continue to pick additional lines through points as required to create additional construction lines radiating from a point.

Options

- Hor/Ver: Draws a construction line parallel to the *x*-axis (Hor) or *y*-axis (Ver). You need only pick single points to define X lines of this type.
- Ang: Draw an *x* line at a specified angle to the *X*-axis by either entering an angle value or by dynamically picking two points.
- Bisect: Creates a construction line that bisects a specified angle. First, pick the angle vertex point and then mark the lines of the angle.
- Offset: Draws a construction line parallel to a selected line object at a specified offset. First, specify the offset by picking two points or entering a numeric value. Then, select a line object and pick a point to indicate the side on which to offset the construction line.

32.24.1.7 Ray

Ray creates the semi-infinite lines that are generally used as construction lines in a drawing. The ray extends from a selected point to infinity.

Command line: Ray

Menu: Draw → Ray

i. Specify start point: specify a start point for the ray.

ii. Specify through point: specify the point through which you want the ray to pass.

iii. Continue to specify points to create multiple rays.

32.24.1.8 Circle

Circle offers several methods for drawing a circle, the default being to choose a center point and enter or pick a diameter or radius.

Command: Circle, C

Menu: Draw Circle Center, Radius / Center, Diameter / 2Points / 3Points / Tan, Tan, Radius / Tan, Tan, Tan

i. Specify center point of the circle or (3P/2P/Ttr): pick a centre point or enter an option.

ii. Specify radius of circle or (Diameter): Provide a radius by dynamically dragging, entering a value or pressing the letter D to display the prompt and then specify the diameter of the circle.

Options

- 3P (3 Points): Allows you to define a circle based on three points. Once you select this option, you are prompted for a first point, second and third point. The circle will pass through these three points.
- 2P (2 Points): Allows you to define a circle based on two points. Once you select this option, you are prompted for a first point and a second point. The two points will be the opposite ends of the diameter.
- Ttr (Tangent, Tangent, Radius): Allows you to define a circle based on two tangent points and a radius. The tangent points can be lines, arcs or circles.

32.24.1.9 Arc

Arc allows you to draw an arc using a variety of methods.

Command line: Arc, A

Menu: Draw → Arc

Options

- Angle: Enters an arc in terms of degrees or current angular units. At the prompt, Specify included angle, you can enter the angle value or use the cursor to select angle points onscreen.
- Center: Enters the location of an arc's center point. At the prompt, Specify center point of arc, enter a coordinate or pick a point with your cursor.
- Direction: Enters a tangent direction from the start point of an arc. At the prompt, Specify tangent direction for the start point of arc, enter a relative coordinate or pick a point with your cursor.
- End: Enters the end point of an arc. At the prompt, Specify end point of arc, enter a coordinate or pick a point with your cursor.
- Chord length: Enters the length of an arc's chord. At the prompt, Specify length of chord, enter a length or drag and pick a length with your cursor.
- Radius: Enters an arc radius. At the prompt, Specify radius of arc, enter a radius or pick a point that defines a radius length.

32.24.1.10 Ellipse

Ellipse draws an ellipse for which you specify the major and minor axes, a center point and two axis points or it draws the center point and the radius or diameter of an isometric circle. It also lets you define a second projection of a 3D circle by using the Rotation option.

Command line: Ellipse, El

Menu: Draw → Ellipse Center/Axis, End/Arc

Options

- Axis Endpoint: Allows you to enter the endpoint of one ellipse axes. Other endpoints of the axis appear after you have defined one point of the ellipse axes. Enter the distance from the center of the ellipse to the second axis point.
- Center: Allows you to pick the center point of ellipse.
- Arc: Creates an elliptical arc. The angle of the first axis determines the angle of the arc.

REVIEW QUESTIONS

1. What is Computer Aided Design? Give a brief description about the importance CAD in engineering.
2. What are the capabilities of CAD?
3. What is CAM? Differentiate between CAD and CAM.
4. Write Short Notes on:
   a. Numerically controlled machine
   b. AutoCAD
   c. AutoCAD LT
5. What is .dwg file? What is its importance in AutoCAD?
6. Discuss briefly about .dxf file format used in CAD applications.
7. What is the importance of DWF in CAD?
8. Explain about the file structure of DXF.
9. What is IGES and what is its importance in CAD?
10. What are the differences between .dxf and .dwg files?
11. What is CAE and what is its importance?
12. What is EDA and what is its importance in circuit design?
13. Write short notes on:
    a. MDO
    b. 3D computer graphics software

index